M.E. O'Connor R.H.C. Bentall
J.C. Monahan
Editors

# Emerging Electromagnetic Medicine

With 107 Illustrations



Springer-Verlag
New York Berlin Heidelberg London
Paris Tokyo Hong Kong Barcelona

Mary Ellen O'Connor, Ph.D., Psychology Department, The University of Tulsa, Tulsa, Oklahoma 74104-3189 USA

Richard H. C. Bentall, F.R.C.S. Ed., Bioelectronics Corporation, The Technical Advancement Program, University of Maryland, College Park, Maryland 20742 USA

John C. Monahan, Center for Devices and Radiological Health, Food and Drug Administration, Rockville, Maryland 20857 USA

Library of Congress Cataloging-in-Publication Data
Emerging electromagnetic medicine / editors, Mary Ellen O'Connor, Richard H. C. Bentall, John C. Monahan.
p. cm.
Proceedings of a conference held May 25-28, 1989, in Tulsa, Okla., sponsored by the University of Tulsa and the Center for Devices and Radiological Health.
Includes bibliographical references.
ISBN-13:978-0-387-97224-4
1. Electromagnetism – Physiological effect – Congresses. 2. Electromagnetism in medicine – Congresses I. O'Connor, Mary Ellen. II. Bentall, Richard H. C. III. Monahan, John C. (John Charles), 1938- . IV. University of Tulsa.
QP82.2.E43E46 1990
615.8'45 – dc20
DNLM/DLC
for Library of Congress 89-26366
CIP

Printed on acid-free paper.



Camera-ready copy provided by the editors.

9 8 7 6 5 4 3 2

ISBN-13:978-0-387-97224-4 e-ISBN-13:978-1-4612-3386-2
DOI: 10.1007/978-1-4612-3386-2

To the memory of my father for his patience, understanding, dedication and love. (MEO'C).

To the hope that patients around the world will one day truly benefit from the technology published in this volume. (RHCB).

To my parents who gave so much of themselves and asked only that I try; and for my wife Gale for her encouragement and patience. (JCM).

# Preface

The contributors to this volume met at an international conference to share their ideas and research on emerging electromagnetic medicine. The purpose of this proceedings is to share their views with a wider audience. Collectively the individual contributors represent many diverse disciplines, with a mutual interest in the use of electromagnetic fields (EMF) in medicine.

The research in this proceedings includes theoretical mechanisms, laboratory experiments, and clinical evidence of beneficial applications. Both diagnostic and therapeutic use of EMF are discussed. The authors accepted the unusual invitation to go on record with their predictions and provide informed projections regarding future medical applications of EMF. The result for those who attended the conference was a lively and exciting scientific interaction. A proceedings can hope to capture only a fraction of that excitement.

Previous research on the bioelectrical effects of radio frequency and other forms of non-ionizing radiation concentrated on the potential hazard that might result from occupational or general environmental exposure to such fields. Prior discussion of the beneficial applications focused primarily on communication and food preparation. Current emphasis is shifting to an array of medical applications. Existing medical applications of EMF (eg, Magnetic Resonance Imaging, pulsed EMF for long bone non-union) have already received considerable attention. The chapters included here focus on future applications. The emphasis on the future demands equal attention to possible risks associated with the technology.

The US Food and Drug Administration (FDA) is preparing for what may develop in the near future. The FDA's involvement signals a growing awareness of the rapid technological developments in this area and the impending array of beneficial applications in medicine that soon will require decisions on effectiveness and safety.

This conference was made possible by the co-sponsorship of the Office of Continuing Education and the Office of Research of The University of Tulsa (TU) and the Center for Devices and Radiological Health of the FDA. Specifically we wish to thank Dr. Milt Jarrett, Dr. Allen Soltow, Dr. Janusz

Beer; the congenial staff of the Westin Hotel in Tulsa, OK; David Wall, Jacqueline Remondet, Tina Clanton and Winnie Stuart of the TU Psychology Department; Diana Beard and John Hamilton of Bioelectronics Corporation; and James Costello of Springer-Verlag.

The thoughtful and insightful comments of all the conference participants were instrumental in the preparation of this volume. We wish to thank the authors for the careful and timely preparation and submission of their chapters. Most importantly we hope that future conferences will expand the international participation, as well as the breadth of the subject matter.

Mary Ellen O'Connor

Richard H. C. Bentall

John C. Monahan

September 9th 1989

Rose Hill Farm, Port Tobacco, Maryland

# CONTENTS

## Section III - Clinical Applications

# CONTRIBUTORS

**VE Amassian**, Chairman
Department of Physiology
SUNY Downstate Medical Center
Brooklyn, NY 11203, USA

**Clyde Barlow**
Professor of Chemistry
Evergreen State College
Olympia, WA 98505, USA

**Richard HC Bentall**
President /Chairman
Bioelectronics Corporation
TAP,University of Maryland
College Park, MD 20742, USA

**N Cheng**
Afdeling Biochemie
Campus Gasthuisberg
Herestraat 49
B-3000 Leuven, Belgium

**CK Chou**
Division of Radiation Oncology
City of Hope Medical Center
Duarte, CA 91010, USA

**Ian A Cook** (on leave)
University of Colorado, Bioelectric
Research Center, Columbia
College Physicians and Surgeons
4200 East 9th Avenue
Denver, CO 80262, USA

**Roger Q Cracco**
Prof. and Chairman of Neurology
SUNY Health Science
Center at Brooklyn
450 Clarkson Avenue
Brooklyn, NY 11203, USA

**Christopher L Davey**
Department of Biological Sciences
University College of Wales
Aberystwyth, Dyfed
SY 23 3DA, UK

**William De Loecker**, Professor
Department of Biochemistry
Katholieke University Te Leuven
B-3000 Leuven, Belgium

**PH Delport**
Afdeling Biochemie
Campus Gasthuisberg
Herestraat 49
B-3000 Leuven, Belgium

**Deborah L Hill**
Research Assistant
City of Faith Hospital
8181 South Lewis
Tulsa, OK 74137, USA

**Douglas B Kell**
Department of Biological Sciences
University College of Wales
Aberystwyth, Dyfed
SY 23 3DA, UK

**Jeffrey Kelly**
Professor of Chemistry
Evergreen State College
Olympia, WA 98505, USA

**Gordon E King**, Professor and
Chairman of Dental Oncology
University of Texas
MD Anderson Cancer Center
Houston, TX 77030, USA

**James C Lin**
Professor and Head of Engineering
University of Illinois at Chicago,
Dept of Bioengineering
851 South Morgan Street
Chicago, IL 60607, USA

**Kenneth H Luk**, Chairman
Division of Radiation Oncology
City of Hope Medical Center
Duarte, CA 91010, USA

**Paul J Maccabee**
Assistant Professor of Neurology
SUNY - Health Science Center
450 Clarkson Avenue
Brooklyn, NY 11203, USA

**Bruce McLeod**
Professor of Electrical Engineering,
Montana State University
Bozeman, MT 59717, USA

**John C Monahan**
Research Psychologist
Food and Drug Administration
Center for Devices and
Radiological Health
5600 Fishers Lane
Rockville, MD 20857, USA

**Mary Ellen O'Connor**
Associate Professor of Psychology
Director, Bioelectromagnetics
Research Laboratory
The University of Tulsa
Tulsa, OK 74104, USA

**Richard G Olsen**
Biomedical Engineer
Naval Aerospace Medical Research
Laboratory
Penacola, FL 32508, USA

**Arun A Patil**
Associate Professor Neurosurgery
Univ of Nebraska Medical Center,
42nd and Dewey
Omaha, NE 68105, USA

**Steven W Phillips**
Research Assistant
City of Faith Hospital
8181 South Lewis
Tulsa, OK 74137, USA

**Jesse Ross**, President
Diapulse Corporation of America
321 East Shore Road
Great Neck, NY 11023, USA

**Seppo J Saksanan**
Associate Professor of Radiology
Central Hospital Dept of Radiology
53130 Lappeenranta
Finland

**Jan Sheetz**
Physical Therapist
University of Texas
MD Anderson Cancer Center
Houston, TX 77030, USA

**Betty Sisken**
Research Associate Professor
Center for Biomedical Engineering
and Department of Anatomy and
Neurobiology
University of Kentucky
Lexington, KY 40506, USA

**Carl Sutton**, Professor
Department of Physical Medicine
Rehabilitation
Medical College of Wisconsin,
Milwaukee, WI 53193, USA

**Paul Tyler**
12604 Stable House Court
Potomac, MD 20854, USA

**Jimmie L Valentine**
Prof. and Chairman Pharmacology
City of Faith Hospital
8181 South Lewis
Tulsa, OK 74137, USA

**Benny D Wagner**
Radiological Technologist
City of Faith Hospital
8181 South Lewis
Tulsa, OK 74137, USA

**James C Weaver**
Associate Director
Biomedical Engineering Center
Harvard MIT Division of Health
Sciences and Technology
Massachusetts Insitute of
Technology
Cambridge, MA 02139, USA

**William S Yamanashi**
Professor of Radiology
Director of NMR Research
Institute
City of Faith Hospital
8181 South Lewis
Tulsa, OK 74137, USA

**Nabil A Yassa**
Dept of Radiology
University of Arkansas for Medical
Sciences
4301 West Markham
Little Rock, AK 72205, USA

# Electromagnetic Energy: A Historical Therapeutic Perspective; its Future

Richard H C Bentall

## INTRODUCTION

To view the historical aspects of any one subject in isolation is to misunderstand how advancement in society is achieved or developed. Frequently society influences historical events by its pervading beliefs, more profoundly than scientific intellect would like to admit or comprehend. This is mainly because a given perception is a function of social attitudes at that particular time; guidelines and beliefs are based upon historical avenues of development. In considering the subject of electromagnetic therapy (EMT) a similar algorithm is at work. Therefore it is important not only to understand the scientific content of what has previously gone before, but also to view that content in the social context of the time.

Culture and politics play an important part in guiding the direction of scientific development. In today's smaller world, as we enter the last decade of this century, communication, ease of transportation, knowledge of social, scientific, and other events, more than ever moderate and influence progress with these same cultural and social attitudes.

Focusing now on medical and particularly therapeutic advancements, these same influential attitudes are very much at play. Science can not significantly create *de novo* therapeutic advancement without addressing, and to some extent understanding, these outside influences. These influences in pure scientific terms can be thought of as co-variants without control or precise capability of quantification, let alone predictive assurity.

It is intended to explore the interaction and influence of these attitudes as they relate to: 1) the history of therapeutic advances; 2) how this history itself influences medical advances; 3) how far the diagnostic and therapeutic use of the electromagnetic spectrum has been effected by such historical considerations; and 4) suggest what course this subject is now embarked upon and how this should be understood in order to maximise the potential value that this subject can bring to the benefit of human health.

## THE HISTORY OF THERAPEUTIC ADVANCES

It thus behooves us to consider carefully how therapeutic advances have been viewed in the past, so that we may benefit from hindsight. From prehistoric man through the Dark Ages a religious undertone pervaded all 'therapy'. Whether it was a solar deity or other gods who were worshipped in the name of therapy, the influence on the patient, and the physician was the same. Some outside force beyond one's daily reach or control was responsible for influencing the outcome of ill health. Sacrificial appeasement of their wrath was an accepted practice.[22]

This healing divinity for instance in the civilized world of the Egyptians, was exemplified by Thoth and Imhotep. The Greek god Ascelpios became the primary healing force in their society.[22] The belief in their healing power led to a combination of mythological incantations with small heed being paid to anything other than what we might now consider local first aid to the lesion being treated. If a topical leaf or herb, or even a systemic potion, was prescribed by the 'physician', it was ordained by these divine influences and religion took on a greater influence as civilization developed over the next few centuries. The 'therapist' invoked the name of the religious beliefs of the sick to add weight, value, and eliminate the risk of responsibility, as a part of the treatment process. An understanding of the value of beliefs, intangibles such as the 'will to live', and other central influences upon immunology or facets of the body's defense system, is only now becoming understood.[20] It is clear that religious therapeutic prescription formulae had a valuable moderating effect on the sick of the middle ages.

### The Dawn of Therapeutic Science

By the arrival of the late 16th century, science was very much on the ascendancy. The earth was firmly established as not being the center of the universe, and by default, religion was forced to relinquish its dominance of scientific discussion in most regions of the world. The publication in 1543 of Andreas Vesalius' 'De Humani Corporis Fabrica' stands as a landmark of the new scientific approach to the anatomy of the human body. The accelerated surge in scientific enquiry seen in the renaissance age of a more identifiable scientific discipline was the background to Michael Faraday's discovery of electricity in the mid 1780's. Here was a monumental discovery, one that for ever changed the world. Electricity, a new 'force', was proving with every advancing year to have powers previously unthinkable, and yet still mystical, and certainly not well understood. Therapeutic medicine, channelled through a new era of enquiring physician, promptly tried this new modality. It took only five years after

Faraday's initial description before the first position as Chief of Medical Electricity at St Thomas' Hospital in London was filled by John Birch.[3] The father of modern scientific surgical enquiry John Hunter (1723-93) availed himself of this treatment for certain specific pathologies occurring in his patients. From the famous experiments of Luigi Galvani (1737-98) came the "Ramsden" type electrogenerator, used for defibrillation, as well as other uses. Victorian engineering ingenuity however, soon outstripped the slow acceptance of physicians, despite the physician still having little proven therapy to offer patients. This was in part because of the resistance to change, entrenched in the medical profession, then as now, and because the concept of medical scientific enquiry into treatments was still nearly half a century away. Even Edward Jenner's 1798 demonstration of the successful development of a smallpox vaccine could hardly be called science as we define it today, but the vaccine worked, and the risk/benefit was self-evident.

When Lord Joseph Lister began his lecture tours in the latter part of the 19th century advising surgeons to wash their hands before surgery, he met with a reluctance to change, and was ostracised by his colleagues.[21] This was despite the investigations into infection pioneered by William Hunter, (1718-83), brother of John, who linked the condition of teeth and gums to obstetrical infection, and also of Sir Fielding Oulds who set a standard for obstetrical cleanliness in the still famous Rotunda Hospital of Dublin, a full century before Lord Lister's more successful arguments. Only after Louis Pasteur's discovery of bacteria, was Lister able to point to why cleanliness in operating rooms was so important, but the uphill struggle took an entire lifetime.[21] Embedded in his colleagues' minds and practices were hundreds of years of not having to be clean in operating rooms, and in believing suppuration was helpful. This embedded opposition is exemplified by Samuel Gross (1805-84) the famous American surgeon's quote when he wrote nine years after Lister's paper "Little if any faith is placed by any enlightened or experienced surgeon on this side of the Atlantic in the so-called carbolic acid treatment of Professor Lister." Even as late as 1901, thirty-five years after Lister's message, surgeons were still operating without gowns and gloves in many areas.[22] Only with continued clinical proof would it become universal. There had yet to develop a good mechanism for keeping surfaces, hands, and theaters clean. For this part industry helped enormously, but only after the demand had been created by Lister's arguments. Two brothers, by the name of Johnson, who listened to Lister's lecture tour in the US started a company to manufacture antiseptic bandages for wounds. This company still makes bandages of many kinds for everyday use with the same principles underlying their rationale, namely having some antiseptic in the dressing covering a wound.

This strong resistance to change has been the hallmark of medical scientific therapeutic endeavour. Physics, electronic engineering, and other basic sciences can move light years faster than medical therapeutic advances. This in part stems from the trust and responsibility inherent within the doctor-patient relationship that was then, and continues today, to be the reason for this reluctance to change. The art of medicine is very real, the placebo therapeutic benefit of this relationship plays a substantial part in treating people as do the latest scientific advances. In the last century this relationship surpassed what could actually be done, but today treatment has become a dichotomy, almost a double standard that physicians have to exhibit. This dichotomy is in part of the physicians' own making because of the success they have achieved in treating patients, but it is also the fault of society, for the reasons of cultural perception which are now undermining this relationship. Society more than ever expects cures from physicians. Their demands are frequently so unrealistic that as Time magazine (1989) recently pointed out, the doctor-patient relationship is deteriorating in the US.[29] In other countries this apparent deterioration is not in evidence primarily because their populations do not view medicine as a consumer product. Both attitudes exemplify these historical and cultural differences, (ie, the expectation of a cure and the perception of how that cure is delivered). Therapies which have proved useful through time, often on an empirical basis, have a more ingrained influence on the management of patients than scientific reality would dare to contemplate.

## HOW HISTORY INFLUENCES MEDICAL ADVANCES

Into this complicated equation of entrenched views and empirical history we now begin to place the other more recent modern over-riding concern for physicians. They now have the ability to offer their patients a far wider range of therapeutic possibilities and therefore one consideration of these beneficial advances are the inherent risks. The physician must now consider the risk/benefit ratio. When Lord Lister introduced his principles of asepsis, the new method had initial problems of unpredicted risk. Clean hands and theaters reduced morbidity and mortality because of a reduction in bacteria; initially something not entirely easy to understand or see, but the benefit was apparent. And yet the rate of conversion continued to be slow. However, the aerosols of carbolic acid used to obtain the standard of asepsis themselves soon found disfavour because of a risk that was not at first fully appreciated. The risks of nausea, rashes, even death for attendants were evident but had to be weighed against the benefit to the patient. Risks were reduced by modification to the disinfecting system and the concept of clean hands itself became entrenched, but a key lesson had been learned. Clear cut benefits to the health of patients are always slow to be accepted by the medical community because with advances come some risks. This concept of risk/benefit is as relevant to us today in regard to the therapeutic and diagnostic use of electromagnetic energy as carbolic acid was at that time.

## THE EFFECT OF HISTORICAL CONSIDERATIONS ON THE DIAGNOSTIC AND THERAPEUTIC USE OF THE ELECTROMAGNETIC SPECTRUM

### A. Diagnostics

Diagnostic advances are easier to introduce for they enhance the capability and reputation of their medical user whose position is in no way jeopardised by the new technique. Advances such as the electrocardiogram and computer assisted tomography (CAT) scans were relatively easy to introduce. The functioning inside their inanimate boxes does not need to be understood, nor does it represent a threat to the physician. On the contrary, the introduction of magnetic resonance imaging (MRI) scanning demonstrated an opposite disciplinary problem. It was two physicists, Freeman and Cope, who began to develop the principle of using physics to investigate living systems. They found that their colleagues obstructed their work being published in the normal channels for that science, so they formed their own Journal of Submolecular Biology. Each edition included something about what they called Fonar, the forerunner of MRI as we know it today. This apparent 'peer' suppression occurred in the mid 1970s and no doubt is in evidence today.

In diagnostic equipment the risk is usually minimal. The diagnostic end-point is frequently an instantaneous finding. Not so with treatments which by definition need time, an important factor regarding the use of electromagnetic treatment. This is not to infer that mis-diagnoses are not very important. The risk/benefit equation will then be the guiding principle. When Röntgen introduced his first x-ray pictures of bones inside the human body, his colleagues' response was less than enthusiastic. Once more the spectra of side-effects reared its head, and again the slow acceptance was followed by a realisation of its benefit. This led to regulation, and thus a clinical balance was struck between the benefits and the side-effects. Compare the ready acceptance by the medical profession of these diagnostic devices with those of therapy. Therapy needs to have a mechanism explained whereas diagnostic equipment will succeed on its instantaneous merits. This is not necessarily a bad attitude, but in the case of the therapeutic use of the electromagnetic spectrum, fundamental mechanisms seem to be a prerequisite of delivering treatment when in fact, full understanding of these mechanisms may be many years away.

B. Treatment

In therapy the wide range of clinical outcomes makes it vital that this risk benefit balance is properly weighed. If the treatment in question can benefit, then some would argue it can also probably do harm. This may be true in some cases of successful treatments where there is no direct harm but only a secondary harmful consequence. For instance, if one uses a therapy to relieve all pain from a patient with a sprained joint for reasons of obvious discomfort, the result may be that the patient re-injures themselves. So in this case the risk/benefit of total pain relief is balanced by the physician who may use the pain itself for some therapeutic value of its own, (ie, because it restricts movement and the use of a limb). This is nature's way of guiding the animal/patient to a protected route to recovery. Frequently the risk comes from the other more obvious direction, namely a side-effect of the therapeutic agent. The post-operative patient's best interest is not necessarily served by a five day course of analgesia consisting of morphine derivatives. The patient almost invariably believes the medical staff are zealots in limiting the prescription of such drugs, but is not cognizant of the respiratory depression and the addictive capability of such a course of treatment. The physician must balance the drug's beneficial effects against the stated risks and this is the prime reason for its careful prescription in the immediate post-operative period.

There is almost always a trade-off in treating patients. This trade-off is where the art and the science of therapeutic medicine work hand in hand. But the ratio of risk to benefit (or trade-off) is sometimes a very difficult equation for the physician to solve, particularly when a patient's long-term health is concerned. Thus physicians will and do prefer to keep to established treatment regimes.

"But, it works in my hands ..." is a frequently heard statement from surgeons, in conversation regarding an established operative technique in comparison to the most recent advances in methods of surgery. From the standpoint of risk/benefit, even if the operative technique used by the protesting surgeon is outmoded or old fashioned, in his hands the patient's best interests may well be served by his familiarity with his established technique. Frequently this argument is in truth the 'best' solution in reality, for each step is well practiced, and a minimal risk/benefit equation is thus at work. A well known surgical example of this is the practice in the US of continuing the procedure of mutilating radical mastectomy, an operation for the most part discarded in Europe where simple mastectomy has found favor. The comparative mortality results for each procedure are the same, and the patients who have a simple mastectomy performed suffer reduced disfigurement and disability by avoiding radical mastectomy. Thus the argument of using what is old fashioned and familiar has inherent dangers, namely a reduced benefit to the patient. No therapeutic advances occur through such a low risk/no risk route. No scientific rigours check these historical dogmas. One of the biggest problems facing the American doctor-patient relationship today is founded around this very point, to some physicians one route is 'best', while another opinion would argue against it, and the patient is placed in a position where he adopts a consumer approach to his healthcare.

The essence of these historical considerations lays the foundation for us to consider and establish the position in which the therapeutic aspects of the electromagnetic spectrum now find themselves, as we enter the last decade of the 20th century. The need for an interdisciplinary approach, and problems arising therefrom, namely communication of each discipline's position, is paramount to the formation of a solid foundation from which this subject can develop. This volume reflects the present struggle to form this solid foundation.

### C. Elements of this Emerging Therapy

#### 1. Engineering

There is no question that electrical engineers know more about components, the workings of circuit configurations, the output of devices, and the restrictions on electrical apparatus than at any time in history. So from the strict engineering standpoint, society is in the very best position it has ever been to understand and manufacture electromagnetic energy based products that are sound, well documented, and fully understood. This is a giant step from similar types of treatment devices in the last century. It is a better situation to be in than pharmacology was at the turn of this century. It is interesting to note that many of the equations forming the basis of today's electromagnetic theory and knowledge were indeed born during the last two hundred years. So Faraday, Hertz, Tesla, Maxwell and others created the very framework from which our present day knowledge is drawn.

#### 2. Biophysical functioning

The advances in knowledge of biophysics today are so far in advance of the last century, there is no doubt that our understanding of the relationship between physics and biology has grown to where it is a subject in its own right. There are an ever increasing number of links being made between one scientific discipline and another. An excellent example of cross disciplinary links is in the developing field of dielectrics of living matter.[1,2,4,5,7,10-14,16-19,25-28]

3. Basic medical disciplines

Anatomy is the same as a thousand years ago; but very much better understood. Likewise physiology is the same, but our understanding of it has advanced to a more sophisticated level. The more questions relating to mechanisms that are asked and answered, the more new questions arise. As with the basic sciences, there begins to be an apparent convergence. The cross-links between the basic sciences of physics and quantum mechanics, with the development of biophysics are becoming so strong and profound, that where the life scientists, and in particular the physiologists, thought of themselves as in a completely different scientific discipline from physicists and electrical engineers, they now realise they are looking at different positions of the same continuum.

4. Treatment in medicine

Treatment however, is the aspect of medicine most restricted by social, cultural and ethical risk/benefits, and only slow progress can be achieved because of these interactive forces. Education is of paramount importance in encouraging a dialogue between the physicians and those who are developing the therapeutic opportunities of electromagnetic fields.

We can now begin to understand the problem that surrounds this therapeutic subject. On the one side of the equation, engineering and biophysics have steam-rollered along until they are now on the verge of actually encroaching on the 'proprietary territory' of therapeutic 'quasi science'. The medical protagonists are suspended in this time warp delay for reasons that science can not alter. The potential therapeutic value of using the electromagnetic spectrum creates huge upheavals that one would naturally expect from a physician's conservative attitude meeting with a therapy whose boundaries are as yet far from clear, and still awaiting scientific definition.

D. <u>The Resolution</u>

Now for the first time we can allow ourselves the cautious luxury of looking into the outcome of this boisterous interaction. Setting the rules for this discussion would be useful. The advancement of scientific knowledge is not going to go backwards, nor is it going to stop. Thus it is must go forward. At what rate, and in what direction? The 'rate', by something as mundane as availability of research funds, and the 'direction', depends on many forces, but unfortunately for science, it is linked also to the first, (ie, money).

The reality of scientific research life, as this century draws to a close, is that new knowledge for the sake of just knowing is becoming an expensive hobby of a few scientists and universities. Funding has been scaled back and for some agencies has disappeared. Most scientific direction is influenced by social perceptions of requirements. In this setting, and in this subject, the requirement could be the investigation of potential hazards related to the use of the electromagnetic spectrum for military programs.[6,14] Some, like the Airforce and Office of Naval Research (ONR), have shifted their emphasis. Power line research is now receiving funding where none existed fifteen years ago.[23] There are scientific questions yet to be asked and answers provided. Will the funds continue? Yes, but probably in reduced amounts, and only if the results are weighted in favour of social development. Whatever the outcome, society is not going to sacrifice electrical power. Modification is expensive so there will be a hesitance to develop scientific knowledge with funds that could result in legal enforcement of expensive solutions. This political and social equation seems therefore to be unstable. Society itself could influence this equation more directly in the name of safety, but has yet to do so with any great fervor.

## ELECTROMAGNETIC FIELDS AND THE THERAPEUTIC FUTURE

The rate of therapeutic development and its direction will depend initially on therapies currently in existence. Bone non-union healing has suffered some setbacks because of diminished numbers of patients requiring this treatment, poor understanding of the necessity of well balanced scientific dialogue for physicians, physicians using the treatment for fresh fractures, and over-zealous marketing. This is in part due to a lack of understanding of how the physician makes choices and what freedom he has to extemporize, and extrapolate in a new prescription area. It is unlikely that the electromagnetic treatment of bone tissue will totally fall from therapeutic grace, particularly as joint surface pathologies may turnout to be amenable to some forms of electromagnetic therapy in conjunction with other treatments.[8] Soft tissue treatments will overtake bone treatments on a broader base in the next fifteen years, because a far greater number of pathologies exist which may benefit from new electromagnetic treatment modalities. Research endeavours in soft tissue must ensure that the same fate does not befall this fast developing research area as happened to bone non-union healing techniques.[9,15] The pathologies of soft tissue injuries, apart from their far greater volume, variation and differing treatment requirements, lend themselves to considerable research effort and the potential to add clinical value.[9] Soft tissue has also both acute and chronic therapeutic possibilities.

Understanding how a treatment works has never been an obstacle for its use, as we have seen for many years with compounds such as aspirin. However, understanding the action of electromagnetic therapy, even in a rudimentary manner, could greatly increase its development and therapeutic use. Thus the rate of dissemination of the therapeutic potential of the electromagnetic spectrum will initially depend upon definitive studies clearly showing its clinical value. Its rate of acceptance will be dependent on the quality of this research but will also be contingent upon a reasonable explanation of its action. Such an explanation does not require a perfect quantum mechanical solution. Physicians practice an art form interpreting a science, so it is not of paramount importance for them to be educated about every innuendo of interaction. They do not need to understand the minutia of mechanisms and to attempt to do so would be a new burden they can well do without. Physicians however, do need to be educated on using these new therapeutic tools which are not as yet a part of their textbook education. Again this brings us back to the historical elements of the development of medicine. Once the subject of electromagnetic therapy is incorporated in medical school textbooks, then the prescribing patterns will change. A fuller understanding will be required as the younger prescribing physician becomes educated in this subject. This may be fifteen to twenty years away, but then and only then, will significant advances be made, as these physicians become knowledgeable and comfortable with the limitations of use and the subject itself become a part of their prescribing history.

One major hurdle will remain for some while. Healing infers the passage of time and occurrence of some established time dependent events. The potential for the therapeutic use of the electromagnetic spectrum to reduce this passage of time is one of the most fascinating aspects of this whole field, whether the reduction is due to greater healing efficiency, improved skin grafting techniques, or pain control. This reduction in time is not within the experience of physicians and therefore will be the greatest barrier to acceptance. Improvements in quality of healing and saving of time are clinically valuable and focus should be placed on those areas where this clinical value can most easily be determined and appreciated. This implies research focused on pathologies where the risk element would be outweighed by the benefits. This does not necessarily mean concentrating on such obvious market demands as chronic leg ulcers or decubitus ulcers, for their clinical reality is so complex, and this difficult area of research asks unending clinical questions which are difficult and expensive to answer. Clinically valuable research will have ensured a foothold for electromagnetic therapy within the next few years. By then there will also be a greater understanding of the physiological events and this will encourage funding from many sources to be channelled into universities and centers involved in this type of research. This in itself will reduce the risks aspect of the physicians'

equation, and lead to diversification of the pathological conditions in which this therapy can be shown to have clinical value.

A. Mechanism of Action

The most frequently asked question relates to the mechanism of action. There is quite an amount that can be said which helps answer this question. As previously emphasized, engineering has been influential in gaining an understanding of how an electrical current might penetrate a tissue, whether it will or will not get into a cell, and what transduction mechanisms are involved. The interface science, while making great strides[5,7,10,11,17-19,25-27] in possible models, has been less successful in quantifying how dosimetry might best be viewed. In terms of physiology, it is beginning to be possible to suggest some parameters which are effected and could be invoked in a logical step by step approach to a mechanism of action.

In clinical practice it is only the physiological events that can be monitored and are therefore of maximum interest to the majority of physicians. Other underlying mechanisms need not enter into clinical considerations. A reduction in interstitial edema allows for an earlier maximisation of the use of oxygen and this concept is readily understandable to clinicians. If the charge across the cell membrane, which is known to be reduced in injury, could be prematurely 'recharged', then this action could be responsible for the cell's earlier return to function. This is understandable from both the life science and engineering standpoints. How does this occur? Cell physiologists and many engineers with life science experience suggest that the cell membrane and/or its surrounding environment is the transducer that enables electrical energy to be converted into biochemical reactions.[10,13,19,25-27] From studies on wound healing, it is clear that this transduction remains within the control of the cell's own regulatory system, because the cells stop migrating at the appropriate stage in wound healing, and do not go out of reproductive control.

B. Dosimetry

One of the most fascinating and puzzling aspects of this area of biological effects is dosimetry. The reason in part is stated above, it is impossible to measure blood levels of an induced electric field. It is clear that dosimetry is highly non-linear[4,5] and this is NOT surprising, although it certainly makes it difficult to understand the coding mechanism of dosimetry. Pharmaceutical action works by an effect that is proportional to the dose given and this is a general effect for all of pharmacology and is thus quantifiable. Bioelectrical interaction, in pure energy terms, does not have these proportional therapeutic characteristics. It is best considered as an all or nothing response that occurs by

way of a threshold mechanism. The converse argument lies in the possibility that biostimulation is in effect working in a type of machine code language of the living system itself. Until one understands those types or levels of instructions, dosimetry will continue to appear to be highly non-linear and subject only to the most gross explanations.

Highly non-linear implies there is a gap in existing knowledge in this subject. Bearing in mind how living systems function, one cannot help but be impressed by the use that the living system makes of time-varying phenomena as evidenced by circadian rhythm, alterations in blood levels of measurable constituents such as hormones, or therapeutic pharmaceutical agents; and also how sensory and other central inputs are interpreted. Time-varying algorithms appear to be used to interpret color, and sensory inputs such as pain or variations in temperature. It is tempting to suggest that <u>nature is using time intervals</u> as the key unit of instruction interpretation. If this is the case, the apparent non-linear dosimetry that we at present suffer from in the discipline of bioelectromagnetics could be clarified by developing equations incorporating time-varying intervals, or harmonics, as a key unit. Thus far such equations have not been fully developed though the concept has, and will be, the subject of a separate communication.

C. <u>Future Products</u>

Products will be dependent on pathology, technology and treatment period, but can be expected to be portable, easy to use, and capable of showing clinical value not only in soft tissue acute injuries (including burns, post-operative wounds and sports injuries) but also able to show benefit in such internal wound healing concerns as cerebral edema, cardiac pathologies, and some abdominal and bowel conditions. Fresh fractures could be shown to benefit from some forms of this type of treatment which will become a major part of normal orthopedic practice. There will come a time when EMT products will be shown to be beneficial in certain types of chronic wounds at time periods during their healing cycle. These products, as a class, are much better considered as <u>adjunctive</u> treatments. This route will allow combination therapies to be developed which will greatly improve the clinical management of chronic skin wounds.

There will be considerable advances in the understanding and capability of this technology to influence the central nervous system. This could easily result in immunological defence enhancement; treatment being wide spread for addictions of various types and such central effects as treatment of depression.

Conceptually there is no reason not to use whole body products, for a given treatment period depending on the pathology, eg, extensive burns.

However, long term application of treatments will be delayed for some years because of risk/benefit considerations and until long term application studies are conducted.

It is very important to understand from these therapeutic possibilities that there will be a range of products for different pathologies. In time their physical appearance and outputs will diverge from one another to such an extent that in a decade their origins will be completely obscure. More fundamental developments in the next ten years will gradually bring better and more efficient treatment regimes. These will include treatment sequencing, frequency-dependent specific pathology treatments, and other more sophisticated regimes. The products themselves will become smaller and 'smarter' and, as their price goes down, become available "over-the-counter" for minor injuries and thus become a familiar treatment concept to the general public. Theoretically at least, there is no reason why combination treatments couldn't be arranged whereby the physician can program the products for the pathology.

Still further into the future, there are endless possibilities as yet unconsidered. In this age of improved communications, there is the possibility of computer diagnosed treatments being arranged via a treatment credit card which is fed into the treatment device in the patient's own home. Perhaps in twenty years one could go into a shopping mall, have an MRI check-up and get a card to put in one's overnight exposure system! This may sound rather whimsical but when one considers the existing availability of standard health monitoring in shopping malls today, there is no reason why this therapy should not also be available in the future. It should be cautioned again that such long term developments and uses will be socially driven. Such a treatment as just outlined might be totally acceptable in Europe or the Far East and accepted a further twenty years later in North America, or vice versa.

In pointing the signpost to such possibilities there is a fundamental rider that should be carefully understood: one product in isolation will not achieve these different effects. It should not be construed that future development of this treatment represents a panacea. Equally, when pharmacology first became a science and moved from cauldrons into the laboratory, it was not one single formula that was developed but a series of many thousands of products each similar in that they were of chemical composition, but as pharmaco-active compounds, very precisely and specifically tailored for different pathologies. A similar science will develop for electromagnetic medicine, but not in the present generation.

D. Therapeutic Boundaries

The boundaries will develop from knowledge, understanding and education. At this point in the history of the science and art of medicine, electromagnetic medicine is not going to disappear. It will develop at a rate which is dependent to a large degree on education. If there is an early resolution to the dosimetric parameter problems, then advancement will be more swift. The emergence of electromagnetic medicine as a clinical speciality will not be significantly held up by side-effects because the treatments function within the control systems of the cells. This does not mean that safety is not a serious consideration. It may well be possible to deliberately, or by chance, make an electromagnetic delivery system which can specifically do harm. For these and other reasons, care in legislation is necessary. One of the most difficult areas to resolve will be the fact that counterfeit treatment systems are very difficult to identify, primarily because the bona fide treatments do not arouse the senses, so neither the physician nor the patient is aware of an immediate physiological response. Criteria for manufacture and monitoring of patient compliance will be required in some instances.

The most harmful way in which the subject can be deleteriously effected is by over-zealous marketing. Nothing will give electromagnetic medicine more problems in gaining clinical recognition than unproven claims.

## CONCLUDING REMARKS

Scientific progress has identified the potential clinical value of using the electromagnetic spectrum for therapeutic applications. It is unclear at this moment just how far reaching this benefit will prove to be. Having established that the potential clinical value does exist, the next 20 years will considerably clarify these unknown variables. The signs are present that the risk to benefit ratio of this therapeutic modality is such that it will slowly gain a clinical usage that would not have been seriously considered 15 years ago.

One of the major cornerstones of this development remains education, based on disseminating high caliber research to physicians, patients, and scientists in university environments. Monitoring and guidelines will need to be developed to police the subject adequately as it broadens and becomes more widespread in its use. Long term treatments will probably require greater care and consideration in view of the potential for unwanted, unknown, and unforeseen side-effects. As with any treatment, the risk/benefit ratio in long-term usage will be different from short-term uses.

The mechanism of action will become elucidated gradually, and not be a one factor development just as the present day development of sophisticated pharmaceutical compounds has taken time. We might expect electromagnetic medicine to take about the same period of time to reach the same level of sophistication and acceptance. The equipment, however, will become much more efficient, inexpensive and will develop into an industry of its own at a far quicker rate. As the mechanism and dosimetry questions become clearer, even if not finally resolved, the equipment will be technologically well placed to keep up with dosimetric monitoring requirements. The possibility of such equipment becoming integrated in computerized links with diagnostic equipment must remain feasible but not in the immediate future.

It is more than likely that our present equipment will seem very antiquated, and our principles outmoded, within a period of 15 years. The Far East can be expected to play a substantial role in the development of this subject for many reasons, including social, cultural, scientific and technical considerations. The diagnostic and therapeutic potential is thus coming of age in the late 1980's and will take its place in the fuller range of mainstream subjects in healthcare over the next generation.

More specific and in depth enquiries into this subject follow this paper.

## ACKNOWLEDGMENTS

The patience of my co-editors in the preparation of this manuscript, Diana for her invaluable support, and the staff of Bioelectronics Corporation who have suffered hours of unavailability both of myself and their computing capabilities.

### References

1. Aaron MW, Grant EH, Young SE. The dielectric properties of some amino-acids, peptides and proteins at decimetre wavelengths. In: Symposium on Molecular Relaxation Processes. Academic Press, UK. 1966; 77-82.

2. Abler RA. Magnets in biological research. In: Biological Effects of Magnetic Fields (Ed. MF Barnothy). Plenum Press, NY. 1969; 1-27.

3. Adams G. An Essay on Electricity, Explaining the Principles of that Useful Science and Describing the Instruments Contrived Either to Illustrate the Theory, or Render the Practice Entertaining (Ed. W Jones). Dillon & Co, London. 1799; 482-575.

4. Adey WR. Ionic nonequilibrium phenomena in tissue interactions with electromagnetic fields. In: Biological Effects of Nonionizing Radiation (Ed. KH Illinger). American Chemical Society, Washington, DC. 1981; 271-297.

5. Adey WR. Nonlinear, nonequilibrium aspects of electromagnetic field interactions at cell membranes. In: Nonlinear Electrodynamics in Biological Systems (Eds. WR Adey, AF Lawrence). Plenum Press, NY. 1984; 3-22.

6. Baker RE Project Sanguine: An overview and status of the Navy's ELF communications system concept. In: Biologic and Clinical Effects of Low-Frequency Magnetic and Electric Fields (Eds. JG Llaurado, A Sances, JH Battocletti). CC Thomas, IL. 1977; 83-90.

7. Behi J, Bone S, Morgan H, Pethig R. Protonic charge transport studies in cyclodextrins. In: Charge and Field Effects in Biosystems (Eds. MJ Allen, PNR Usherwood). Abacus Press, Tunbridge Wells, UK. 1984; 139-146.

8. Bassett CA. Fundamental and Practical Aspects of Therapeutic Uses of Pulsed Electromagnetic Fields (PEMFS). CRC Press, Boca Raton, FL, 1989 (in press).

9. Bentall RHC. Low-level pulsed radiofrequency fields and the treatment of soft-tissue injuries. Bioelectrochem and Bioenergetics 1986;16:531-548.

10. Cope FW. Overvoltage and solid state kinetics of reaction at biological interfaces. In: Bioelectrochemistry (Eds. H Keyzer, F Gutman). Plenum Press, NY. 1980; 297-330.

11. Drissler F, Santo L. Coherent excitations and raman effect. In: Coherent Excitations in Biological Systems (Eds. H Fröhlich, F Kremer). Springer Verlag, Heidelberg. 1983; 6-9.

12. Edelberg R. Electrical properties of skin. In: Biophysical Properties of the Skin (Ed. HR Elden). Academic Press, NY. 1971; 513-550.

13. Eley DD. Semiconductivity in biological molecules. In: Horizons of Biochemistry (Eds. M Kasha, B Pullman). Academic Press, NY. 1962; 341-380.

14. Index of Publications on Biological Effects of Electromagnetic Radiation (0-100 GHz) (Eds. JB Kinn, E Postow). United States Environmental Protection Agency, NC, 1981.

15. Frank CB, Szeto AYJ. A review of electromagnetically enhanced soft tissue healing. IEEE Eng Med Biol 1983; 12:27-32.

16. Fröhlich H. Organisation and long range selective interaction in biological and other pumped systems. In: Synergetics (Ed. H Haken). Teubner, Stuttgart. 1973; 241-245.

17. Grant EH, Nightingale NRV, Sheppard RJ, Gough SR. Dielectric properties of water in myoglobin solution. In: Biological Effects of Nonionizing Radiation (Ed. KH Illinger). American Symposium Series 157, Washington, DC. 1981; 57-62.

18. Kanagy JR. Sorption of water by collagen. In: Biophysical Properties of Skin (Ed. HR Elden). Academic Press, NY. 1971; 373-391.

19. Kell DB. Non-thermally excited modes and free energy transduction in proteins and biological membranes. In: Energy Transfer Dynamics (Eds. TW Barrett, HA Pohl). Springer Verlag, Heidelberg. 1987, 237-246.

20. Levy SM. Behavior as a biological response modifier: The psychoimmuno-endocrine network and tumor immunology. Behav Med Abstr 1985; 6:1-4.

21. Lister Baron Joseph. The Collected Papers. Vols. I & II. Clarendon Press, Oxford, Special Edition, 1979.

22. Lyons AS, Petrucelli RJ. Medicine: An Illustrated History. Abradale Press, NY, 1987.

23. Miller DA. Electric and magnetic fields produced by commercial power systems. In: Biologic and Clinical Effects of Low Frequency Magnetic and Electric Fields (Eds. JG Llaurado, A Sances, JH Battocletti). CC Thomas, Springfield, IL. 1974; 62-70

24. Patterson MA. Getting Off the Hook. Harold Shaw Publishers, IL, 1983.

25. Pethig R. Dielectric and Electronic Properties of Biological Materials. J Wiley, Chichester, UK.1979.

26. Pethig R, Szent-Gyorgi A. Bioelectrochemistry, the living state, and electronic conduction in proteins. In: Bioelectrochemistry (Eds. H Keyzer, F Gutmann). Plenum Press, NY. 1980; 227-252.

27. Pilla AA. Electrochemical information transfer at cell surfaces and junctions - application to the study and manipulation of cell regulation. In: Bioelectrochemistry (Eds. H Keyzer, F Gutmann). Plenum Press, NY. 1980; 353-396.

28. Szent-Gorygi A. Electronic Biology and Cancer: A New Theory of Cancer. Marcel Dekker, NY, 1976.

29. Time Magazine: Doctors and Patients (Cover Story). July 31st, 1989; 48.

# SECTION I

# THEORY AND MECHANISMS

# THE DIELECTRIC PROPERTIES OF CELLS AND TISSUES: WHAT CAN THEY TELL US ABOUT THE MECHANISMS OF FIELD/CELL INTERACTIONS?

Christopher L. Davey and Douglas B. Kell

## INTRODUCTION

"A knowledge of the passive electrical properties of biological systems must underpin any significant understanding of the nature, role and mechanisms of bioelectrical phenomena."[42] In other words, and not least because of the possible physiological effects exerted on tissues following their absorption of non-ionising electromagnetic radiation,[1,2,11,18,33,34,35,51,56,57,58,59,60 65,69,72, 77,78,83,89,90,91,102] there is an increasing awareness that if we are to exploit electric fields in medical technology, and in other forms of diagnosis and therapy, the first thing we must do is to measure and then to understand how and why such fields actually are absorbed by the target tissues or cells. In this sense, we treat tissues as concentrated suspensions of cells, and, recognising that the magnetic susceptibility of virtually all tissues is essentially identical to that of water, we consider only the electrical component of any imposed electromagnetic field.

In the present article, therefore, we shall: 1) discuss in elementary terms how we measure the (predominantly RF) passive electrical properties of living systems; 2) describe the so-called $\alpha$– and $\beta$–dispersions as examples of linear dielectric behaviour; 3) suggest some future uses that we may make of our knowledge of such properties; 4) introduce readers to the ideas and existence of the non-linear dielectric behaviour of living systems; and finally 5) suggest some novel means by which these nonlinear properties may be exploited in emerging electromagnetic devices.

## MEASUREMENT OF THE DIELECTRIC PROPERTIES OF LIVING SYSTEMS

As reviewed elsewhere extensively,[1,39,56,64,87,88,94,95,96,97,99] living systems possess dielectric (ie, passive electrical) properties very different from those generally found in inanimate matter. One of the chief characteristics of living systems, from an electrical point of view, is that they are typically rather conductive or 'lossy', ie, they contain a substantial concentration of small, mobile ions. **Conductivity** ($\sigma'$) is a measure of the ease with which free charges can migrate through the material under the influence of an electrical field. By contrast, if an exogenous electrical field can induce or modulate significant charge separations (ie, polarisations), the material will have a high **permittivity** (ε').

Except for unusual cases (such as nerve axons, not discussed here) in which inductive reactances and negative resistances are present,[3,21,31,32,111] the passive electrical properties of living systems are completely characterised by their frequency-dependent conductivity and permittivity. These are related to the macroscopic conductance and capacitance of the material held between two or more electrodes by a geometric factor, the cell constant (which has units of $cm^{-1}$). The vector sum of the conductance and the (angular frequency times the) capacitance is known as the admittance. Thus, to determine the dielectric properties of cell suspensions at frequencies up to say 30 MHz or so, one applies a sinusoidally modulated current and measures the resulting voltage (at the frequency of excitation) and phase angle (Fig. 1 a and b).

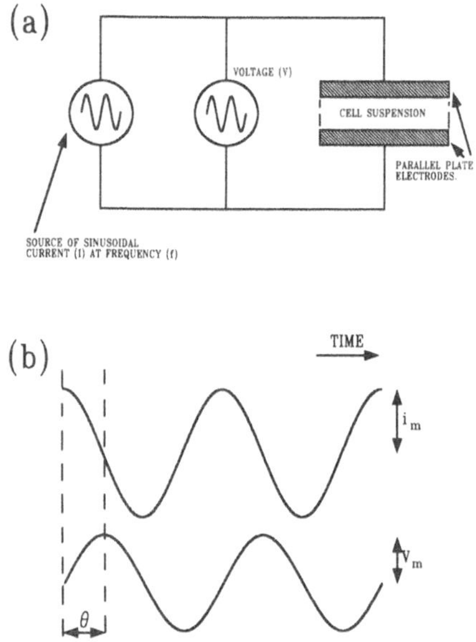


**Figure 1.** A dielectric measurement, in which a sinusoidally modulated AC current is applied to the system of interest, in this case a cell suspension. The result of this is that a voltage of the same frequency, phase shifted by an amount (θ radians) which reflects the macroscopic capacitance and conductance of the suspension, is generated across the system. For a pure capacitor the current leads to voltage by $\pi/2$ radians, whilst for a pure resistor the current and voltage are exactly in phase. (a) generalised measurement system. (b) waveforms of voltage and current: the overall admittance is the ratio $\mathbf{i_m / V_m}$.

A particular disadvantage of this arrangement is that at 'low' frequencies (< 5 MHz) there is a significant contribution to the measured capacitance and conductance from reactions occurring at the electrode-solution interfaces, a phenomenon usually referred to as "electrode polarisation." To minimise this problem, one may use a 4-terminal system in which the outer electrodes are the source of current, whilst the inner pair of electrodes are used to determine the voltage drop. By connecting the inner electrodes to a voltmeter of high input impedance, one may ensure that electrode polarisation phenomena (which are caused by current flow across the electrode-solution interfaces) do not contribute to the voltage drop (and hence admittance) measured.[56,62,98]

A. Dielectric Dispersions

When one measures the permittivity and conductivity of a cell suspension, it is found that permittivity falls and conductivity rises as the frequency is increased, in a series of steps known as dispersions (Fig. 2a). In the frequency range of interest (say DC-100 MHz), two or three main dispersions may usually be discerned, known respectively as the α-, β- and δ-dispersions (Fig. 2b).

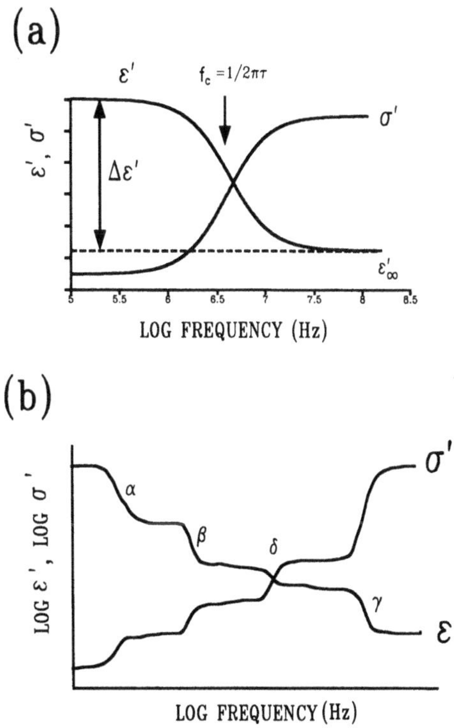


**Figure 2.** Dielectric dispersions. (a) In a dielectric dispersion, the permittivity falls and the conductivity rises as the frequency of measurement is increased. The dielectric increment (Δε') describes the magnitude of the change in permittivity, whilst the characteristic frequency ($f_c$) is the frequency at which the change is half-completed. Similarly one may describe a conductivity increment.

(b) The classical dispersions typically observed in living tissues. The frequency range and dielectric increments are somewhat arbitrary, but the characteristic frequencies and dielectric increments are respectively of the order of 1 kHz and $10^8$ (α), 1 MHz and $10^4$ (β), 100 MHz and 20 (δ) and 10 GHz and 70 (γ). The dispersions are less sharp than those of a single Debye dispersion.

The δ-dispersion is due mainly to the rotation of the side-chains of amino acids in proteins and of bound water molecules (whilst the γ-dispersion, occurring at somewhat higher frequencies still, is dominated by the rotation of the dipoles of 'free' water molecules).[39] Whilst our present interest is focussed upon the α- and β-dispersions occurring at the lower audio and radio frequencies, it is convenient first to consider the simplest type of dielectric dispersion, that due to dipole rotation.[25]

1. The Rotating Dipole

Figure 3 shows the simplest type of molecule suitable for our present purposes: the dipolar billiard ball. Such a molecule contains a permanent dipole moment due to the fact that it has two (or more) charges of opposite sign separated in space. If the (two) charges are of magnitude **+q** and **-q** Coulombs, and they are separated by a distance **s** metres, the molecule has a permanent dipole moment of **m = qs (in C.m)**. Though not an SI Unit, dipole moments are often quoted in Debye units, where **1 D = 3.33 x $10^{-30}$ Cm**, and it is convenient to note that the displacement of 1 electronic charge through $10^{-10}$ m gives a dipole moment of 4.8D, ie, 1 "charge-Ångstrøm". Pure water has a dipole moment of some 1.8D, whilst typical proteins have dipole moments of some hundreds of Debyes. As one may expect, the dielectric increments observed, ie, the change in permittivity as one passes between two plateau regions as one increases the frequency, is related both to the intrinsic dipole moment of the relaxing dipoles and to their concentration. The frequency (in Hz) at which the above transition is half-completed is known as the characteristic frequency (**$f_c$**) and is related to the relaxation time (τ) by the relation **$f_c = 1/(2\pi\tau)$** (Fig. 2a). The relaxation times to be expected for the rotation of a molecular dipole are those for the spheroidal molecules whose rotation is opposed by frictional interaction with the surrounding viscous medium. Thus in this case the energy that the system absorbs by the field is transduced into heat (by virtue of the frictional interactions between the rotating dipole and the solvent). We shall find later in this paper that this type of mechanism is in contrast to certain other mechanisms by which biological systems may interact with exogenous fields.

In the situations considered here, only a miniscule fraction of the charges and dipoles in the ensemble present are effectively moving in response to the applied field, such that doubling the voltage (field) doubles the current flowing such that their vector ratio, the admittance, is voltage-independent. This is why the electrical properties referred to here are called 'passive' or 'linear'. Nonlinear properties are discussed later.

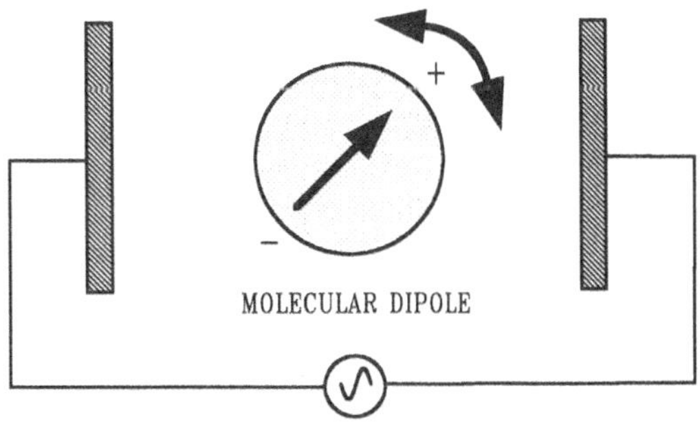


**Figure 3**. The rotating dipolar billiard ball. In this model system of dielectric relaxation, the system of interest is a hard sphere possessing unit charges at opposite poles. The alternating nature of the electrical field means that the dipole seeks to rotate to an orientation of minimum energy with respect to the field.

If we ask what sort of frequency ranges are characteristic of this sort of dipole rotation, we find that for water rotating in water, $f_c = (2\pi\tau)^{-1}$ is approximately 25 GHz at room temperature[39,46,87] whilst concentrated protein solutions have an observable dispersion due to dipolar rotation centred at 1 MHz or so.[85,109,110] In practice this is essentially invisible, since it is dominated by the β -dispersion typically occurring in this frequency range.

2. The β–Dispersion

Biological membranes have conductances of the order of perhaps $10^{-3}$ mS.cm$^{-2}$ and may be regarded (with respect to the extracellular and intracellular phases) as essentially nonconductors. On each side of this insulator are conducting ionic solutions (cell cytoplasm and suspending medium) and so a cell membrane is analogous to a classical electric capacitor. This means that when an exciting electrical voltage is applied across a cell suspension, the membrane capacitance ($C_m$) is charged up by ions moving under the influence of the electrical field. Because of the essentially nonconducting nature of the membrane, the membrane has the effect (Fig. 4) of strongly amplifying the exciting field generated between the electrodes.[113]

However, as the frequency rises, fewer and fewer ions have time to charge up the membrane(s) before the field changes direction. Thus the electrical charge stored by the suspension for a given exciting voltage falls, and the capacitance (permittivity) of the suspension drops. At low frequencies, the admittance (ie, the conductance to alternating current) of the cell membranes is very low, such that they behave as nonconductors suspended in a conducting medium and most of the current flowing in the suspension must flow round the cells. As the frequency increases, the membrane admittance rises and an increasing proportion

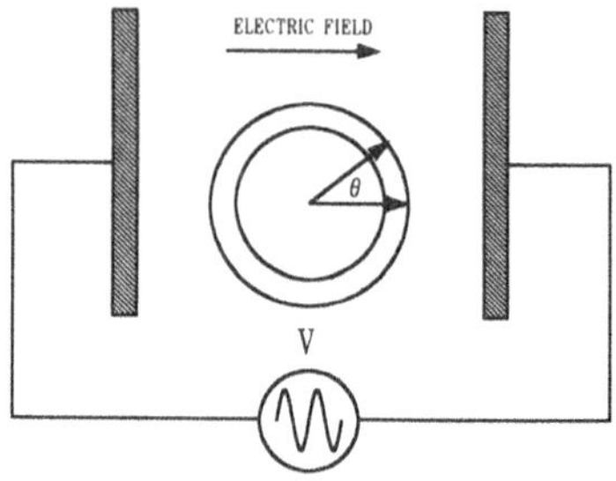


**Figure 4.** Amplification by a spherical shell membrane of an exogenous electrical field. In the present case, the (maximum) exogenous field strength $\mathbf{E_0}$ (neglecting electrode polarisation) experienced by the spherical shell is $\mathbf{V_{(max)}}$ **/d**, where **d** is the distance between the electrodes (typically of the order of 1 cm). At the moment shown, the right-hand electrode is negative. However, the potential induced across the membrane is $\psi_m = 1.5rE_0 \cos\theta / [1 + (f/f_c)^2]^{1/2}$, where **r** is the sphere's radius and **f** and $\mathbf{f_c}$ respectively the excitation frequency and the characteristic frequency of the β-dispersion. This potential will superimpose vectorially upon any pre-existing transmembrane potential [61]. The field strength across the membrane is then $\psi_m$ divided by the thickness of the membrane (typically 5 nm). Thus the membrane amplifies the field by an amount depending in particular upon the cell radius and the frequency of the field.

of the current can flow through the membrane and via the conductive cytoplasm of the cells. Thus the conductivity of the suspension increases. These features are illustrated in Fig. 5, whilst an electrical equivalent circuit for the shell membrane system characteristic of cell suspensions is given in Fig. 6, and a more complete overview of the β-dispersion.[23,64]

Since only living cells have these properties (in that cells with leaky membranes are dead or moribund,[43] we have been able to exploit this fact to devise a biomass probe for use in laboratory and industrial fermentations, based on the 4-terminal measurement of the RF dielectric properties of fermentor broths.[24,45,66,67] This device, the βugmeter (Fig. 7), is now being produced commercially. Most recently, in addition to its use in fermentor broths, we and others have shown that it may be used 1) in toxicological studies to follow cell death subsequent to a challenge with toxic xenobiotics,[104] 2) in the control of yeast pitching in breweries,[14] and 3) in solid-substrate fermentations by the mould Rhizopus oligosporus used in the production of Tempeh (personal communication Davey, Penaloza, Kell, Hedger).

### 3. The α-Dispersion

Cell surfaces are normally negatively charged, the cell surface charge density depending upon the cell type.[9,75] The presence of this cell surface

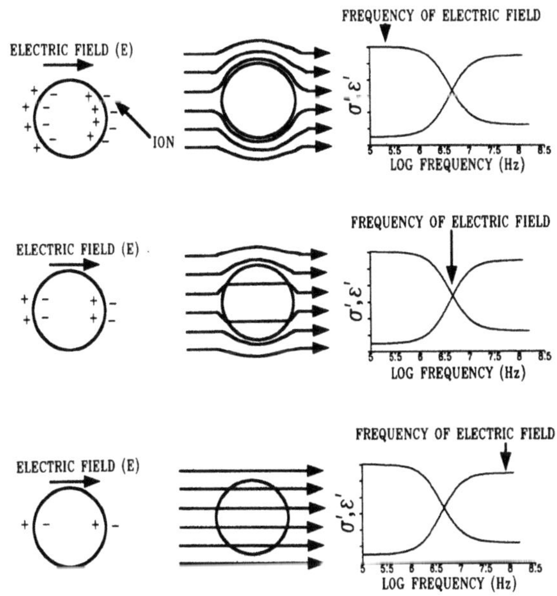


**Figure 5.** Effect of frequency upon the polarisation, current flow and dielectric properties of spherical shell suspensions. The frequency increases from top to bottom. It is assumed that the electrodes are to the left and right of the cells illustrated, and are observed at an instant when the right-hand electrode is negative. The left-hand portion shows the relative polarisation of the plasma membrane, the middle portion the flow of current around and/or through the cells, and the right-hand portion the approximate frequency (relative to $\mathbf{f_c}$) at which the behaviour indicated would be observed. For further discussion, see the text.

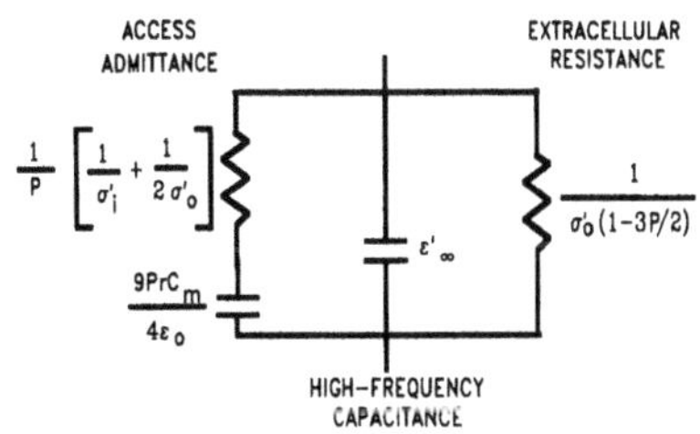


**Figure 6.** An electrical equivalent circuit often used to describe the β-dispersion. The values of the components are dimensionally inaccurate but are used to illustrate the physical contributions of the different subsystems. In the absence of cells, the 2 right-hand components are the only ones present, whilst the membrane capacitance (lower left) is charged up via the 'access admittance' consisting of internal and external conductivities. The symbols are for volume fraction (**P**), internal ( $\sigma_i'$ )and external ($\sigma_0'$ ) conductivities, the cell radius (r) and membrane capacitance per unit area ($\mathbf{C_m}$ ). $\varepsilon_0$ is the permittivity of free space.

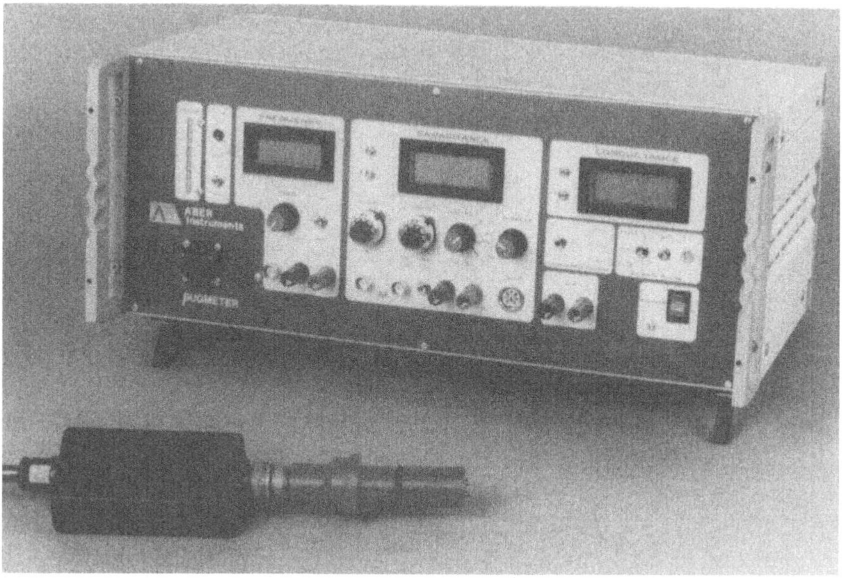

**Figure 7.** The βugmeter, an instrument for the real-time estimation of cellular biomass and viability, based upon the measurement of the radio frequency dielectric properties ($\Delta\varepsilon'$ for the β-dispersion). The probe is a standard 25 mm fermentor probe. Figure courtesy of Aber Instruments.

charge results in a diffuse counterion layer around the cells. When an electric field is applied to this cell, the counterions move tangentially along the cell surface, so that an induced dipole is formed along the length of the cell (Fig. 8). As it takes a finite time for the counterions on the cell surfaces to reach the ends of the cell, the number doing so, and hence the measured capacitance, will again increase with decreasing frequency. As before, the conductivity rises with increasing frequency. This is because the energy in the exciting field must either be stored (as reflected in $\varepsilon'$) or dissipated (as reflected in $\sigma'$), and therefore for a linear system a permittivity fall must be accompanied by a rise in conductivity (the Kronig-Kramers relationship). It is not improbable that other factors, such as the field-induced gating of transmembrane ion transfers, may also be involved in the α-dispersion, and it is certainly true to say that the classical explanations[28,84] do not alone account for the independence of the magnitude of the α-dispersion from the number and valency of counterions in bacterial chromatophores.[55] None the less, the low-frequency dielectric properties of extended, charged macromolecules such as DNA are caused by counterion relaxation mechanisms of this type.[76,112] Further experimental developments in this area would benefit from the adoption of simple experimental systems such as HPLC resins.[73]

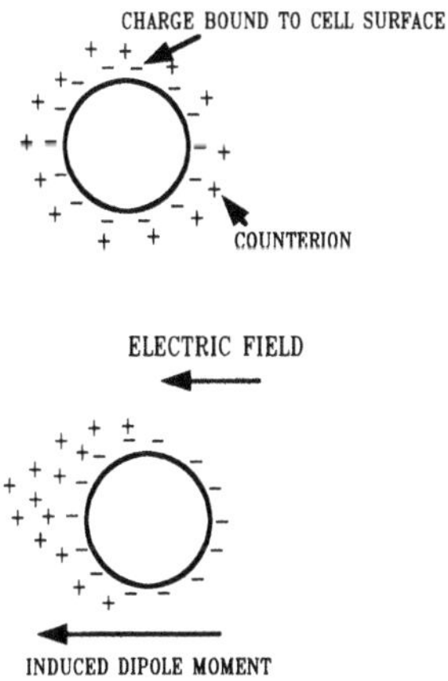


**Figure 8.** The production of a field-induced dipole along the length of a cell as a mechanism of dielectric dispersion. The upper half of the figure illustrates the counterion distribution in the absence of the field, whilst the lower part indicates what happens when a (low-frequency) electric field is present.

4. Other Dispersions

In essence, any field-induced relaxation or motion of a charge or dipole will result in a dielectric dispersion, those of particular interest including the passage of 'gating' charges,[3] the lateral motions of charged components in the plane of biological membranes (the μ-dispersion[44,55,63,64]) and the hydration-dependent changes in flexibility and in protonic conductivity exhibited by protein colloids (for lysozyme see[12,13,17,30]). Space does not permit a detailed discussion of these, save to mention that they will always be present and are likely at least to contribute to some of the more classical dielectric dispersions observed in tissues.

B. The Effect of a Distribution of Relaxation Times (τ)

The electrical properties of individual cells in a suspension are not identical. This means that the distribution of cell sizes (and of other properties) inevitably present must result in a distribution of relaxation times, since the τ value of the β- dispersion is proportional to the cell radius (**r**), while that of the α-dispersion is proportional to $\mathbf{r}^2$. If such a distribution in τ (and hence $\mathbf{f_c}$) values exists then the fall of permittivity and rise in conductivity with increasing frequency will be less steep than in the case of a system exhibiting but a single relaxation time. These effects are normally discussed in terms of an empirical parameter, the Cole-Cole α[22] (Fig. 9), which may take a value between 0 (no distribution of relaxation times) and 1 (infinite distribution), though there are many reasons to doubt that this apparent spread of relaxation times is ascribable solely to a spread in the properties of non-interacting subsystems.[26,27,49,50,54,59,63,64,70,96,119] Notwithstanding, and as pointed out by Schwan[96] an enormous number of possible distributions of relaxation times will give behaviour that is experimentally indistinguishable from 'true' Cole-Cole

**Figure 9**. The effect of the magnitude of the Cole-Cole α in smoothing out a dielectric dispersion. The curves are for changes in α in steps of 0.1.

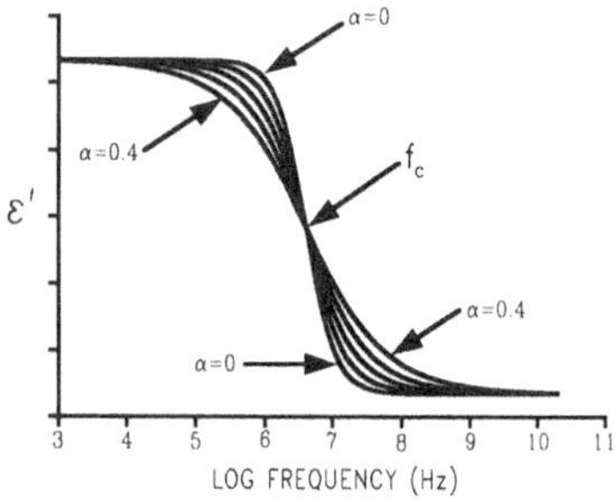


behaviour; in view of this most investigators use the Cole-Cole formalism for the analysis and condensed description of their data.

## DECONVOLUTING DIELECTRIC SPECTRA

Especially given the above-mentioned spread of apparent relaxation times, one of the major problems encountered in dielectric studies is that each dispersion can take several decades of frequency to run to completion. This means that if two dispersive mechanisms occur in the same frequency range, as is generally the case with the α- and β-dispersions, they will seriously overlap each other. A further problem encountered at low frequencies, especially when 2-terminal approaches are used, is that the admittance due to reactions occurring at the electrodes can dominate measurements in the sub-MHz region. To deconvolute such data one must minimise the contribution due to this electrode polarisation; in our experience[23] this can be done by subtracting the polarisation control data (obtained in a cell-free medium of identical low-frequency conductivity) from the actual cell data. Once this compensation has been carried out one fits the Cole-Cole equation for two dispersions to the resulting data points. Free parameters in the fit are the dielectric increments, the characteristic frequencies and the Cole-Cole α for the high- and low-frequency dispersions, and the permittivity at frequencies that are high relative to the frequencies of interest.

We have devised two computer programs (for use with IBM-PCs and compatibles) to deconvolute permittivity data. The first program (COLE.WKS) is written on a LOTUS 1-2-3 -type spreadsheet (VP-Planner) and allows us first to correct the data for polarisation. Estimated values for the variables of interest may then be entered and the computer then uses them to calculate the locus of the points generated, in the form of a 'fit' to the data. Graphs of the permittivity data and the fit can then be presented on the computer screen in various ways (eg, of log permittivity vs. log frequency (**f**), permittivity vs. log **f**, permittivity and conductivity vs. log **f**, error vs. log **f**, imaginary vs. real part of the permittivity) by using the function keys. Our experience is that manual iteration, comparing the effects of changes in the estimates on the goodness of fit as judged by the overall (modulus of the) percentage error, allows one rapidly to obtain an excellent fit to the data of interest.

Alternatively, the data compensated for polarisation may then be downloaded into a BASIC program (COLE.BAS) together with the estimates of the fit. The data are then fitted iteratively to (the double) Cole-Cole equation by the program, using the initial estimates. Because of the bias towards the low-frequency end, where the permittivity can become very high, the fit is not a nonlinear least-squares type of fit, as used for instance by MacDonald[74] and Grant,[39] but, as in the program COLE.WKS, is judged by the overall modulus of the percentage error in the permittivity domain. This program has been run using both an interpreted BASIC (GWBASIC) and a compiled BASIC (Borland Turbo BASIC), the latter running some 20-fold faster, especially in PCs possessing a maths coprocessor (80x87) chip. Together, these programs provide a convenient and accurate means of registering the linear dielectric properties of biological systems and discerning the mechanisms underlying such properties. Figure 10 gives an example of the type of fit that one may obtain, using experimental data from a suspension of erythrocytes.

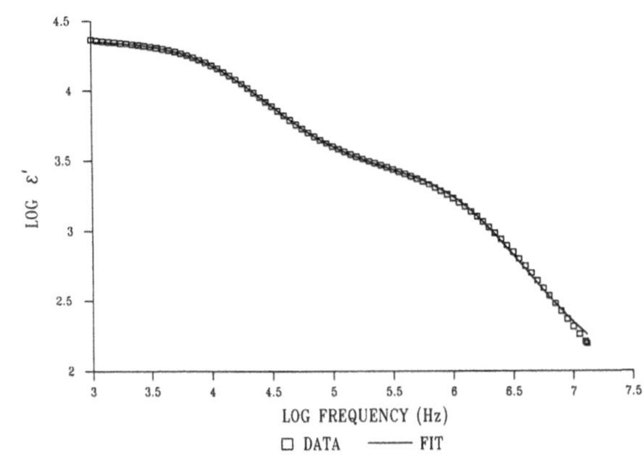


**Figure 10.** Fitting of dielectric data to 2 dispersions using the programs COLE.WKS and COLE.BAS. Dielectric data were obtained using an HP 4192A Impedance Analyser, as described by Harris and Kell (1983). Human erythrocytes were obtained locally, washed three times and resuspended in 1 mM NaCl, 290 mM sorbitol, 10 mM tris chloride, pH 6. The 84 data points are illustrated and the fit to the double Cole equation with the parameters $\varepsilon_{inf}$ = 70, $\Delta\varepsilon'_{high}$ = 2400, $f_{c(high)}$ = 1.41 MHz, $\alpha_{high}$ = 0.11, $\Delta\varepsilon'_{low}$ = 20900, $f_{c(low)}$ = 14 kHz, $\alpha_{low}$ = 0.16. The 'low' and 'high' frequency dispersions correspond to the α- and β-dispersions respectively.

## EXPLOITATION OF LINEAR DIELECTRIC MEASUREMENTS IN MEDICAL TECHNOLOGY

The electronic biomass probe that we have devised for the registration of the biomass in fermentors has been alluded to above. The present instruments determine the dielectric properties of the suspension of interest in the frequency range 0.1 - 10 MHz, and use the permittivity value obtained at an appropriate frequency to estimate the amount (and in favourable cases the nature) of the biomass present. It is important to note that these properties correlate with viable biomass, and not simply with cell number.[14,104]

Given that, as has been known for many years, the dielectric properties of living systems at both audio and radio frequencies change dramatically after death,[96,106,107,108] it seems reasonable to propose that the measurement of such properties might provide a novel, non-invasive and useful approach to the estimation of the time of death of a subject in forensic medicine, an estimation which is still subject to many uncertainties.

The distinction between life and non-life is of course but one extreme subset of the variety of physiological states that a tissue may adopt. Diagnostic techniques that rely on dielectric measurements[40,52] include impedance plethysmography[4,15,48,80,86,121] and pneumography[7,47] and electrical impedance tomography.[5,8,29,81,82,92] Various pathological states of muscle may also be accompanied by changes in dielectric properties.[71] Further, the electrical impedance of the meridians, and especially the needle points, recognised in the science of acupuncture are significantly lower than that of the surrounding tissue,[11,53,93] a fact that seems more than coincidental. The dielectric properties at microwave frequencies of tumours and of adipose tissue are greatly different from those of other tissues,[103] such that we may expect to be able to distinguish the former in the presence of the latter. The hydration-dependencies of the microwave dielectric properties of cells are discussed from a more biological point of view by Clegg et al.[19,20]

More generally, the interaction of electromagnetic energy with tissues is also of importance in RF and microwave hyperthermia,[41,105] in the gentle thawing of cryogenically preserved tissue,[16] in the use of pulsed electromagnetic fields to aid tissue and bone regeneration and healing.[10,11,89] Measurement of this interaction, additional to the simple application of these electrical fields, can only improve the quality and reproducibility of these regimes. From this point of view, 'dielectric diagnosis' may (and may be expected to) be exploited in a variety of existing or projected biosensing devices.[56,57,62]

## NONLINEAR INTERACTIONS OF CELLS WITH ELECTRICAL FIELDS

We saw above that the simplest type of dipolar billiard ball rotation resulted ultimately in the simple transduction of exogenous electrical energy via frictional forces into heat. However, based on recent work of Tsong and colleagues,[100,101] Westerhoff, Astumian and co-workers have shown that the properties required of an enzyme to 'harvest' energy from an exogenous electrical field are common to all enzymes, viz. the possession of conformational states which possess different dipole moments and which interconvert hierarchically in a fashion that is coupled to their chemical environment.[6,114,115,118,119,120] Importantly, it was shown that such energy converters must act nonlinearly, and it was proposed that this should be visible as a conversion of the frequency of an exciting electrical field to another frequency.[59,70,119] This would be true at field strengths in which traditional measurements (of a voltage-independent impedance) would suggest that a purely linear system was being observed.

In essence, this turns out to be a case of rose-coloured spectacles: because of the fact that impedimetric devices assume linearity, such measuring systems are normally so organised (electronically) that they reject currents at frequencies other than that of the exciting voltage. Thus a system may appear linear (in that the observable current is linear with the exciting voltage) but be nonlinear (in that the system causes what may be a constant fraction of the exciting voltage to be tranformed into currents at other than the exciting frequency). We discuss one type of system explicitly.

## A 4-STATE ENZYME CAPABLE OF HARVESTING ELECTRICAL ENERGY FOR THE PERFORMANCE OF USEFUL CHEMICAL WORK

We note (and would stress) that the average potential in a sinusoidally modulated field is zero, that in most cases where it interacts with an aqueous system, it merely produces heat under macroscopically isothermal conditions. However, it certainly constitutes a source of free energy, as is clear from its action as such across every electrical power point or wall socket. To provide a simple illustration of the properties of a system which does not merely turn this electric field energy into heat and which can therefore conserve this free energy as useful (electro)chemical work,[61,116] we consider (Fig. 10) an enzyme with the following properties. We assume that the protein is a membrane-located pump (which does not therefore rotate), capable of transporting an uncharged molecule (S) which is present at a higher concentration inside the cell than that outside in a direction oppposite to its (electro)chemical potential.

The protein possesses a negatively charged binding site for S which can flip-flop between the inner and outer surfaces of the membrane. (Of course it does not cross the whole membrane, merely the region where the potential drop is the greatest.) The protein possesses, and cycles between, 4 conformational states (1 to 4), which represent the combination of bound or non-bound substrate with the negatively charged binding site facing inwards or outwards. States 4 and 2 have lower basic free energies[48,120] than states 1 and 3, and thus are more stable (ie, highly populated). Cycling of the protein in a clockwise direction would have the effect of 'pumping' S against its chemical potential, such that to effect this an exogenous source of free energy is necessary. In the absence of any field, therefore, we assume that, because of a higher affinity for its substrate than any of the other states, state 4 is the most highly populated (Fig. 11a) in the protein ensemble.

If we apply a low-frequency alternating field (whose frequency $\mathbf{f} < \mathbf{f_c}$ for the β-dispersion), during the first half cycle it will be oriented (say) in a direction that attracts the negatively charged binding site towards the inner face of the membrane (Fig. 11b). This effectively causes transitions from states 4 --> 1 and state 3 --> 2. Thus state 1 has become the most highly populated state and there will have been a net translocation of S from outside to inside. However, as state 2 also has its negative charge in the energetically favoured inner position, and as it has a lower free energy than state 1, there will be a re-equilibration between (the populations in) states 1 and 2 (Fig. 11c). Thus state 2 becomes more highly populated than the other states, and because the transition 1 --> 2 releases S on the inside, there will be a net release of S to the cytoplasm.

When the second half of the ac field is applied, (ie, the applied electric field is now of the opposite polarity), the favoured position of the negatively charged binding site is now at the outer face of the membrane. This causes transitions from 2 --> 3 and from 1 --> 4. The immediate result now is that state 3 becomes more populated than the other states (Fig. 11d). In other words, there has been a net movement of empty binding sites from the inside to the outside of the membrane. However, since state 4 also has its negative charge in the energetically more favoured outside position, and has a lower free energy than does state 3, there is another re-equilibration (Fig. 11e) in the direction 3 --> 4.

The net result of this is that, from the protein's point of view, we have returned to our starting position (Fig. 11a). In a sense this is to be expected, since the net potential of the field was indeed zero. Yet despite this fact the field has done work on the system, since the clockwise cycling of the protein caused it to pump S against its electrochemical gradient and under macroscopically

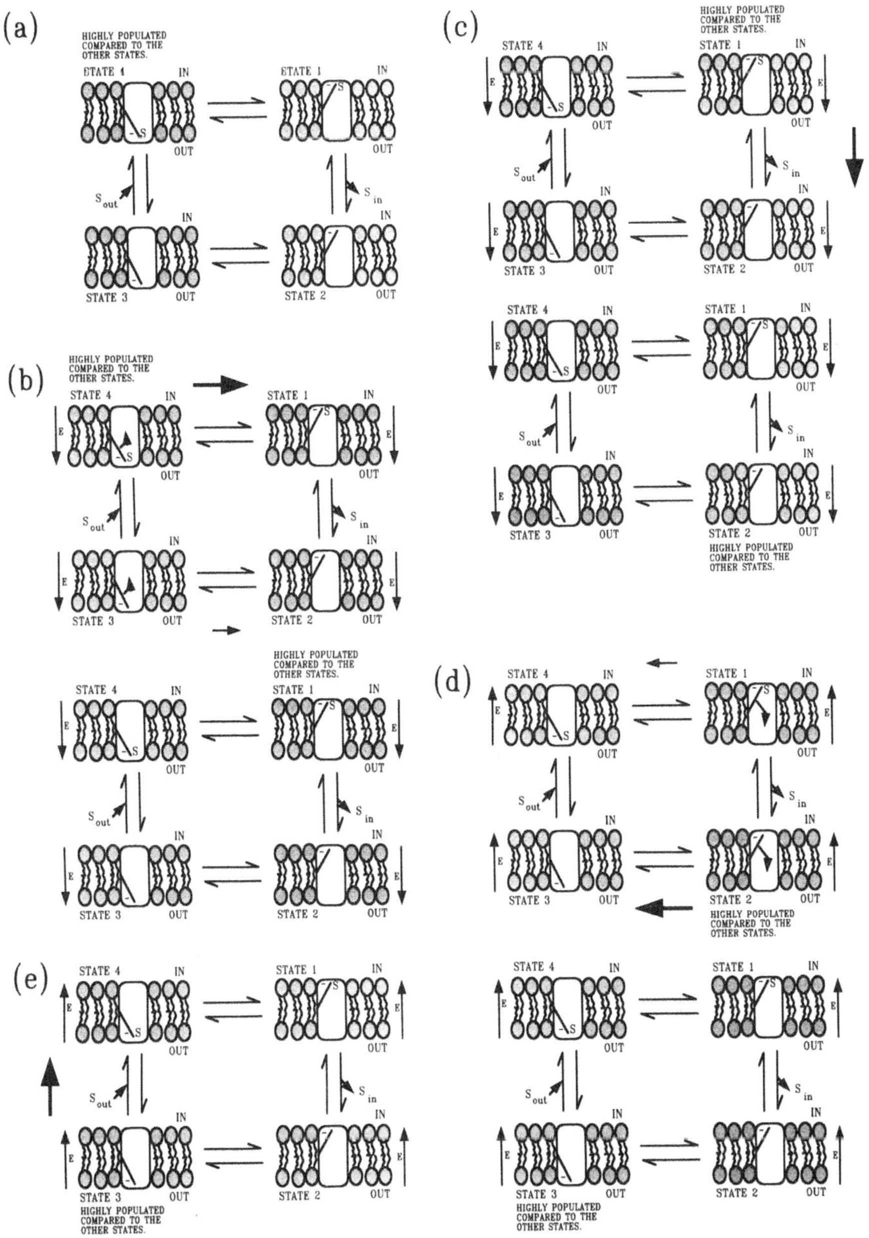


**Figure 11.** A 4-state enzyme capable of transducing electrical energy in the form of a non-stationary electrical field into useful (electro)chemical work. For discussion, see the text.

isothermal conditions. This does not violate the Second Law,[116] but merely indicates that proteins are not simple dipolar billiard balls.

Clearly, the properties required of our model protein here are common to all enzymes, viz. the possession of conformational states of different dipole moments which are coupled to each other in a hierarchical fashion, which possess different free energies and the transitions between which are coupled to electrochemical reactions. Whilst this does not of itself indicate which enzymes one should seek to consider as the most likely or suitable 'targets' for low-energy exogenous electrical fields, the metabolic control analysis[68,117] indicates that there is rarely a unique target. The amplifying effects of membranes on electrical fields suggest that initial attention might most fruitfully be directed at membranous ones. Similarly, the fact that the exogenous field causes the re-equilibration of protein conformational states which have different dipole moments indicates that one should expect to see field-induced currents at frequencies related to the (pseudo-first-order) rate constants of the normal rate constants for such transitions.

On the assumption that the fluctuation-dissipation theorem has at least some validity in this type of system,[61] these frequencies would mainly in fact be the inverse of the pseudo-first-order rate constants. In this regard, we may mention that harmonics have been observed experimentally (under conditions in which the fundamental appears linear) in artificial polymers[36,37] and in nerve axons,[79] consistent with the arguments developed herein and elsewhere.[70,119] Whilst we can at this stage say no more about the appropriate frequencies, we may state that the design of optimal waveforms for interacting with particular targets does not in principle differ from the design of drugs aimed at selective interaction with appropriate targets or receptors.[58] What sort of considerations might we apply?

## DIELECTRIC FINGERPRINTING

The answer to the previous question is to be found in the opening quotation, and requires that we jettison the idea that the dielectric properties of living systems, even at 'low' field strengths, are linear *sensu stricto*. This means that we need properly to characterise the nonlinear dielectric properties of target enzymes and tissues. The idea behind the dielectric fingerprinting approach[56] is that because of the properties of real enzymes alluded to above, and their modulation by interaction with ligands, the imposition of a high-strength field will cause time- and field-dependent changes in the dielectric properties of the target enzyme (as measured at 'low' field strengths), which may therefore be used to construct a time-, field- and frequency-dependent dielectric spectrum or 'dielectric fingerprint' of the target of interest. One way of viewing this approach, is to regard its relationship to conventional (linear) dielectric spectroscopy as in some ways analogous to that between 2-dimensional and 1-dimensional NMR spectroscopies. We are presently developing the hardware necessary for the construction of such a system. Thus, and as fore-shadowed in the opening statement (although one might well argue (and without doubt we shall soon do so) whether the word 'passive' is wholly appropriate), one should then be able to base the design of a more specific waveform on the dielectric fingerprint of the target enzyme or tissue.

## AN OVERVIEW OF FUTURE USES

From the foregoing, we may anticipate future developments in this area to lie in 3 directions in particular:

1). **Tissue Diagnostics.** Here it is easy to envisage (linear) RF-dielectric probes for studying the physiological state of tissues, including, for instance, their adiposity, and, on a somewhat macabre note, the possibility of a means for estimating the time after death. Non-invasive 'probe' methods[16,38] provide a particularly convenient approach.

2). **Biosensing.** Especially in the nonlinear regime, the possibility of studying the dynamics of proteins and of DNA by dielectric means, and how they may be modulated by interactions with ligands, provides a novel and powerful generic biosensing technology.

3). **Therapeutics.** If fields can affect enzymes and cells, there is no reason of principle why one should not expect to be able to tailor a waveform as a therapeutic agent in much the same way as one now modulates chemical structures to obtain pharmacological selectivity,[58] and perhaps without many of the side-effects common to pharmaceutical substances.

## ACKNOWLEDGMENTS

We are grateful to the Biotechnology Directorate of the Science and Engineering Research Council, U.K., the Wolfson Foundation and Aber Instruments for financial support. DBK wishes to acknowledge many stimulating discussions of these and other topics with Dean Astumian and Hans Westerhoff.

## References

1. Adey WR. Tissue interactions with non-ionizing electromagnetic fields. Physiol Rev 1981; 61:435-514.

2. Adey WR, Lawrence AF (Eds.). Non-linear Electrodynamics of Biological Systems. Plenum Press, New York, 1984.

3. Almers W. Gating currents and charge movements in excitable membranes. Rev Physiol Biochem Pharmacol 1978; 82:96-190.

4. Anderson FA, Jr. Impedance plethysmography in the diagnosis of arterial and venous disease. Ann Biomed Eng 1984; 12:79-102.

5. Anon. Electrical impedance tomography-applied potential tomography: Bibliography. Clin Phys Physiol Meas 1987; 8,Suppl. A:181-184.

6. Astumian RD, Chock PB, Tsong TY, Chen Y, Westerhoff HV. Can free energy be transduced from electric noise? Proc Natl Acad Sci USA 1987; 84:434-438.

7. Baker LE. Electrical impedance plethysmography. In: Non-invasive Physiological Measurements (Ed. P Rolfe). Academic Press, London, 1979.

8. Barber DC, Brown BH. Applied potential tomography. J Phys E, Sci Instr 1984; 17:723-733.

9. Barber J. Influence of surface charges on thylakoid structure and function. Ann Rev Pl Physiol 1982; 33:261-295.

10. Barker AT, Lunt MJ. The effects of pulsed magnetic fields of the type used in the stimulation of bone fracture healing. Clin Phys Physiol Meas 1983; 4:1-27.

11. Becker RO, Marino AA. Electromagnetism and Life. State University of New York Press, Albany, NY, 1982.

12. Bone S, Pethig R. Dielectric studies of protein hydration and hydration-induced flexibility. J Mol Biol 1985; 181:323-326.

13. Bone S. Electrostatic interactions in protein structures: the influence of water on enzyme fluctuations. J Electrostatics 1988; 21:245-256.

14. Boulton CA, Maryan PS, Loveridge D. The application of a novel biomass sensor to the control of yeast pitching rate. Proc European Brewing Convention Congress, Zürich (in press).

15. Brown BH. Tissue impedance methods. In: Imaging With Non-ionising Radiation (Ed. DF Jackson). Surrey University Press, Guildford, England. 1983; 85-110.

16. Burdette EC, Cain FL, Seals J. In vivo probe measurement technique for determining dielectric properties of VHF through microwave frequencies. IEEE Trans Microwave Theory Techn 1980; MTT-28:414-427.

17. Careri G, Giansanti A, Rupley JA. Proton percolation on hydrated lysozyme powders. Proc Natl Acad Sci 1986; 83:6810-6814.

18. Chiabrera A, Nicolini C, Schwan HP (Eds). Interactions Between Electromagnetic Fields and Cells. Plenum Press, New York, 1985.

19. Clegg JS, McClean VER, Szwarnowski S, Sheppard RJ. Microwave dielectric measurements (0.8 - 70 GHz) on Artemiarcysts at variable water content. Phys Med Biol 1984; 29:1409-1419.

20. Clegg JS, Szwarnowski S, McClean VER, Sheppard RJ, Grant EH. Interrelationships between water and cell metabolism in Artemia cysts. X: Microwave dielectric studies. Biochim Biophys Acta 1982; 721:458-468.

21. Cole KS. Membranes, Ions and Impulses. University of California Press, Berkeley, CA, 1972.

22. Cole KS, Cole RH. Dispersion and absorption in dielectrics. 1: Alternating current characteristics. J Chem Phys 1941; 9:341-351.

23. Davey CL, Kell DB. Low-frequency dielectric properties of cell suspensions. In: Electric Field Phenomena in Biological Systems (Ed. R Paris). IOP Short Meetings Ser, IOP, London. 1989; 21:51-62.

24. Davey CL, Kell DB, Kemp RB, Meredith RWJ. On the audio- and radio-frequency dielectric behaviour of anchorage-independent, mouse L929-derived LS fibroblasts. Bioelectrochem Bioenerg 1988; 20:83-98.

25. Debye P. Polar Molecules. Dover, New York, 1929.

26. Dissado LA, Hill RM. A cluster approach to imperfect materials and their relaxation spectroscopy. Proc R Soc 1983; A390:131-180.

27. Dissado LA, Hill RM. Small-signal ac frequency response functions. Solid-state Ionics 1987; 22:331-336.

28. Dukhin SS, Shilov VN. Dielectric Phenomena and the Double Layer in Disperse Systems and Polyelectrolytes. John Wiley, Chichester, England, 1974.

29. Eisenberg RS, Mathias RT. Structural analysis of electrical properties of cells and tissues. CRC Crit Rev Biophys Bioeng 1980; 4:203-232.

30. Finney JL, Poole PL. Protein hydration and enzyme activity: the role of hydration-induced conformation and dynamic changes in the activity of lysozyme. Comments Mol Cell Biophys 1984; 2:129-151.

31. Fishman HM. Relaxations, fluctuations and ion transfer across membranes. Progr Biophys Mol Biol 1985; 46:127-162.

32. Fishman HM, Moore LE, Poussart D. Squid axon K conduction: Admittance and noise during short- versus long-duration step clamps. In: The Biophysical Approach to Excitable Membranes (Eds. WJ Adelman, DE Goldman). Plenum Press, New York. 1981; 65-95.

33. Fröhlich H. The biological effects of microwaves and related questions. Adv Electron Electron Phys 1980; 53:85-152.

34. Fröhlich H (Ed.). Biological Coherence and Response to External Stimuli. Springer-Verlag, Heidelberg, 1988.

35. Fröhlich H, Kremer F (Eds.). Coherent Excitations in Biological Systems. Springer-Verlag, Heidelberg, 1983.

36. Furukawa T, Nakajima K, Koizumi T, Date M. Measurements of nonlinear dielectricity in ferroelectric polymers. Jap J Appl Phys 1987; 26:1039-1045.

37. Furukawa T, Tada M, Nakajima K, Seo I. Nonlinear dielectric relaxation in a vinylidene cyanide / vinyl acetate copolymer. Jap J Appl Phys 1988; 27:200-204.

38. Gabriel C, Grant EH, Young IR. Use of time-domain spectroscopy for measuring dielectric properties with a coaxial probe. J Phys E, Sci Instr 1986; 19:843-846.

39. Grant EH, Sheppard RJ, South GP. Dielectric Behaviour of Biological Molecules in Solution. Clarendon Press, Oxford, 1978.

40. Grant JP, Spyrou NM. Complex permittivity differences between normal and pathological tissues: Mechanisms and medical significance. J Bioelectricity 1985; 4:419-458.

41. Hahn G M. Hyperthermia and Cancer. Plenum Press, New York, 1982.

42. Harris CM, Kell DB. The radio-frequency dielectric properties of yeast cells measured with a rapid, automated, frequency-domain dielectric spectrometer. Bioelectrochem Bioenerg 1983; 11:15-28.

43. Harris CM, Kell DB. The estimation of microbial biomass. Biosensors 1985a; 1:17-84.

44. Harris CM, Kell DB. On the dielectrically observable consequences of the diffusional motions of lipids and proteins in membranes. 2: Experiments with microbial cells, protoplasts and membrane vesicles. Eur Biophys J 1985b; 13:11-24.

45. Harris CM, Todd RW, Bungard SJ, Lovitt RW, Morris JG, Kell DB. The dielectric permittivity of microbial suspensions at radio frequencies: A novel method for the real-time estimation of microbial biomass. Enzyme Microbial Technol 1987; 9:181-186.

46. Hasted JB. Aqueous Dielectrics. Chapman Hall, London, 1973.

47. Henderson RP, Webster JG. An impedance camera for spatially specific measurements of the thorax. IEEE Trans Biomed Eng 1978; BME-25:250-254.

48. Hill DW. The role of electrical impedance methods for the monitoring of central and peripheral blood flow changes. In: Non-invasive Physiological Measurements (Ed. P Rolfe). Academic Press, London, 1979.

49. Hill RM, Dissado LA. Debye and non-Debye relaxation. J Phys C: Solid-state Phys 1985; 18:3829-3836.

50. Hill RM, Jonscher AK. The dielectric behaviour of condensed matter and its many-body interpretation. Contemp Phys 1983; 24:75-110.

51. Illinger KH. (Ed.) Biological Effects of Non-ionizing Radiation. American Chemical Society, Washington, DC, 1981.

52. Iskander MF, Durney CH. Electromagnetic techniques for medical diagnosis: A review. Proc IEEE 1980; 68:126-132.

53. Jakoubek B, Rohlicek V. Changes of electrodermal properties in the 'acupuncture points' in man and rats. Physiol Biochem 1982; 31:143-149.

54. Jonscher AK. Dielectric Relaxation in Solids. Chelsea Dielectrics Press, London, 1983.

55. Kell DB. Dielectric properties of bacterial chromatophores. Bioelectrochem Bioenerg 1983; 11:405-415.

56. Kell DB. The principles and potential of electrical admittance spectroscopy: An introduction. In: Biosensors: Fundamentals and Applications (Eds. APF Turner, I Karube, GS Wilson). Oxford University Press, Oxford. 1987; 429-470.

57. Kell DB. Non-thermally excited modes and free energy transduction in proteins and biological membranes. In: Energy Transfer Dynamics (Eds. TW Barrett, HA Pohl). Springer-Verlag, Heidelberg. 1987; 237-246.

58. Kell DB. Bioelectrochemical phenomena: their role and exploitation in science and technology. Univ Wales Rev Sci Technol 1987; 1:64-71.

59. Kell DB. Dielectric spectroscopy of biological systems. ISI Atlas of Science: Biochemistry 1988; 1:25-29.

60. Kell DB. Coherent properties of energy coupling membrane systems. In: Biological Coherence and Response to External Stimuli (Ed. H Fröhlich) Springer-Verlag, Heidelberg. 1988; 233-241.

61. Kell DB. Protonmotive energy-transducing systems: some physical principles and experimental approaches. In: Bacterial Energy Transduction (Ed. CJ Anthony). Academic Press, London. 1988; 429-490.

62. Kell DB, Davey CL. Conductimetric and impedimetric devices. In: Biosensors: A Practical Approach (Ed. AEG Cass). Oxford University Press, Oxford (in press).

63. Kell DB, Harris CM. On the dielectrically observable consequences of the diffusional motions of lipids and proteins in membranes. 1: Theory and overview. Eur Biophys J 1985; 12:181-197.

64. Kell DB, Harris CM. Dielectric spectroscopy and membrane organisation. J Bioelectricity 1985; 4:317-348.

65. Kell DB, Hitchens GD. Coherent properties of the membranous systems of electron transport phosphorylation. In: Coherent Excitations in Biological Systems (Eds. H Fröhlich, F Kremer) Springer-Verlag, Heidelberg. 1983; 178-198.

66. Kell DB, Samworth CM, Todd RW, Bungard SJ, Morris JG. Real-time estimation of microbial biomass during fermentations, using a dielectric probe. Studia Biophys 1987; 119:153-156.

67. Kell DB, Todd RW. Determination of Biomass. U.S. Patent 4,810,650. 1989.

68. Kell DB, van Dam K, Westerhoff HV. Control analysis of microbial growth and productivity. Symp Soc Gen Microbiol 1989; 44:61-93.

69. Kell DB, Westerhoff HV. Catalytic facilitation and membrane bioenergetics. In: Organized Multienzyme Systems: Catalytic Properties (Ed. GR Welch). Academic Press, New York. 1985; 63-139.

70. Kell DB, Astumian RD, Westerhoff HV. Mechanisms for the interactions between nonstationary electrical fields and biological systems. I. Linear dielectric theory and its limitations. Ferroelectrics 1988; 86:59-78.

71. Kleibel A, Pfützner H, Krause E. Measurement of dielectric loss factor: A routine method of recognising PSE muscle. Fleischwirtschaft 1983; 63:1183-1185.

72. Konig HL, Krueger AP, Lane S, Sonnig W. Biologic Effects of Environmental Electromagnetism. Springer-Verlag, Heidelberg, 1981.

73. Lim K, Franses EI. Electrical properties of aqueous dispersions of polymer microspheres. J Coll Interf Sci 1986; 110:201-213.

74. Macdonald JR. Impedance Spectroscopy. Wiley, New York, 1987.

75. McLaughlin S. Electrostatic potentials at membrane-solution interfaces. Curr Top Membr Trans 1977; 9:71-144.

76. Mandel M. Dielectric properties of charged linear macromolecules with particular reference to DNA. Ann NY Acad Sci 1977; 74:303-311.

77. Marino AA. (Ed.). Modern Bioelectricity. Marcel Dekker, New York, 1988.

78. Marino AA, Ray J. The Electric Wilderness. University of San Francisco Press, San Francisco, CA, 1986.

79. Miyamoto S, Fishman HM. $Na^+$ conductance kinetics in the low-frequency impedance of isolated snail neurons. IEEE Trans Biomed Eng 1986; BME-33:644-653.

80. Mohapatra SN. Non-invasive Cardiovascular Monitoring by Electrical Impedance Technique. Pitman Medical, London, 1981.

81. Murai T, Kagawa Y. Electrical impedance computed tomography based on a finite element model. IEEE Trans Biomed Eng 1985; BME-32:177-184.

82. Newell JC, Gisser DG, Isaacson D. An electric current tomograph IEEE Trans Biomed Eng 1988; 35:828-833.

83. Nuccitelli R (Ed.). Ionic Currents in Development. Alan R Liss, New York, 1986.

84. O'Brien RW. The response of a colloidal suspension to an alternating electrical field. Adv Coll Interf Sci 1982; 16:281- 320.

85. Oncley JL. The electrical moments and relaxation times of proteins as measured from their influence on the dielectric constant of solutions. In: Proteins, Amino Acids and Peptides (Eds. EJ Cohn, JT Edsall). Reinhold, New York. 1943; 543-568.

86. Penney DC. Theory and cardiac applications of electrical impedance measurements. CRC Crit Rev Biomed Eng 1986; 13:227-281.

87. Pethig R. Dielectric and Electronic Properties of Biological Materials. Wiley, Chichester, England, 1979.

88. Pethig R. Dielectric properties of biological materials: Biophysical and medical applications. IEEE Trans Electr Insul 1984; EI-19:453-474.

89. Pilla AA, Sechaud P, McLeod B. Electrochemical and electrical aspects of low-frequency electromagnetic current induction in biological systems. J Biol Phys 1983; 11:51-58.

90. Polk C, Postow E. (Eds.). CRC Handbook of Biological Effects of Electromagnetic Fields. CRC Press, Boca Raton, FL, 1986.

91. Presman AS. Electromagnetic Fields and Life. Plenum Press, New York, 1970.

92. Pryce LR. Electrical impedance tomography (ICT): A new CT imaging technique. IEEE Trans Nucl Sci 1979; NS-26:2736-2739.

93. Reichmanis M, Marino AA, Becker RO. Electrical correlates of acupuncture points. IEEE Trans Biomed Eng 1975; BME-22:533-535.

94. Salter C. Quantifying skin disease and healing in vivo using electrical impedance measurements. In: Non-invasive Physiological Measurements (Ed. P Rolfe). Academic Press, London. 1979; 1:21-68.

95. Schanne OF, Ceretti ERP. Impedance Measurements in Biological Cells. Wiley, New York, 1978.

96. Schwan HP. Electrical properties of tissue and cell suspensions. Adv Biol Med Phys 1957; 5:147-209.

97. Schwan HP. Determination of biological impedances. In: Physical Techniques in Biological Research (Ed. WL Nastuk). Academic Press, New York. 1963; VIB:323-407.

98. Schwan HP, Ferris CD. Four-electrode null techniques for impedance measurement with high resolution. Rev Sci Instr 1968; 39:481-485.

99. Schwan HP, Foster KR. RF-field interactions with biological systems: Electrical properties and biophysical mechanisms. Proc IEEE 1980; 68:104-113.

100. Sepersu EH, Tsong TY. Stimulation of an ouabain- sensitive $Rb^+$ uptake in human erythrocytes with an external electric field. J Membr.Biol 1983; 74:191-201.

101. Sepersu EH, Tsong TY. Activation of electrogenic $Rb^+$ transport of $(Na^+,K^+)$-ATPase by an electric field. J Biol Chem 1984; 259:7155-7162.

102. Sheppard AR, Eisenbud M. (Eds.). Biological Effects of Electric and Magnetic Fields of Extremely Low Frequency. New York University Press, New York, 1977.

103. Smith S, Foster KR. Dielectric properties of low-water-content tissues. Phys Med Biol 1985; 30:965-973.

104. Stoicheva NG, Davey CL, Markx GH, Kell DB. Dielectric spectroscopy: A rapid method for the determination of solvent biocompatibility during biotransformations. Biocatalysis 1989; 3 (in press).

105. Storm FK (Ed.). Hyperthermia in Cancer. GK Hall, Boston, MA, 1983.

106. Surowiec A, Stuchly SS, Swarup A. Radiofrequency dielectric properties of animal tissues as a function of time following death. Phys Med Biol 1985; 30:1131-1141.

107. Surowiec A, Stuchly SS, Swarup A. Postmortem changes of the dielectric properties of bovine tissue at low radiofrequencies. Bioelectromagnetics 1986; 7:31-43.

108. Swatland HJ. Postmortem changes in electrical capacitance and resistivity of pork. J Animal Sci 1980; 51:1108- 1112.

109. Takashima S. Dielectric properties of proteins. I. Dielectric relaxation. In: Physical Principles and Techniques of Protein Chemistry, Part A (Ed. AJ Leach). Academic Press, New York. 1969; 291-333.

110. Takashima S, Minakata A. Dielectric behaviour of biological macromolecules. In: Digest of Dielectric Literature. National Research Council, Washington, DC. 1975; 37:602-653.

111. Takashima S, Schwan HP. Passive electrical properties of squid axon membrane. J Membr Biol 1974; 17:51-68.

112. Takashima S, Gabriel C, Sheppard RJ, Grant EH. Dielectric behaviour of DNA solution at radio and microwave frequencies (at 20°C). Biophys J 1984; 46:29-34.

113. Tsong TY, Astumian RD. Absorption and conversion of electric field energy by membrane-bound ATPases. Bioelectrochem Bioenerg 1986; 15:457-476.

114. Tsong TY, Astumian RD. Electroconformational coupling and membrane function. Progr Biophys Mol Biol 1987; 50:1- 45.

115. Tsong TY, Astumian RD. Electroconformational coupling: How membrane-bound ATPase transduces energy from dynamic electric fields. Ann Rev Physiol 1988; 50:273-290.

116. Welch GR, Kell DB. Not just catalysts: The bioenergetics of molecular machines. In: The Fluctuating Enzyme (Ed. GR Welch). Wiley, New York. 1986; 451-492.

117. Westerhoff HV, van Dam K. Thermodynamics and Control of Biological Free-energy Transduction. Elsevier, Amsterdam, 1987.

118. Westerhoff HV, Tsong TY, Chock PB, Chen Y, Astumian RD. How enzymes can capture and transmit free energy from an oscillating electric field. Proc Natl Acad Sci USA 1986; 83:4734-4738.

119. Westerhoff HV, Astumian RD, Kell DB. Mechanisms for the interactions between nonstationary electrical fields and biological systems. II. Nonlinear dielectric theory and free-energy transduction. Ferroelectrics 1988; 86:79-101.

120. Westerhoff HV, Kell DB, Astumian RD. The dynamics of electrostatic interactions between membrane proteins. J Electrostatics 1988; 21:257-298.

121. Wheeler HB, Penney C. Impedance plethysmography: theoretical and experimental basis. In: Non-invasive Diagnostic Techniques in Vascular Disease (Ed. EF Bernstein). CV Mosby, St Louis. 1982; 104-116.

# EFFECTS OF PULSED ELECTROMAGNETIC FIELDS ON MEMBRANE TRANSPORT

W. De Loecker, N. Cheng and P.H. Delport

## METABOLIC EFFECTS OF PULSED ELECTROMAGNETIC FIELDS

The fundamental research into the mechanism of action [2,3,5,11,18,25,31] of electromagnetic field stimulation was initiated and stimulated by the successful clinical treatments of pseudarthrosis, non-union, bone fracture, and nerve and ligament regeneration. Furthermore, treatment with pulsed electromagnetic fields (PEMF) has facilitated the healing of skin ulcers and reduced the pain of hip prostheses while simultaneously stimulating bone growth.[26,41] Bone deposition in relation to orthodontic problems has been stimulated when animals were treated with electromagnetic fields.[20,45] It has been argued that, due to a reduced release of lysosomal enzymes observed in cultured fibroblasts, PEMF may favorably affect chronic inflammatory processes.[34] In specific circumstances PEMF may affect the activity of enzyme systems as observed in murine melanoma cells.[27]

### A. Effects on Membrane Transport

Modifications in membrane transport activity by which movements of molecules and ions across the cell membrane are affected lead to alterations in metabolic activity. Small changes in ionic fluxes may result in unexpected and disturbing metabolic effects.[7] Electric fields are observed to decrease calcium release from cerebral tissue. Calcium movements also have been further affected by interactions of electromagnetic fields with $H^+$ and $HCO_3^-$ probably involving a competitive $H^+$-$Ca^{2+}$ binding at the cell membrane surface.[6] Depending on the frequency and amplitude of inductively coupled PEMF magnetic fields, calcium uptake by avian chondrocytes could be either stimulated or inhibited.[4] The effects of PEMF on calcium uptake by osteoporotic bone of rats have been suggested to be due to stimulation of bone formation or a reduction in bone resorption. Exposure of cultured embryonic bone cells to inductively coupled electromagnetic fields and capacitively coupled electric fields may increase calcium uptake.[6] It has been suggested that ($Na^+$-$K^+$) ATPase activity may be affected by inductively coupled electromagnetic fields increasing $Na^+$ efflux.[14] In vitro experiments on rat skin demonstrated the stimulatory effect of PEMF on the transport of alpha-amino-isobutyric (AIB) acid. This amino acid analogue is actively transported as an amino acid by the cell membrane but not further metabolized, thus allowing evaluation of membrane transport activity.[13,39,40] The increased amino acid transport is directly related to an increased protein synthesizing activity.

B. Effects on Protein Synthesis

Specific combinations of physical parameters characterizing the PEMF are capable of stimulating protein synthesis in rat skin.[13] PEMF treatment of primary cultures of embryonic chicken tendon fibroblasts resulted in a stimulatory effect that depended on the application of a specific wave form. PEMF treatment also stimulated collagen synthesis in slowly growing chondroblasts.[33,36] Stimulatory effects were equally observed in cultures of embryonic bone cells and fibroblasts.[17] Specific parameters characterizing the PEMF were able to stimulate the rate of collagen synthesis in bone tissue after experimentally induced osteoporosis.[10] The stimulatory effects on collagen synthesis in cultured fibroblast have been related to possible changes occurring in cyclic AMP metabolism.[33] On murine embryonal carcinoma cells, PEMF induced growth stimulation.[1] The specific activities of messenger RNA in dipteran salivary gland cells were increased after exposure to specific field sources.[22] Analysis of autoradiographs indicated an increased mRNA synthesis during PEMF stimulation affecting protein synthesis in salivary gland cells and hamster ovary cells.[21,23] After partial hepatectomy the rate of rat liver regeneration was increased by treatment with PEMF.[38]

Nerve regeneration in vitro has been stimulated as well as limb regeneration in specific organisms.[43,44] PEMF treatment also has been observed to stimulate the addition of anionic constituents to existing oligosaccharide chains in human fibroblasts and rat osteosarcoma cell lines.[16] DNA synthesis of cultured rapidly growing chondroblasts was stimulated by PEMF, while slowly growing chondroblasts were not affected. Thymidine incorporation into DNA of embryonic chondrocytes and fibroblasts was enhanced under influence of PEMF.[42] By modifying the field intensity, opposite effects could be induced in Chinese hamster cells.[46]

C. Effects on Cellular cAMP

As a second messenger, intracellular concentrations of cAMP condition the plasma membrane responsiveness. cAMP concentrations were increased after stimulation of epiphyseal cartilage from chicken embryo when orientation of the long axis of the bone was parallel to the oscillating electric field.[37] Exposure of embryonic calvaria to capacitively coupled electric fields resulted in a reduced or increased cAMP concentration depending on the field strength applied.[28,30]

Inductively coupled electromagnetic fields were found to enhance cAMP levels in cultured embryonic tibia cells and osteoblast-like bone cell lines.[9,15,24] Electromagnetic fields affect the activation of adenylate cyclase that controls cAMP levels. This suggests the translation of a physical signal by the plasma membrane into a chemical effect that initiates metabolic responses.

## INTRODUCTION

Most of the effects of PEMF on biological systems appear to depend on specific physical parameters governing the electromagnetic field. It appears that only a limited number of specific combinations of driving voltages, pulse frequencies, pulse width and waveform are capable of stimulating metabolic reactions. Modifications in the field characteristics may even result in completely opposite effects. It has been argued that ionic transport through cell membranes may be affected by externally applied magnetic fields greatly affecting subsequent metabolic reactions.[7,12,32]

Since the cell membrane is the first barrier to be negotiated to induce metabolic modifications, we developed a tissue model by which we specifically analyzed the effects of PEMF, characterized by different physical parameters, on the cell membrane transport mechanism. Based on the experimental data we intend to explain the mechanism of action by which PEMF may affect metabolism in rat skin tissue.

## MATERIALS AND METHODS

The skin of the back of locally inbred male Wistar R rats was removed at 21 days of age when the animals just finished their first hair cycle. The hair was plucked easily and skin flaps measuring 5 cm by 6 cm and 0.5 mm thickness were halved longitudinally. Excessive subcutaneous fat and connective tissue was removed with a scalpel. One half of each skin flap was used as a non-treated control, while the other half was treated with PEMF.[13] The skin segments were fixed on a Perspex frame with wooden plugs, placed in a plastic container and submerged in either a buffer medium at pH 7.4 characterized by an extracellular high $Na^+$, low $K^+$ or an intracellular low $Na^+$, high $K^+$ electrolyte composition.[12] One hundred ml of incubation medium further contained 100,000 units of penicillin-K (Sigma, St. Louis, MO), 100 mg of streptomycin sulfate (Sigma), 20 mg of gentamycin sulfate (Sigma) and 1 ml of an amino acid supplement (Vamin, Vitrum, Stockholm 12, Sweden). Forty micro-Ci of AIB (specific radioactivity 60 mCi/mmol; Amersham International) were added to 100 ml of incubation buffer to follow the amino acid transport activity through the cell membrane. PEMF was applied during incubation periods from 15 min up to 360 min in a 37°C waterbath to one skin half while the other half was identically incubated outside the electromagnetic field.

Ouabain (Sigma) to block $(Na^{+},K^{+})$-ATPase and 2,4-dinitrophenol (DNP, Sigma) to uncouple oxidative phosphorylation were added to specific experiments in final concentrations of $10^{-3}$ molar. The electromagnetic field source (ESAT, Faculty of Engineering, University of Leuven) allowed modifications of five variable parameters: driving voltage from 5 to 20 V, pulse frequency from 0.5 to 10 kHz, pulse widths from 10 to 150 microsec., pulse-burst frequency from 1 to 70 Hz and number of pulses per pulse burst from 1 to 100. The waveform applied to the coils was rectangular and the field was applied perpendicularly on the skin surface.[13] After incubation with AIB, the tissues were washed four times with 10 ml fractions of ice-cold physiological saline containing 30 mmolar of non-radioactive AIB (Sigma) and placed in 3 ml of Lumasolve (Lumac Systems AG, Basel, Switzerland) to be dissolved overnight in an oven at 60°C. After the addition of 10 ml of Lipoluma (Lumac Systems AG) the radioactivity in the samples was counted in a liquid scintillation spectrometer (Rack Beta, LKB, Wallac, OY, Turku, Finland). The radioactivity taken up by the cells was expressed as disintegrations per minute (DPM) ± S.E.M. Every experiment was carried out on six different rats. Statistical evaluation of the significance was carried out according to Student's t-test.

## RESULTS

Inductively coupled electromagnetic fields characterized by specific physical parameters when applied perpendicularly onto the skin were able to affect membrane transport. PEMF with driving voltages between 5 and 20 V, with 20 pulses per pulse burst, with pulse frequencies between 1 and 5 kHz, pulse widths of 10 and 20 microsec. and pulse burst frequencies between 10 and 30 Hz resulted in metabolic effects that were of identical magnitude. Instead of pulse bursts, continually running pulses gave analogous results. Modifications of these parameters outside the described limits reduced or blocked the metabolic effects.

These electromagnetic fields applied at 37°C during incubation of the skin in an extracellular buffer medium resulted in the AIB uptake being increased over the experimental period of 6 hours. During the first hour of incubation the stimulatory effect amounted to around 45% while upon further incubation stimulation progressively declined but tracer uptake remained significantly above control levels at the end of the incubation period (Fig. 1). During PEMF stimulation of skin tissue incubated for 2 hours in an extracellular medium, the

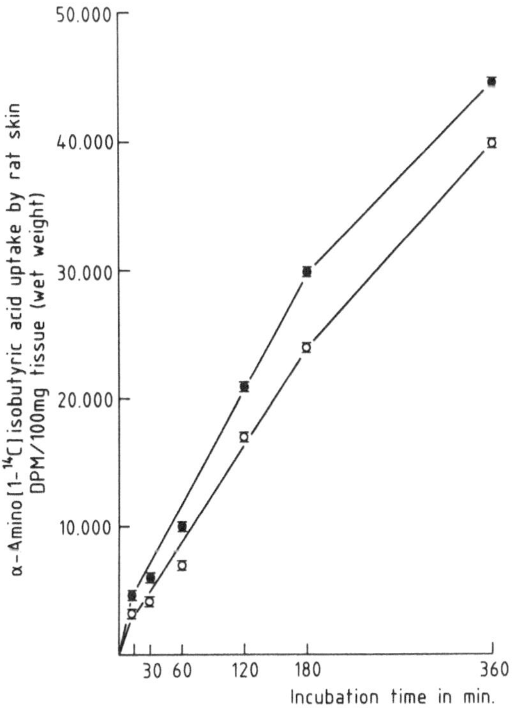


**Figure 1.** The effects of PEMF on membrane transport. Skin tissue flaps were incubated for 15, 30, 60, 120, 180 and 360 min in 100 ml of buffer characterized by extracellular electrolyte composition at 37°C containing 40 micro-Ci of alpha-amino-[1-14C]-isobutyric acid (radioactivity 60 mCi/mmol). One half of the skin served as control (o) while the other half was electromagnetically stimulated (•) with continuously running pulses, driving voltage of 10 V, pulse frequency of 5 kHz and pulse widths of 20 microsec. After incubation 100 mg skin tissue portions were taken and prepared for liquid scintillation counting. Each value represents the mean of 8 experiments + SEM.

amino acid analogue uptake increased by 20% ($P < 0.01$) while in an intracellular medium the AIB uptake became reduced by 10% ($P < 0.05$) compared to non-stimulated control skin (Fig. 2).

Without the electromagnetic field, the transfer of skin tissue from an extracellular medium, where it was preincubated for 30 minutes, into an intracellular medium containing AIB reduced the label uptake by 30% ($P < 0.01$) after 60 min when compared to tissue further incubated in extracellular buffer. After preincubation in an intracellular buffer followed by incubation in an extra-cellular medium containing the radioactive amino acid analogue, transport stimulation again occurred (Fig. 3). Addition of $10^{-3}$ M ouabain to the extra-

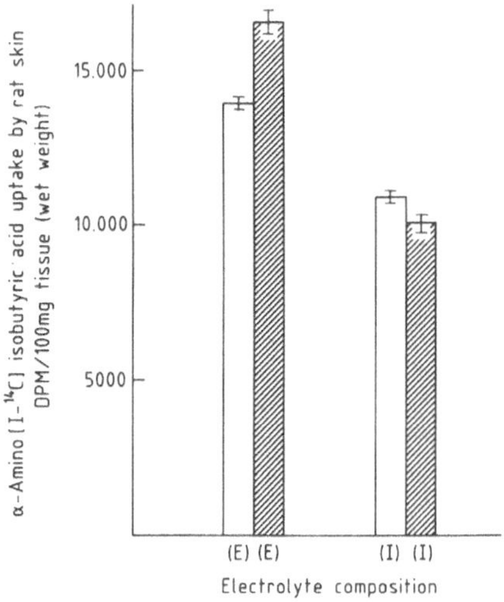


**Figure 2.** Interaction of buffer composition on PEMF effects: alpha-amino-[1-14C]-isobutyric acid uptake. Skin tissue was prepared and incubated in incubation buffers with different electrolyte composition, either with an extracellular (E) electrolyte composition ($Na^+$,115 meq; $K^+$, 10 meq; $Cl^-$,15 meq; $HCO_3^-$, 10 meq; $HPO_4^{2-}$ 85 meq; $K_2PO_4^-$, 15 meq, glucose, 194 mM) or an intracellular (I) electrolyte composition ($Na^+$, 10 meq; $K^+$,115 meq; $Cl^-$, 15 meq; $HCO_3^-$, 10 meq; $HPO_4^{2-}$, 85 meq; $H_2PO_4^-$, 15 meq, glucose, 194 mM) in the presence of AIB. Preparation of the samples for counting was identical to Fig. 1. Open and shaded blocks represent average uptake values of 8 control and PEMF treated samples + SEM.

cellular buffer solution considerably reduced transport activity (Fig. 4). Upon PEMF stimulation of skin in the presence of ouabain, AIB uptake could not be stimulated further or otherwise affected (Fig. 5).

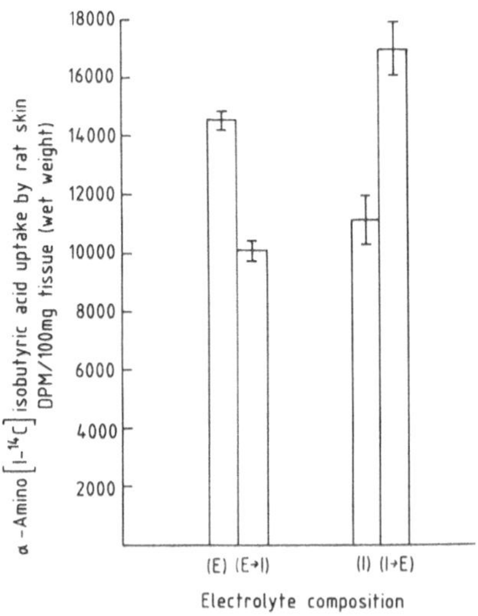


**Figure 3.** Effects of electrolyte composition on membrane transport. Skin samples as prepared in Fig. 1, were preincubated for 30 minutes either in buffer medium with extracellular (E) electrolyte composition or in buffer with intracellular (I) characteristics. After preincubation each tissue was transferred respectively into intra- and extracellular buffer containing alpha-amino-[1-$^{14}$C]-isobutyric acid for a further incubation during 60 min to follow the amino acid analogue transport. Each value represents the mean of 8 experiments ± SEM.

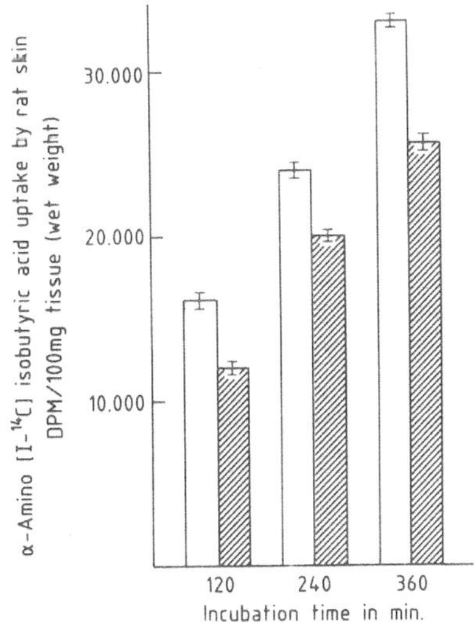


**Figure 4.** Effect of ouabain on alpha-amino-[1-$^{14}$C]-isobutyric acid transport. Skin tissue was incubated for 120, 240, and 360 min in 100 ml of pure extracellular electrolyte buffer (open blocks) or in identical buffer containing $10^{-3}$ M of ouabain (shaded blocks) with 40 micro-Ci of alpha-amino-[1-$^{14}$C]-isobutyric acid. Each value represents the mean of 8 experiments ± SEM.

In the presence of 2.4-dinitrophenol, blocking the oxidative phosphorylation and reducing protein synthesis, AIB uptake was reduced by approximately 10% (Fig. 6). Exposed to a PEMF, membrane transport could still be stimulated although at a much reduced significance (Fig. 7).

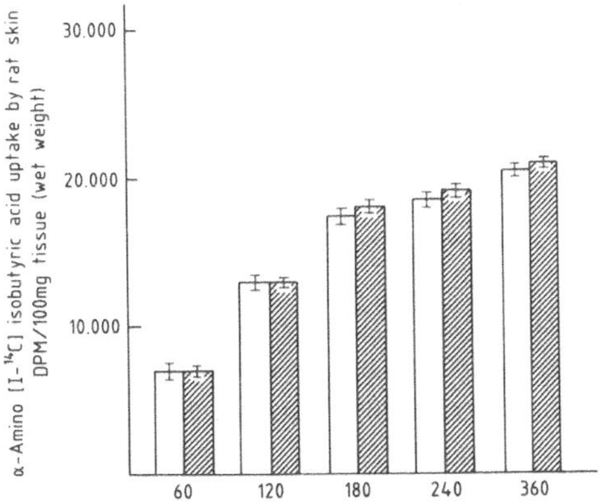


**Figure 5.** Effect of PEMF on membrane transport in the presence of ouabain. Ouabain in a concentration of $10^{-3}$ M was added to the extracellular incubation buffer in which skin tissue was incubated for 60, 120, 180, 240 and 360 min with (shaded blocks) or without (open blocks) PEMF treatment in the presence of alpha-amino-[1-$^{14}$C]-isobutyric acid. Each value represents the mean of 8 experiments ± SEM.

**Figure 6.** Effect of DNP on alpha-amino-[1-$^{14}$C]-isobutyric acid transport. Skin tissue was incubated in fresh extracellular buffer (open blocks) or in buffer containing $10^{-3}$ M of DNP (shaded blocks) for 120, 240, and 360 min. 100 ml of buffer equally contained 40 micro-Ci of alpha-amino-[1-$^{14}$C]-isobutyric acid. Each value is the mean of 8 experiments ± SEM.

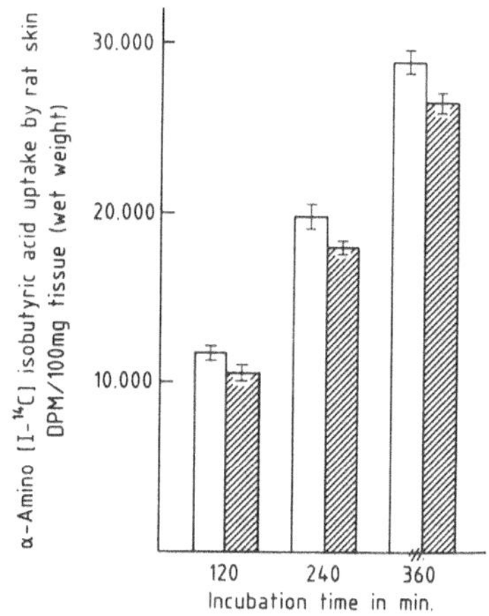


**Figure 7.** Effect of PEMF on membrane transport in the presence of DNP. Skin tissue was incubated in extracellular buffer containing $10^{-3}$ M of DNP, without (open blocks) or with (shaded blocks) PEMF treatment for 60, 120, 180, 240 and 360 min in the presence of 40 micro-Ci of alpha-amino-[1-$^{14}$C]-isobutyric acid. Each value represents the mean of 8 experiments ± SEM.

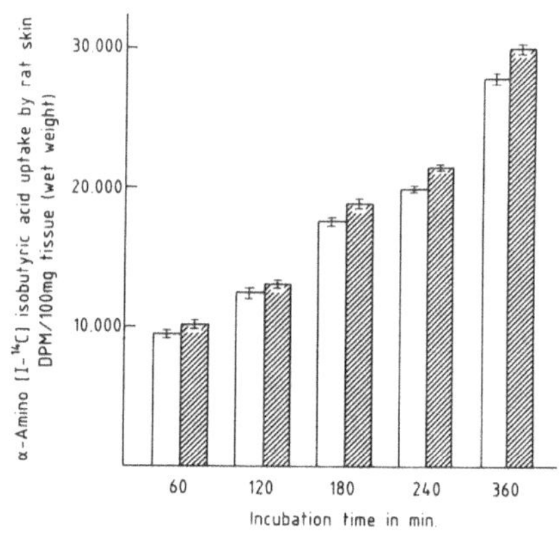


## DISCUSSION

PEMF characterized by specific electromagnetic parameters have been observed to be able to affect transport of amino acids through cell membranes.[12] Changes in lipid and protein molecules of the membranes attract counter ions resulting in a surface transmembrane potential that arises from differences in membrane permeability for various ions. An electric field in the vicinity of a membrane may be able to affect and modify the endogenously produced potentials. Numerous biological molecules are known to present various conformational states with different electric properties that may interact with externally applied electric fields.[49] The dipole moments of proteins thus vary with the conformation of the molecules. Induced conformational changes of biological molecules have been observed under the influence of electric fields and specific relationships seem to exist between conformational and electrical energy.[35] Opening and closing of membrane channels as well as ion pump mechanisms dependent on enzyme systems imply conformational changes that are dependent on transmembrane electric fields.[19,29,50] Although ATP synthesis obtains its energy from the proton electrochemical potential energy according to the chemiosmotic theory, external electric fields have been observed to induce ATP synthesis in energy-transducing membranes.[8,50] Stimulation of ion pump activity against a concentration gradient has been achieved by an externally applied electric field without excess consumption of ATP.[47,48]

According to the electroconformational coupling theory developed by Tsong and Astumian,[51] one cycle of a sinusoidal electric field, in principle, induces an enzyme to turnover once. However, effective energy coupling only takes place when the biological system is in resonance with the external field.[49,51] Transmembrane enzymes such as ($Na^+$-$K^+$)-ATPase, are capable of absorbing free energy from oscillating electric fields to be transduced into chemical or transport work.[52] The electric field as a thermodynamic quantity may thus affect the equilibrium between enzyme states with oscillations inducing enzyme cycling. As PEMF in the vicinity of biological membranes induce alternating currents, our results of PEMF effects on membrane transport activity provides illustration and confirmation of the electroconformational coupling theory. By increasing the pulse frequency, the metabolic stimulation disappears. Rapid changes in polarity may prevent electroconformational coupling, thus blocking the intended enzyme cycle by which substrates are transported through the cell membrane.[12]

The effect of PEMF on membrane transport seems conditioned by the electrolyte composition of the incubation medium. Modifications of the electrolyte composition of the buffer solutions directly affect the ($Na^+$-$K^+$)-ATPase activity, resulting in modified PEMF effects. In an intracellular environment PEMF inhibits the amino acid transport activity, which in an extracellular medium is stimulated considerably. These modifications in membrane transport can be simulated even outside an electromagnetic field simply by changing the ionic composition of the incubation buffers. In the presence of ouabain, AIB transport is reduced by inhibition of the ($Na^+$-$K^+$)-ATPase system, thus rendering PEMF treatment without effect. After uncoupling of the oxidative phosphorylation by DNP addition to the incubation medium, PEMF treatment remains active in stimulating the amino acid transport mechanism that is independent of oxidative phosphorylation. The effects of

PEMF on membrane transport may be understood by the effects of an oscillating electric field on transport enzyme systems through electroconformational coupling.[12]

## ACKNOWLEDGMENTS

The authors are indebted to the Belgian National Foundation for Medical Research (FGWO) for a grant to the laboratory and to Professor W. Sansen and E.S.A.T., Faculty of Engineering, University of Leuven for the development of the electromagnetic field source.

### References

1. Akamine T, Muramatsu H, Hamada H, Sakou T. Effects of pulsed electromagnetic field on growth and differentiation of embryonal carcinoma cells. J Cell Physiol 1985; 124:247-254.

2. Bassett CAL. The development and application of PEMFs for ununited fractures and arthrodesis. Orthop Clin N Am 1984; 15:61-88.

3. Bassett CAL. The development and application of pulsed electromagnetic fields (PEMFs) for ununited fractures and arthrodesis. Clin Plast Surg 1985; 12:259-277.

4. Bassett CAL, Chokshi HR, Hernandez E, Pawluk RJ, Strop M. The effect of pulsing electromagnetic fields on cellular calcium and calcification of non-unions. In: Electrical Properties of Bone and Cartilage: Experimental Effects and Clinical Applications (Eds. CT Brighton, J Black, SR Pollack). Grune and Stratton, New York. 1979; 427-441.

5. Bassett CAL, Pilla AA, Pawluk RJ. A non-operative salvage of surgically resistant pseudarthroses and non-unions by pulsing electromagnetic fields: A preliminary report. Clin Orthop 1977; 124:128-143.

6. Bawin SM, Adey WR, Sabbot IM. Ionic factors in release of $^{45}Ca^{2+}$ from chicken cerebral tissue by electromagnetic fields. Proc Natl Acad Sci USA 1978; 75:6314-6318.

7. Blank M. Recent developments in the theory of ion flow across membranes under imposed electric fields. In: Modern Bioelectricity (Ed. AA Marino). Marcel Dekker, New York. 1988; 345-364.

8. Boyer PD, Chance B, Ernster L, Mitchell P, Racker E, Slater EC. Oxidative phosphorylation and photophosphorylation. Ann Rev Biochem 1977; 46:955-1026.

9. Cain CD, Adey WR, Luben RA. Evidence that pulsed electromagnetic fields inhibit coupling of adenylate cyclase by parathyroid hormone in bone cells. J Bone Miner Res 1987; 2:437-441.

10. Cruess RL, Kan K, Bassett CAL. The effects of pulsing electromagnetic fields on bone metabolism in experimental disuse osteoporosis. Clin Orthop 1983; 173:245-250.

11. Dal Monte A, Fontanesi G, Cadossi R, Poli G, Giancecchi F. Pulsed electromagnetic field therapy in the treatment of congenital and acquired pseudarthrosis. In: Modern Bioelectricity (Ed. AA Marino). Marcel Dekker, New York. 1988; 711-756.

12. De Loecker W, Delport PH, Cheng N. Effects of pulsed electromagnetic fields on rat skin metabolism. Biochim Biophys Acta 1989; 982:9-14.

13. Delport PH, Cheng N, Mulier JC, Sansen W, De Loecker W. The effects of pulsed electromagnetic fields on metabolism in rat skin. Bioelectrochem Bioenerg 1985; 14:93-98.

14. Farndale RW, Maroudas. Low frequency pulsed magnetic fields do not modify several aspects of ion transport in biological materials. Bioelectrochem Bioenerg 1985; 14:105-114.

15. Farndale RW, Murray JC. The action of pulsed electromagnetic fields on cyclic AMP levels in cultured fibroblasts. Biochim Biophys Acta 1986; 881:46-53.

16. Fisher SJ, Dulling J, Smith SD. Effect of a pulsed electromagnetic field on plasma membrane protein glycosylation. J Bioelectricity 1986; 5:253-267.

17. Fitton-Jackson S, Jones DB, Murray JC, Farndale RW. (abstract) The response of connective and skeletal tissues to pulsed magnetic fields. Trans BRAGS 1981; 1:86.

18. Frank C, Schachar N, Dittrich D, Shrine N, De Haas W, Edwards G. Electromagnetic stimulation of ligament healing in rabbits. Clin Orthop 1983; 175:263-272.

19. Furman RE, Tanaka JC, Mueller P, Barchi RL. Voltage-dependent activation in purified reconstituted sodium channels from rabbit T-tubular membranes. Proc Natl Acad Sci USA 1986; 83:488-492.

20. Gerling JA, Sinclair PM, Roa RL. The effect of pulsating electromagnetic fields on condylar growth in guinea pigs. Am J Orthod 1985; 87:211-223.

21. Goodman R, Abbott J, Krim A, Henderson AS. Nucleic acid and protein synthesis in cultured chinese hamster ovary (CHO) cells exposed to the pulsed electromagnetic fields. J Bioelectricity 1985; 4:565-575.

22. Goodman R, Bassett CAL, Henderson AS. Pulsing electromagnetic fields induce cellular transcription. Science 1983; 220:1283-1285.

23. Goodman R, Henderson AS. Stimulation of RNA synthesis in the salivary gland cells of Sciara Coprophila by an electromagnetic signal used for treatment of skeletal problems in horses. J Bioelectricity 1987; 6:37-47.

24. Hiraki Y, Endo N, Takigawa M, Asada A, Takahashi H, Suzuki F. Enhanced responsiveness to parathyroid hormone and induction of functional

differentiation of cultured rabbit costal chondrocytes by a pulsed electromagnetic field. Biochim Biophys Acta 1987; 931:94-100.

25. Ito H, Bassett CAL. Effect of weak, pulsing electromagnetic fields on neural regeneration in the rat. Clin Orthop 1983; 181:283-290.

26. Jeran M, Zaffuto S, Moratti A, Bagnacani M, Cadossi R. PEMF stimulation of skin ulcers of venous origin in humans: Preliminary report of a double blind study. J Bioelectricity 1987; 6:181-188.

27. Jones DB, Pedley RB, Ryaby JT. The effects of pulsating electromagnetic fields on differentiation and growth in Cloudman S91 murine melanoma cells in vitro. J Bioelectricity 1986; 5:145-169.

28. Korenstein R, Somjen D, Danon A, Fischler H, Binderman I. (abstract) Pulsed capacitive electric induction of cyclic AMP changes, $Ca^{45}$ uptake and DNA synthesis in bone cells. Trans BRAGS 1981; 1:34.

29. Lauger P. Ionic channels with conformational substrates. Biophys J 1985; 47:581-591.

30. Luben RA, Cain CD, Chen MCY, Rosen DM, Adey WR. Effects of electromagnetic stimuli on bone and bone cells in vitro: Inhibition of responses to parathyroid hormone by low-energy low-frequency fields. Proc Natl Acad Sci USA 1982; 79:4180-4184.

31. Marcer M, Musatti G, Bassett CAL. Results of pulsed electromagnetic fields (PEMFs) in ununited fractures after external skeletal fixation. Clin Orthop 1984; 190:260-265.

32. McLeod BR, Smith SD, Liboff AR. Calcium and potassium cyclotron resonance curves and harmonics in diatoms (A Coffeaeformis). J Bioelectricity 1987; 6:153-168.

33. Murray JC, Farndale RW. Modulation of collagen production in cultured fibroblasts by a low-frequency pulsed magnetic field. Biochim Biophys Acta 1985; 838:98-105.

34. Murray JC, Lacy M, Fitton-Jackson S. Degradative pathways in cultured synovial fibroblasts: Selective effects of pulsed electromagnetic fields. J Orthop Res 1988; 6:24-31.

35. Neumann E, Katchalsky A. Long-lived conformational changes induced by electric impulses in biopolymers. Proc Natl Acad Sci USA 1972; 69:993-997.

36. Norton LA. Effects of pulsed electromagnetic fields on a mixed chondroblastic tissue culture. Clin Orthop 1982; 167:280-290.

37. Norton LA, Rodan GA, Bourret LA. Epiphyseal cartilage cAMP changes produced by electrical and mechanical perturbations. Clin Orthop 1977; 24:59-68.

38. Ottani V, Monti MG, Morocutti M, Ferri M, Stracchi R, Ruggeri A, Barbiroli B. Influence of pulsed electromagnetic fields on regenerating rat liver after partial hepatectomy. J Anat 1984; 139:253-263.

39. Oxender DL, Christensen HN. Distinct mediating system for the transport of neutral amino acids by the Ehrlich cell. J Biol Chem 1963; 238:3686-3699.

40. Riggs TR, McKirahan KJ. Action of insulin on transport of L-alanine into rat diaphragm in vitro. J Biol Chem 1973; 248:6450-6455.

41. Rispoli FP, Corolla FM, Mussner R. The use of low frequency pulsing electromagnetic fields in patients with painful hip prostheses. J Bioelectricity 1988; 7:181-187.

42. Rodan GA, Bourret LA, Norton LA. DNA synthesis in cartilage cells is stimulated by oscillating electric fields. Science 1978; 199:690-692.

43. Sisken BF. Effects of electromagnetic fields on nerve regeneration. In: Modern Bioelectricity (Ed. AA Marino). Marcel Dekker, New York. 1988; 497-527.

44. Smith SD. Limb regeneration. In: Modern Bioelectricity (Ed. AA Marino). Marcel Dekker, New York. 1988; 529-555.

45. Stark TM, Sinclair PM. Effect of pulsed electromagnetic fields on orthodontic tooth movement. Am J Orthod Dentofacial Orthop 1987; 91:91-104.

46. Takahashi K, Kanedo I, Date M, Fukada E. Effect of pulsing electromagnetic fields on DNA synthesis in mammalian cells in culture. Experientia 1986; 42:185-186.

47. Teissie J, Knox BE, Tsong TY, Wehrle J. Synthesis of ATP in respiratory-inhibited submitochondrial particles induced by microsecond electric pulses. Proc Natl Acad Sci USA 1981; 78:7473-7477.

48. Teissie J, Tsong TY. Evidence of voltage-induced channel opening in Na/K ATPase of human erythrocyte membrane. J Membr Biol 1980; 55:133-140.

49. Tsong TY. Electroconformational coupling: How membrane-bound ATPase transduces energy from dynamic electric fields. Ann Rev Physiol 1988; 50:273-290.

50. Tsong TY, Astumian RD. Absorption and conversion of electric field energy by membrane-bound ATPases. Bioelectrochem Bioenerg 1986; 15:457-476.

51. Tsong TY, Astumian RD. Electroconformational coupling and membrane protein function. Prog Biophys Molec Biol 1987; 50:1-45.

52. Westerhoff HV, Tsong TY, Chock PB, Chen YD, Astumian RD. How enzymes can capture and transmit free energy from an oscillating electric field. Proc Natl Acad Sci USA 1986; 83:4734-4738.

# CELLULAR PHYSIOLOGY AND ELECTROMAGNETIC FIELDS: CELLULAR COMMUNICATION, HUMAN COMMUNICATION

Ian A. Cook

## INTRODUCTION

The editors have charged us with presenting material to help scientists, clinicians, and regulators alike to examine critically the growing body of knowledge concerning electromagnetic medical technologies. I would like to suggest a central proposition as a guide to such an examination, by proposing the following:

1) subtle and specific changes in cellular function may be achieved by applying specific electromagnetic fields; and

2) in evaluating emerging electromagnetic medical technologies, it is sometimes appropriate to consider not just the <u>energetics</u> of the applied field, but also its <u>informational content</u>, where this content may reflect both frequency components and phase relationships and other characteristics of the signal as well.

## SCIENTIFIC BASIS FOR AN INFORMATION-BASED PERSPECTIVE

### A. <u>Changes in Cellular Function</u>

This central proposition ([1] above) may, at first blush, appear to be an alien concept, but it should not. Much of clinical medicine and biological research is predicated on manipulations which are fundamentally concerned with information flow. Specific changes in a cell's activity are brought about when the information entering a cell is modified, for example, by using a pharmacologic agent which changes the activity of a membrane receptor or channel. Calcium-channel blockers, insulin, and L-DOPA are very important drugs which can rest on their therapeutic laurels because of just such interactions. There is a growing body of evidence which suggests the value of taking a similar, information-based perspective when examining the alterations in cell physiology produced by specific electromagnetic fields.

An underlying supposition here is that cells can distinguish fields on the basis of their information content. The basis for this lies in experimental work, much of which is reviewed at great length in a forth-coming volume,[2] and need not be repeated here exhaustively. For this discussion, several key threads of evidence can be followed to illustrate the value of an information-based approach to cellular communication.

1. RNA Transcription

One set of observations derives from the work performed with fruit-fly salivary-gland models using Sciara larvae.[15-17] RNA transcription patterns were examined in one set of experiments using autoradiography, nick translation methods, and tritiated uridine incorporation assays. These experiments revealed that different translation patterns were found in comparisons between specimens exposed to no applied field, those exposed to one particular clinically-used pulse pattern ("single pulse"), and those exposed to another, different clinical signal ("pulse burst"). While these fields are somewhat different in terms of their energetics, they are markedly different in terms of the distribution of energy as a function of frequency, as assessed by Fourier transform methods.[7] Further work with the same model system has shown that these cells are able to discriminate between clinical pulses and sine waves of different frequency as well, and that different polypeptide products ultimately appear to be associated with the different field patterns. Since the expression of genetic information via mRNA is critical to higher-order biological processes, electromagnetic manipulation of information can become an increasingly important tool in basic biomedical research as well as human therapeutics.

2. Biosynthesis in Tissue Culture

Another set of observations has been made using tissue culture models such as chick limb rudiments and explants.[10-14,18-19] Here, again, cells have been shown to alter their production of macromolecules (eg, glycosaminoglycans, collagen) in response to the particular characteristics of the applied field (eg, repetition rate of a repeating single pulse). Cells evidently are able to discriminate between signals and alter their synthetic and degradative behavior in response.

3. Calcium Flux

Yet another group of evidence comes from work with amplitude modulated fields.[1,5,6,20] Transport of calcium in CNS tissues has been shown to be highly dependent upon the modulating frequency impressed upon a very high frequency (VHF) or microwave carrier wave. Variation of a few hertz was associated with large changes in calcium flux, again consistent with a very selective detection mechanism.

B. Information Content and Constraints of the Communication Channel

This and other work suggests that cells can distinguish between different applied fields. One must consider, however, that the information delivered to the cell depends not simply on the characteristics of the applied field, but also upon the local electrical environment in which the cell is situated. Previous work[7-9] revealed that the voltage waveforms detected in a specimen of bone were not the same as those detected in size-matched specimens of cartilage or tendon which were exposed under identical conditions to the same field. The local field which is presented to a cell, therefore, depends upon the type of tissue in which it is situated. Consequently, the final message which is available to the cell reflects both the characteristics of the applied field (the message that is sent) and the characteristics of the tissue matrix in which the cell resides (which forms a transmission path).

C. Similarity to Drug Interactions

The body of scientific evidence suggested that it is prudent for experimentalists to examine electromagnetic interactions with living systems as potentially being similar to drug interactions. It is clearly recognized that one must control the chemical environment in an experiment to avoid spurious chemical influences. One does not include small doses of the active drug under investigation in pills considered "placebo" in doing a double-blind comparison of the active agent vs. no treatment unless one really wishes to conduct a dose-response test. Likewise, it is imperative to be cognizant of the electromagnetic environment in which an experiment takes place, and to control or identify other sources of electromagnetic radiation which may have biological activity.

D. Economy of Means in Living Systems

As my long-term colleague Dr. Andrew Bassett has often said, one must learn to "think like a cell." This maxim can serve both as guidance for experimental work and for the ways in which bench research is translated into effective clinical therapies. Living organisms tend to be remarkably adept at recycling solutions to problems, and one sees comparatively few differences in, for example, the inflammatory response at various locations within the body. We must learn, then, to evaluate electromagnetic technologies with regard to their biological setting, and recognize that a therapeutic system that has, for example, demonstrated its efficacy (in a meticulously controlled study) in controlling an inflammatory response in the skin is likely to be efficacious for inflammation in the colon or in the pericardium. The policies of science and regulation should work in concert with the laws of nature.

## FUTURE APPLICATIONS AND CONCERNS

My crystal ball is colored by experience as a surgeon, as a biomedical scientist, as an engineer, and as a promulgator of policy. With that caveat, I will offer my projections for future applications. I believe we can lay the groundwork for a future in which clinicians will be able to accept electromagnetic ***therapies*** (emphasis on the plurality) as another family of mainstream tools, just as drugs form a diverse family, diagnostic imaging techniques are a family, and surgical procedures are yet another. Each may have specific indications, contra-indications, and mechanisms of action, just as selection of a drug or operation depends upon clinical setting. To achieve this, there must be a continued growth of the scientifically sound, rational basis for these tools. An understanding of application at the clinical level ultimately is most potent when coupled with an understanding of mechanism at the tissue, cellular, and molecular levels. But an impressive body of science is neither wholly necessary nor sufficient to convince clinicians to use electromagnetic therapies. Clear benefits need to be demonstrated in carefully conducted animal and clinical trials. Benefits will likely be measured not only in terms of salvage rates for otherwise recalcitrant problems, as has been the case for non-unions[3] and avascular necrosis of the femoral head,[4] but also in terms of reduced morbidity, improved clinical outcome, and cost effectiveness for more common, less desperate, situations. Pilot studies already suggest that significant advances are near in osteoporosis, neurological disorders, the complications of diabetes, some cardiovascular diseases, and perhaps malignancies.

In recent years, our national economy has changed from one largely dependent upon moving physical objects (capital goods, currency, raw materials) to one in which operations take place with the transfer of information (electronic bank deposit, FAX's). Medicine may follow suit by replacing some instances of rearrangement of physical elements (pharmacologically or surgically) with appropriate alterations of the informational content in those physical elements by electromagnetic fields.

To encourage wider exploration of the uses of electromagnetic technologies, we need an environment which is conducive to prudent innovation. Funding from public and private sources will need to be deployed and regulations promulgated, from the viewpoint that this is a field with an increasingly sound scientific basis and potentially bright future, not a black box technology based on quackery. This, in turn, calls for concerted efforts on the part of the scientists working at this frontier to educate fellow researchers, clinicians, the lay public, and regulators alike, about the true state of the art.

Safety concerns loom large for the public, who unfortunately are sometimes exposed to incomplete reports and un-informed interpretations. A single ill-conceived story in the popular media can undo years of solid science by engendering a negative perception. Researchers must play a role in disseminating what is known, while honestly representing that all the answers are not yet in.

This certainly is a tall order. But those who work at the frontiers of science can see, perhaps most clearly, the bright future that may await. Like the frontier explorers who struck out in unknown directions and found themselves looking on the greatness of the north American continent, we must be both explorers of the new territory and trail-blazers who enable others to follow and derive benefit from the discoveries.

Orthopedics has benefitted the most from electromagnetic therapies so far. However, the human body has a great amount of soft tissues as well, and these tissues hold great interest for a large number of physicians and patients alike. Realizing the promise of electromagnetic therapeutic technologies calls for a great deal of hard work. A tall order, yes; but a bright future as well.

## References


1. Adey WR, Bawin SM, Lawrence AF. Effects of weak amplitude-modulated microwave fields on calcium efflux from awake cat cerebral cortex. Bioelectromagnetics 1982; 3:295-307.

2. Bassett CA. Fundamental and Practical Aspects of Therapeutic Uses of Pulsed Electromagnetic Fields (PEMFs). CRC Critical Reviews in Biomedical Engineering. CRC Press, Boca Raton, FL, 1989 (in press).

3. Bassett CA, Pilla AA, Pawluk RJ. A non-surgical salvage of surgically-resistant pseudarthroses and non-unions by pulsing electromagnetic fields: a preliminary report. Clin Orthop 1977; 124:128-143.

4. Bassett CA, Schink MM, Mitchell SN. Treatment of osteonecrosis of the hip with specific, pulsed electromagnetic fields (PEMFs), a preliminary clinical report. In: Bone Circulation (Eds. J Arlet, RP Ficat, DS Hungerford). Williams & Wilkins, Baltimore, MD. 1984; 343-354.

5. Bawin SM, Adey WR, Sabbot IM. Ionic factors in release of 45-$Ca^{+2}$ from chicken cerebral tissue by electromagnetic fields. Proc Natl Acad Sci USA 1978; 75:6314-6318.

6. Bawin SM, Kaczmarek LK, Adey WR. Effects of modulated VHF fields on the central nervous system. Ann NY Acad Sci 1975; 247:74-81.

7. Cook IA. Techniques in Cellular Communication. Thesis, B.S.E. Princeton University, School of Engineering/Applied Science. Princeton, New Jersey. 1982.

8. Cook IA, Bassett CA. (abstract) Effects of tissue type and orientation on electromagnetically-induced voltage. J Bone Joint Surg Orthop Trans 1983; 7:361.

9. Cook IA, Bassett CA. (abstract) Electromagnetically-induced pulse waveforms are dependent upon tissue factors. Proc Bioelectric Repair Growth Soc 1983; 3:70.

10. Farndale RW, Murray JC. Low frequency pulsed magnetic fields enhance collagen production in connective tissue cultures. Bioelectrochem Bioenerg 1985; 14:83-91.

11. Farndale RW, Murray JC. Pulsed electromagnetic fields promote collagen production in bone marrow fibroblasts via athermal mechanisms. Calcif Tiss Int 1985; 37:178-182.

12. Fitton-Jackson S. Biophysical studies of pulsing magnetic field interaction with biological systems. I. Biophysical interactions. In: Interactions Between Electromagnetic Fields and Cells (Eds. A Chiabrera, C Nicolini, HP Schwan). NATO ASI, Series A, Life Sciences 97. Plenum Press, New York. 1985;537-545.

13. Fitton-Jackson S, Bassett CA. The response of skeletal tissues to pulsed magnetic fields. In:. Tissue Culture in Medical Research (Eds. RJ Richards, KT Rajan). Pergamon Press, Oxford. 1980;21-28.

14. Fitton-Jackson S, Bord S. (abstract) The effect of PEMFs on connective tissue repair. Proc 11th Ann Bioelectromagnetics Soc, 1989; 11:16.

15. Goodman R, Abbott J, Henderson AS. Transcription patterns in the X chromosome of Sciara coprophila following exposure to magnetic fields. Bioelectromagnetics 1987; 8:1-7.

16. Goodman R, Bassett CA, Henderson AS. Pulsing electromagnetic fields induce cellular transcription. Science, 1983; 220:1283-1285.

17. Goodman R, Henderson AS. Exposure of salivary gland cells to low-frequency electromagnetic fields alters polypeptide synthesis. Proc Natl Acad Sci USA 1988; 85:3928-3932.

18. Murray JC, Farndale RW. Modulation of collagen production in cultured fibroblasts by a low-frequency, pulsed magnetic field. Biochim Biophys Acta 1985; 838:98-105.

19. Murray JC, Lacy M, Fitton-Jackson S. Degradative pathways in cultured synovial fibroblasts: selective effects of pulsed magnetic fields. J Orthop Res,1988; 6:24-31.

20. Lin-Liu S, Adey WR. Low frequency amplitude modulated microwave fields change calcium efflux rates from synaptosomes. Bioelectromagnetics 1982; 3:309-322.

# LOW FREQUENCY ELECTROMAGNETIC ENERGY: ITS EMERGENCE AS A MEDICAL TREATMENT

Bruce McLeod

## INTRODUCTION

Electromagnetic field (EMF) technology in medicine has been emerging, or attempting to emerge, for over 25 years. In fact, one could look back several times 25 years, and find references in history to the use of, or the study of, electric or magnetic fields applied to various organs or parts of the human body. In many ways, this long history of use, or abuse, has been a factor in preventing widespread application of low frequency EMF energy in the medical arena.

Unfortunately, during the 1920's and 1930's, anything electric or magnetic was viewed with some awe, perhaps because electricity and electromagnetic fields were fairly new technologies. There was an almost mystical power associated with these terms and what the general public perceived or understood about this energy. Catalogs of that period were filled with electric or magnetic rings, belts, hats, pads, etc., for which wonderful healing properties were claimed, (see almost any mail-order catalog from that period). Those claims were for the most part properly debunked. However, the echoes of those false claims were still present when early researchers such as Robert Becker and Andrew Bassett began to publish their work about 25 years ago.[3] At this time, the use of electric power (sixty cycle power) to light homes, streets, buildings, and to power industry had become as accepted as breathing. The general public gave no thought to the fact that EM fields were generated by this power. The fields did not arouse any of our typical sensory modalities. They could not be seen, felt, smelled, or tasted. Therefore, it was assumed that they had no effect on the animals, including humans, who lived in an environment filled with these fields.

Thus, a scientist reporting biological effects involving EM fields had at least two battles to fight, independent of science. On the one hand, many fraudulent claims of specific effects were still ghosting in the background, and on the other, much of the world's energy was being converted into EM energy in order to provide what had become the expected standard of living. The public clearly viewed this EM energy as harmless, and indeed, most of the public probably did not identify the lights and electrical appliances in their home as being connected with a form of EM energy. A third barrier to electromagnetic energy emerging into the medical field came with the advent of widespread concerns associated with the word "radiation". Attention to nuclear radiation spawned instant fear and suspicion in the mind of the public when the word "radiation" was in any way connected with the word "electromagnetic". If it was perceived that the EM fields "radiate" into any space near the transducer producing the fields, there was little hope that such a device would gain public acceptance. Thus, any scientist announcing to the world, even the scientific or medical world, that EM fields could alter biological systems was treading in some rather deep water. He or she had at least these three obstacles with which to be concerned, as well as that of obtaining solid scientific proof that could be verified by colleagues.

In spite of this background, a few scientists began to report that cells and systems of cells not only exhibited electrical and magnetic properties, but that some of these properties could be duplicated and fed back to the cells. Robert Becker and Andrew Bassett[3] published some of the early reports in the United States of changing cell properties with EM fields. Previously, reports had been published in several Eastern European countries, where, it seems, minds were a bit more open. Gradually, a handful of other scientists began to do work in the field. It is not possible to name all of them, but names like Brighton, Friedenberg, Marino, Pilla, Pollack, and Spadaro, come to mind. This field of research grew very slowly due, in part, to the problems mentioned above, and in part, due to two new factors. As the reports of biological effects attributable to EM fields grew in numbers, so did the reports of failed efforts to repeat some of these experiments.[8] Non-repeatability of experiments became so prevalent that at one point even the researchers in the field could be heard referring to the fact that the entire field appeared to be suffering from the "Cheshire Cat" effect. Another factor crept in with reports that non-ionizing radio frequency (RF) fields could change the course of the life of a biological system (eg, Bioelectromagnetics Society [BEMS] Annual Meeting abstracts from 1980 to 1986). This seemed to split the research into two separate worlds, almost competing with each other. One world promoted the idea that there were no real effects while the other suggested the effects were not only real, but also of vital importance.

Thus, the movement toward using EM fields in medicine seemed to pause for a considerable period of time while various debates were carried out. One group of scientists were convinced that EM fields could alter cell processes, while another group insisted that the whole area was nonsense. Yet a third group felt that public health was the real issue and that better electromagnetic energy exposure or emission standards should be the primary focus of the research. At least four separate scientific societies came into existence to serve as focal points for the research and the debates. The Bioelectrical Repair and Growth Society (BRAGS) was formed in 1980 with an emphasis in clinical uses of low frequency EM fields. The Bioelectromagnetics Society (BEMS) had been formed a year earlier (1979), with its emphasis more directed toward high frequency EM fields and possible health hazards. The Bioelectrochemical Society (BES) was formally incorporated in 1979, but it had started regular meetings in 1971. Its emphasis was primarily directed at cellular research with parallels looked for between EM fields and chemical reactions. Another society, the Institute of Electrical and Electronic Engineers (IEEE) formed a special group called the IEEE Engineering in Medicine and Biology (EMBS). This group, as the name implies, was made up primarily of electrical engineers who were interested in how electricity and magnetism could play a role in cell or biological system regulation. The membership in these societies was not mutually exclusive. Many researchers were members of at least two and sometimes all of these societies. The reason for multiple memberships was simply that these seemingly diverse groups provided forums for the debates on the most recent research. Initially, meetings of these groups seemed to involve only discussions in the areas of emphasis of each group. However, as the research field matured and as the data began to show an unquestionable link between the various groups, these distinctions began to blur. This was also due, in part, to the multiple memberships held by scientist with widely diverse backgrounds. A very positive result came about because of these societies. The mixing of many ideas and many scientists in very different fields gradually pushed the research into promising areas. More science and less speculation was encouraged which began to eliminate data that could not be repeated. It is perhaps now possible to suggest that there is a need for only one society that can serve as the clearinghouse and focal point for research in this area, since the emphasis in all of the societies has now shifted toward finding the mechanism(s) of interaction.

One low frequency EM technology did begin to make its way out of the laboratory and into the clinic. Orthopedic surgeons faced with patients with broken bones that refused to heal (non-unions) began to use pulsed electromagnetic fields (PEMF) as a last resort treatment before amputation of the limb. The results were good, with a claimed healing success rate of about 80% in these rather desperate cases. However, patient, physician, and regulatory agency acceptance of these early clinical devices was slow. No theory of how the device worked was accepted by the scientific community, and a great debate arose about whether the EM device really altered bone cell growth or whether a superior job of casting the broken bone would accomplish the same end result. For several years, a double blind study was needed to establish the efficacy of the PEMF devices. At least one such study has now been done and reported by Sharrard in England.[25] His data is being widely quoted at this point in that he showed there was a statistically significant increase in the healing of the non-union bone fracture when pulsed electromagnetic fields were applied to the limbs.

This brings us to the question of whether low energy electromagnetic fields are about to emerge into the field of medicine as a positive force for healing, or will they be written off as they were between 1920 and 1950. It is perhaps too early to state that everyone believes devices are about to emerge, but it is probably safe to say that little doubt remains that EM fields can and do effect biological systems. A fair percentage of scientists working in the field report that certain devices can be used for healing, and seem to believe that these devices will make their way into clinical medicine in the next two to five years. As more evidence accumulates that certain fields, frequencies and amplitudes can control cellular functions, the potential for electromagnetic medicine becomes greater.

In this chapter, no attempt will be made to discuss all the clinical success of EM fields in medicine, but an attempt is made to convince the reader that the technology really is about to emerge in some areas. In addition, succeeding portions of the paper are devoted to discussion of risks and benefits, the formation of a specific model, and some views on the position of the Food and Drug Administration (FDA) with respect to these devices.

## DISCUSSION

Without preparing an extensive review of all of the available literature it is easy to identify some recent trends by looking at abstracts from two recent meetings of the Bioelectrical Repair and Growth Society (BRAGS)[29] and the Bioelectromagnetics Society (BEMS).[30] These two sources are particularly appropriate when considering devices using low energy level EM fields and addressing the question of whether they will emerge into medical technology. Both societies are approximately 10 years old although satellite meetings with other primarily engineering oriented groups have been occurring for nearly 15-20 years. A major portion of the research done in the United States in this field has appeared in the meeting abstracts of these two societies. Two other very good sources for recent information are the Journal of Bioelectricity and Bioelectromagnetics.

In the 1988 BRAGS meeting[29] one full day was devoted to clinical applications or clinical studies involving low energy level EM fields. A glance at the abstracts suggests the extent to which the research has progressed. One immediately sees the following applications:

1. Ununited fractures[22]

2. Fresh fracture repair[9]

3. Treatment of osteonecrosis of the femoral head[24]

4. Lumbar spinal fusions[5]

5. Enhancement of bone maturation[6]

6. Nerve regeneration[7]

7. The study of PEMF on mice undergoing bone marrow transplantation[32]

8. Wound healing in rat skin flaps[8]

9. Treatment of loosened hip prosthesis with PEMF;[11] and

10. Healing of skin ulcers of venous origin in humans.[10]

If one examines the program for the 1989 BEMS meeting,[30] it again becomes clear that considerable meeting time was devoted to reports of potential clinical applications of these EM fields. The topics include:

1. Symposium of ELF fields and neuroendocrine function[31]

2. Calcium effects and ELF-in vitro studies;[4,16] and

3. Symposium on application of time-varying fields for tissue healing, (three separate sessions).[2]

The 1989 BRAGS meeting proposes to feature device usage, special sessions on clinical applications, reports of double blind studies on the efficacy of low energy level EM fields used in the clinic and information about material being developed to increase the amount of communication between the regulatory agencies and the scientific community.[31]

While this paper presents only a limited look at clinical uses of this technology, it certainly sets the tone of what is happening to the field. Each year the use of these devices in clinical medicine is increasing. The continuation of this trend depends on how well the devices actually work. Two other key criteria must be met for the EMF emergence to gain momentum. A testable theory about the mechanism involved in EMF cell interaction must be established, and regulatory agencies such as the Food and Drug Administration (FDA) must become aware that electromagnetic medicine may be just around the corner. This will require additional education of their current staff, and the addition of new staff that can perform timely evaluations of the increasing flow of requests to use EM fields on human subjects.

## THE FDA POSITION

The FDA currently is in the midst of the controversy regarding EMF in medicine. On one side they have various commercial groups that are requesting exemptions in order to obtain clinical data on their particular device. Part of the public who are aware of these devices are asking why the technology is not more available. Other public groups are pressuring the agency to not allow any use of EM fields on humans due to the perceived connection with negative effects of radiation. Yet another point of view suggests there is no effect at all from these devices, and that the FDA should therefore disallow any use of them in human medicine. The FDA maintains a position based on policies developed in response to their legislative mandate regarding determination of the safety and efficacy of devices. In addition, FDA suggests that a conservative position is necessary until sufficient data are available showing the risk/benefit factors for human exposure to these low energy EM fields. The key then, is the flow of information, with the flow being required in both directions. The FDA has set up advisory panels made up of clinicians and scientists who provide input to the decision making process for new devices. This can help in the flow of information into the FDA. It would be a great help if the four professional societies mentioned above would extend themselves to increase the flow of information into the FDA. Conversely, the FDA needs to make a clear effort to inform the scientific and commercial communities about the type of data that is needed in order that informed, reasonably rapid decisions can be made by the FDA about allowing more of this technology into the clinic. Obviously, the technology will emerge in the United States only as rapidly as is allowed by the FDA.

## THE FORMATION OF A MODEL

After almost thirty years of research, there is still no accepted explanation about how non-ionizing EM fields interact with living systems to cause an alteration in some life process. However, a model was proposed in 1984 by Liboff[12] during a NATO Advanced Research Workshop in Erice, Italy. He suggested that ion cyclotron resonance could occur in the membrane of individual cells and, as such, increased transport of ions such as calcium, magnesium, potassium, and many others would occur if specific magnetic fields were applied to the cells. Since 1984, Liboff, McLeod and Smith have published a number of papers, both experimental and theoretical, that support this model[13,14,17,18,19,21,24,26,27] Researchers in other laboratories are also beginning to report cell responses using cyclotron resonance conditions for specific ions.[16,23] The model has not been accepted generally since unanswered questions still exist about how certain of the features of the model can exist in a biological membrane. These will be pointed out below. However, many researchers have come to believe that the model is on the correct track. If this is the case, an acceptable working model will emerge in the next two or three years.

The concept of cyclotron resonance is explained rather easily for a charged particle moving in free space in the presence of a static magnetic field. If one imagines one wall of a room as being a north pole of a magnet, and the opposite wall as being the south pole, most people can "see", in their mind's eye, the lines of magnetic flux going from one wall to the other. If one now throws a tennis ball that has charge on its surface into the lines of flux (assume the velocity of the ball to be at right angles to the flux for simplicity) the path of the ball will be a circle about several lines of flux with the radius of the circle being proportional to the strength of the static magnetic field. One could use a stop watch and find out how much time it takes for the tennis ball to complete one revolution about the lines of flux. If one now applies a second, time varying (ac) field to the tennis ball such that the ac frequency is exactly one over the time it takes for the ball to complete a single revolution, resonance will occur. That is, energy will be efficiently transferred between the time varying field and the tennis ball. If the frequency is gradually changed from the resonance frequency, the transfer of energy to the tennis ball will become less and less efficient. One could easily add energy to the ball at resonance (like a child pumping a swing), or one could subtract energy from the ball, eventually bringing it to rest. Liboff and McLeod[13] concluded that even though the charged particles in biological systems (ions) were not in free space, it was possible for resonance to occur if one considered the channels in the cell membranes through which the ions must pass and where these ions would be effectively isolated. If resonance did occur, one would expect to see modification of the biological processes of the cell upon application of the externally applied, time varying electromagnetic fields since the calcium ion, for instance, is a necessary precursor to almost every biological action. If one could move more (or less) calcium across the cell membrane by adding or subtracting energy from the ions, it should be possible to control the cell.

Some of the problems that exists with the model are pointed out by scientists that are familiar with charged particle motion in a sparse plasma in free space. If one calculates the radius of an ion such as calcium ($Ca^{++}$) at a resonant dc magnetic field of about 50 microTesla (approximately the earth's geomagnetic field of strength), that radius is much too large when compared to the radius of a membrane channel. Biologists point out that the ion should be hydrated (water attached to the ion) inside the channel and, since this would greatly increase the ion mass, the cyclotron resonance should occur only at parts of a cycle per second rather than at the resonant frequency of several cycles per second for a non-hydrated ion. Another criticism of the model is that there has been no conclusive evidence that indicates that more of a specific ion is being moved across the cell membrane at resonance. Each of these points have been addressed in several ways, but to date, none of the explanations have been proven.

An extension of the theory by McLeod and Liboff[13,18] was done recently and first discussed at this meeting on emerging technologies. More of this theory was presented at the 1989 BEMS meeting.[20] Very briefly, the theoretical model has been extended to try and account for the ion being constrained to stay within the walls of the membrane channel as it is exposed to the magnetic fields. The results of this work indicate that adding the effects of the boundary within which the ion must move (the channel wall) seems to answer the question of why the calculated resonant radius was so far in error. The theory indicates that the ion moves in a helical path through the channel, bounded by the channel wall, and that the radius of the helical path is effectively modulated by the low frequency, low energy magnetic fields. The theory suggests the possibility that the applied fields may be tuned to make the path of the ion fit the shape of the inside of the channel (or conversely make the path a poor fit to the shape of the channel) and, thus, to allow the applied fields to control ion conduction by controlling the proximity of the ion to the channel wall. This new theory will be published soon and is bound to stir further controversy. The theory answers a number of questions, predicts a large portion of the experimental data, and raises new questions. It does, however, represent another step in the road to a model that can be tested.

## RISK AND BENEFIT

One could speculate on events that might increase or decrease the speed with which these low level EMF devices will emerge into medical practice. There seem to be three points, or questions on which this emergence hinges. First, if cells can, using these fields, be made to speed up their growth (such as bone healing), what is the potential that cells will go into uncontrolled growth when exposed to these fields? Second, how can new devices be developed without a real understanding of the mechanisms of interaction? And, third, can the public or government perception of these devices be swayed in such a fashion that the devices are never allowed to reach their full potential in clinical medicine?

The first question is of great importance from a safety standpoint as well as from a scientific standpoint. There is an abundance of data that indicate that most cells that are in growth equilibrium do not respond to the application of these low energy level EM fields. The PEMF devices have been in widespread use for a number of years now, and there is no report of cells near an exposure site going out of control. There are however, numerous reports that these EM fields can influence the growth of certain cancer cells. There are also reports from both the United States and abroad that embryonic cells can be affected with these fields.[27] This suggests that a cell in transition is not exactly the same cell after it reaches equilibrium. Until an understanding of this phenomenon is obtained, the emergence of the technology must be carefully controlled. Risks must be weighed against benefits, and safety data must be obtained for each new device. The general public must be made aware of the facts also. It is quite possible that misinformation about what these fields can do to plants or animals would lead to long delays in the acceptance of any medical devices. The power companies should be actively involved in this research since many of the fields these companies generate fall into the range where resonance could occur given the earth's geomagnetic field. A firm position that these fields have no biological effect will expose the power companies to litigation, as has been the case in New York, Texas, and Florida. Such a position will also impede the acceptance of beneficial EM medical devices.

The second question concerns the lack of understanding of the underlying mechanisms. The development of devices without a theory or model from which to work is rather like constructing a complex electronic device in a dark room. There are many ways to fit the pieces together, some of which may work, but most of which will not work. The probability of discovering the optimum design in the dark is very low. The question of credibility is fuelled by the lack of an acceptable model. The problem is, of course, that a biological system is enormously complex, and its built-in feed-back mechanisms make it almost impossible to obtain a clear cause and effect relationship that will lead to a model. If one attempts to model an entire biological system (an animal, or plant for instance) it is soon obvious that there is simply no way to account properly for all the various outcomes, given a specific input to the system. If one attempts to model a specific biological event or pathway that leads to a known outcome, the model often is too restricted to be of general use. A compromise between these two extremes is the attempt to formulate a model at the cell level. If one can develop a model that accurately predicts how a single cell (or group of the same kind of cells) responds to EM signals, there is some hope of scaling the model to predict the reaction of whole animals or plants to the EM signals. The only way of ascertaining the accuracy of the scaling is by testing, much the same as has been done in the development of pharmaceuticals. In a sense then, the focus of the work of trying to develop the model is forced to the cellular level. Even here, however, the system is not simple. Perhaps the best that can be done is a black box representation of the cell. However, it is imperative that enough understanding be obtained to predict the parameters of the EM signal that will cause a response from the cell. Without at least that level of a model, one is working with an infinity of possibilities in the design of new devices.

The last point may be the most critical. Public opinion governs the eventual use of any of these devices that emerge into the clinic as well as how the uses of EM fields are regulated. There is a real danger that public opinion will be swayed toward the idea that everything causes cancer. Publicity involving power line sites and the publicity surrounding studies indicating that levels of childhood leukemia may increase in areas near power lines are bound to create doubts about these devices. Care must be taken, particularly by the scientific community, to avoid giving the impression that these EM fields have been thoroughly studied, and that no possible adverse effects can be attributed to the fields. The interaction of low energy EM fields with biological systems must be treated with caution since all the answers are not known. Both good and bad can come from the exposure to these fields, but the good will outweigh the bad if caution is used in their further development.

## CONCLUSIONS

Is this technology emerging? It appears that the answer is yes. The major role that low energy, low frequency electromagnetic fields will play in the clinic in the next five years will still be primarily in the area of bone healing and bone treatment. That role will expand, however, to cover many new aspects of bone treatment, including osteoporosis, joint replacements, and joint diseases. Further away, but a possible clinical application of the technology, are devices that may assist the body to grow or regenerate nervous system cells. Perhaps at the far end of the tunnel, there is a glimmer of data that suggests that cell growth enhancement or growth reduction is possible. In plain words that means regeneration of organs or cancer cell control. It is this glimmer of hope for true cell control using externally applied EM fields that causes increasing numbers of scientists to devote their careers to this field. The potential exists for EM fields to emerge as a replacement for or adjunct of pharmacologic therapies and to open new ideas of treatment that do not exist today.

## References

1. Aaron RK, Lennox D, Bunce G, Ebert T. (abstract) The conservative treatment of osteonecrosis of the femoral head: A comparison of pulsed electromagnetic fields and core decompression. Bioelectric Repair and Growth Society Meeting Abstracts 1988; 24.

2. Bassett CAL. (abstract) Do PEMF'S have a legitimate role in medicine and biology. Bioelectromagnetics Eleventh Annual Meeting Abstracts 1989; 14.

3. Becker RO, Sleden G. The Body Electric. Quill, New York, 1985.

4. Blackman, CF. (abstract) Response of an in vitro brain-tissue preparation to extremely low-intensity electric and magnetic fields. Bioelectromagnetics Eleventh Annual Meeting Abstracts 1989; 13-14.

5. Brodsky AE, Khalil MA. (abstract) Update of experience with electrical stimulation for enhancement of lumbar spine fusions (both pulsing electromagnetic fields therapy and implantable batteries). Bioelectric Repair and Growth Society Meeting Abstracts 1988; 29.

6. Ciombor DM, Aaron K. (abstract) Enhancement of bone maturation in experimental endochondral ossification by stimulation of chondrogenesis with pulsing electromagnetic fields. Bioelectric Repair and Growth Society Meeting Abstracts 1988; 29.

7. Feinberg JH, Khouri RK, Spielholz N, Harper A, Chiu DTW. (abstract) Nerve regeneration through an autologus venous nerve-conduit: The effects of conduit diameter and of neurite promoting factors. Bioelectric Repair and Growth Society Meeting Abstracts 1988; 34.

8. Herbst E, Sisken BF, Wang HZ. (abstract) Assessment of vascular network in rat skin flaps subjected to sinusoidal EMFs using image analysis techniques. Bioelectric Repair and Growth Society Meeting Abstracts 1988; 40.

9. Hisenkamp M. (abstract) Consequences of faster ossification induced by electromagnetic stimulation on the biomechanics of fresh fracture repair. Bioelectric Repair and Growth Society Meeting Abstracts 1988; 23.

10. Jeran M, Moratti A, Zaffuto S, Bagnacani M, Annovi M, Cadossi R. (abstract) Effect of PEMF on the healing of skin ulcers of venous origin in humans: A double blind study. Bioelectric Repair and Growth Society Meeting Abstracts 1988; 47.

11. Kennedy WF, Roberts CG. (abstract) Treatment of loosened hip prostheses with pulsed electromagnetic fields. Bioelectric Repair and Growth Society Meeting Abstracts 1988; 45.

12. Liboff AR., Cyclotron resonance in membrane transport. In: Interactions Between Electromagnetic Fields and Cells (Eds. A Chiabrera, C Nicolini, HP Schwan). Plenum, London. 1985; 281-296.

13. Liboff AR, McLeod BR. Kinetics of channelized membrane ions in magnetic fields. Bioelectromagnetics 1987; 9:39-51.

14. Liboff AR, McLeod BR. (abstract) Cyclotron resonance in ion channel proteins with 3-fold cylindrical symmetry. Bioelectromagnetics Tenth Annual Meeting Abstracts 1988; 31.

15. Liboff AR, Smith SD, McLeod BR, Experimental evidence for ion cyclotron resonance mediation of membrane transport. In: Mechanistic Approaches to Interactions of Electric and Electromagnetic Fields With Living Systems.(Eds. M Blank, E Findl). Plenum Press, NY. 1987; 109-132.

16. Lyle DB, Ayotte RD, Wang Z, Sheppard AR, Adey WR. (abstract) Activation and proliferation of normal and leukemia T-lymphocytes exposed to magnetic fields under calcium cyclotron resonance conditions. Bioelectromagnetics Eleventh Annual Meeting Abstracts 1989; 13.

17. McLeod BR, Liboff AR. (abstract) Electromagnetically induced osteogenesis: Cyclotron resonance as a testable hypothesis. Fifth Annual Meeting of the Bioelectrical Repair and Growth Society 1985; 16.

18. McLeod BR, Liboff AR. Dynamical characteristics of membrane ions in multi-field configurations at low frequencies. Bioelectromagnetics 1986; 7:177-189.

19. McLeod BR, Liboff AR. Cyclotron resonance in cell membranes; The theory of the mechanism. In: Mechanistic Approaches to Interactions of Electromagnetic Fields With Living Systems (Eds. M Blank, E Findl). Plenum Press, NY. 1987.

20. McLeod BR, Liboff AR, Smith SD. (abstract) A theoretical model that predicts frequency, amplitude and harmonic resonances in cell channels. Bioelectromagnetics Eleventh Annual Meeting Abstracts 1989; 33.

21. McLeod BR, Liboff AR, Smith SD, Cooksey K. (abstract) Harmonic response patterns of biosystems exposed to weak EM fields. Bioelectromagnetics Ninth Annual Meeting Abstracts 1987; 23.

22. Pethica BA, Brownell J. (abstract) The dose-response relationship in PEMF therapy on ununited fractures. Bioelectric Repair and Growth Society Meeting Abstracts 1988; 22.

23. Reese JA, Frazier ME, Morse JE, Miller DL. (abstract) A confirmation of Diatom mobility at 16-Hz electromagnetic fields. Bioelectromagnetics Eleventh Annual Meeting Abstracts 1989; 59.

24.Rozek RJ, Sherman ML, Liboff AR, McLeod BR, Smith SD. Nifedipine is an antagonist to cyclotron resonance enhancement of $^{-45}Ca$ incorporation in human lymphocytes. Cell Calcium 1987; 8:413-427.

25. Sharrard WJW. A double blind trial of pulsed electromagnetic fields for delayed fracture union. Orthopedic Transactions 1988; 12:555-556.

26. Smith SD, McLeod BR, Liboff AR. Cyclotron resonance control of explanted chick femurs. Eighth Annual Meeting of the Bioelectrical Repair and Growth Society 1988; 6.

27. Smith SD, McLeod BR, and Liboff AR. Control of embryonic chick bone growth. (under editorial review).

28. Transactions of the Second Annual Meeting of the Bioelectrical Repair and Growth Society 1982; 2.

29. Transactions of the Eighth Annual Meeting of the Bioelectrical Repair and Growth Society 1988.

30. Bioelectromagnetics Eleventh Annual Meeting Abstracts 1989.

31. Transactions of the Ninth Annual Meeting of the Bioelectrical Repair and Growth Society 1989.

32. Winters WD, Liboff AR, McLeod BR. (abstract) Enhancement and inhibition of protein synthesis in human chondrosarcoma cells exposed to 60 Hz magnetic fields. Seventh Annual Meeting of the Bioelectrical Repair and Growth Society 1987; 86.

# ELECTROPORATION: A NEW PHENOMENON TO CONSIDER IN MEDICAL TECHNOLOGY

James C. Weaver

## INTRODUCTION

Medical procedures and devices which cause electric fields or currents within the body may cause electroporation in exposed cells and tissues. In summary form, electroporation is the occurrence of a significant size and number of transient pores in cell membranes because of an elevated transmembrane potential [U(t)]. Several reversible and irreversible phenomena are now believed to be a consequence of these pores, but whether reversible or irreversible behavior occurs at a microscopic level is probably not the most important question. Instead, whether or not physiologically significant damage occurs, and whether or not the extent of any damage is acceptable in comparison to the benefits of the procedure, is much more important. With this in mind, we review what is presently known about electroporation, and suggest questions that might be asked in considering medical technology which involves exposure of cells and tissue to electrical fields and currents.

## BACKGROUND

Early experimental reports of phenomena now attributed to electroporation include the observation of a reversible electrical breakdown in the nodes of Ranvier,[37] non-thermal killing of microorganisms by electric fields,[35] dramatic permeability changes in biological vesicles[28] and permeability changes and lysis of red blood cells (RBCs).[26,49] With the exception of references,[37,49] primary electrical behavior was not sought in early work, but instead emphasized altered transport of ions and molecules, or overall cellular effects such as cell killing. The emphasis on molecular transport has continued, with much of the present work concerned with uptake of genetic material by cells.[30,31,47] Further, almost all of the studies to date utilized isolated cells, and only a few recent papers reported tissue electroporation.[10,25,26,33]

Electroporation involves the occurrence of transient pores in membranes which contain phospholipid bilayers, and is therefore believed to occur universally in cell membranes.[30,31,47] The pores are caused by a large transmembrane potential, U(t), that is due to the combined effects of a cell's resting potential, $U_0$, and a change, $\Delta U(t)$, caused by the applied electric field,

$\vec{E}_e$. Alternatively, one can regard an applied current density, $\vec{J}_e$ as causing $\Delta U(t)$, as the relationship between current density and electric field is general, $\vec{E}_e = \vec{J}_e/\sigma_e$. If $\vec{J}_e$ is considered to be the applied quantity, this basic relationship emphasizes that it also is important to know the electrical conductivity of the medium, $\sigma_e$. Thus, in considering whether or not electroporation occurs as the result of a medical device or procedure, it is imperative to estimate $\vec{E}_e$, either directly, or indirectly through estimates of $\vec{J}_e$ and $\sigma_e$.

The mechanism for electroporation is not yet well understood. Although the detailed dynamic process of pore formation has not yet been described, it is generally believed the pores form because of the electrical energy density of the field, $E_m$, across the membrane. As $U(t)$ increases, the fact that electrical energy density within the membrane is proportional to $E_m^2$ results in more favorable conditions for openings (pores) with a large dielectric constant ($K_W \approx 80$) rather than the continuous sheet of lipid-rich membrane ($K_L \approx 2$).[1,2,34,38,44] Whether or not transient pores exist at zero or normal physiologic transmembrane potentials is not yet known. However, it appears that if $U(t)$ reaches values in excess of about 200 mV, then a large number of pores with a distribution of sizes appears in the membrane.[2,34,45] The pores result in a number of phenomena, and include both reversible and irreversible effects.

## SIGNIFICANCE OF ELECTROPORATION

Electroporation is an important topic of study because of: 1) potential insights into fundamental membrane behavior, 2) potential applications in biology and medicine, and 3) potential damage to cells and tissue incidental to the use of biomedical technology. In the first case, it is significant that electroporation appears to occur in all bilayer-containing membranes, (ie electroporation seems to be a universal cell membrane phenomenon). For this reason, the dramatic consequences of electroporation cannot help but be important to fundamental understanding of bilayer-containing membranes. In the second case, electroporation is significant because a wide variety of applications can be expected from a phenomenon which allows electrical processes to transiently alter cell membrane permeability. Existing electrical phenomena in biology and medicine are generally important, because of the relative ease (compared to chemical phenomena) with which electrical phenomena can be measured and caused to occur. Important naturally occurring bioelectric phenomena include neural action potentials, electrocardiac potentials, myoelectric action potentials and other active cellular potentials. Other, non-bioelectric phenomena include electrophoresis, microelectrodes, electrochemical sensors and membrane potential fluorescent dyes. A tremendous effort has gone into the investigation and application of these electrical phenomena. In contrast, the effort directed towards the study of electroporation has been much smaller, with the result that both empirical and theoretical understanding is at an early stage. As further empirical and theoretical understanding is gained, it is likely that widespread technical applications will result, extending well beyond the present use[31] for introducing genetic material into cells. The third case is of primary interest here. Although most knowledge of electroporation was obtained from investigations of isolated cells, a few recent studies explicitly considered tissue electroporation.[10,25,26,33] Given the paucity of direct studies with tissue, hints concerning tissue electroporation can be sought from studies on isolated cells. The cited literature provides illustrations of what is presently known, but it should be understood that ongoing and future studies can be expected to significantly improve our understanding of electroporation.

## SEQUENCE OF EVENTS ASSOCIATED WITH ELECTROPORATION

Figure 1 illustrates some of the main events that are believed to occur during electroporation of an isolated spherical cell. A more detailed, but still brief, discussion is provided below.

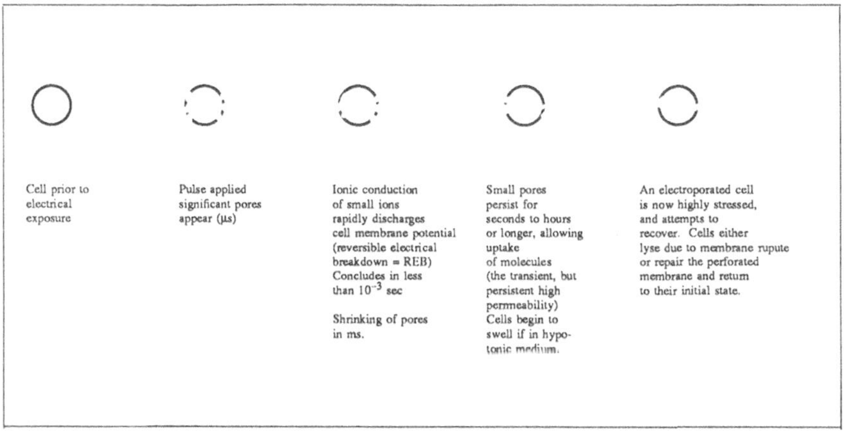


**Figure 1.** Illustration of qualitative sequence of electroporation events. A brief description of these events is given in the text. An isolated spherical cell is assumed, as this is a good approximation to the usual experimental arrangement of a cell suspension.

A high voltage pulse of magnitude $\Delta\phi$ is applied across two electrodes (assumed to be the usual configuration of parallel planes of separation L) that contain a cell suspension. The resulting nominal electric field is $E_{nominal} = \Delta\phi / L$, but this is not necessarily the field actually experienced by the cells.[11,43] Instead an electric current, I(t), flows within the aqueous electrolyte comprising the suspending medium. The corresponding current density is $j(t) = I(t)/A_{electrode}$, where $A_{electrode}$ is the macroscopic surface area of one of the two electrodes. There is an associated electric field, $E_e = j/\sigma_e$, where $\sigma_e$ is the electrical conductivity of the medium. Generally $E_e \leq E_{nominal}$ because significant potential drops can occur at the electrode/electrolyte interface.[21]

The transmembrane potential, U(t), increases rapidly, with U(t) having different values at different sites on the cell membrane. For a spherical cell at low values of U(t), the relationship between the applied field and the change in transmembrane potential, $\Delta U(t\theta)$ at different sites on the membrane is given by the well-known.[20]

$$\Delta U(t\theta) \approx 1.5E_e\,(t)R_{cell}\,\cos\theta$$

Here $R_{cell}$ is the cell's radius, and $\theta$ is the angle between the direction of $E_e$ and the site on the membrane under consideration. The maximum change occurs at the poles ($\cos\theta \pm 1$), yielding the widely used estimate

$$\Delta U(t)_{max} \approx 1.5E_e\,(t)R_{cell}$$

If the cell's resting transmembrane potential prior to application of $E_e$ was $U_0$, then $U(t) \approx U_0 + \Delta U(t)$, an estimate which neglects the perturbation of the resting potential by the changes in permeability caused by electroporation. Dramatic electroporation effects are believed to occur if $\Delta U(t)_{max}$ reaches values of about 200 mV to 1500 mV.

Some type of membrane openings (pores) rapidly develop in the sense that large numbers and/or large sizes of transient aqueous pores appear if $200\ mV < U(t) < 1500\ mV$. Such pores are generally capable of passing both ions and molecules, but the specificity and size cutoff change with time. In addition to small ions and molecules, macromolecules such as proteins and DNA, or small particles such as latex beads or viruses, may begin to cross the membrane at this stage.

If U(t) reaches a larger value, often approximately 500 mV, reversible electrical breakdown (REB) occurs by massive ionic conduction through pores.

A very rapid decrease in membrane resistance, **R(t)** therefore occurs, but the membrane capacitance, **C**, hardly changes. As a result the membrane is rapidly discharged, with an approximate **RC** time constant governed by the changing **R(t)**. Essentially complete discharge is usually obtained within $10^{-3}$secs, often within $10^{-6}$ sec to $10^{-5}$ sec, or even faster.[2,4,32,45]

The temperature, **T**, of the medium rises at a tremendous rate (eg, of order $10^5$ degrees $sec^{-1}$). However the actual temperature change is often negligible (eg, $1^{o}C$), because of the short duration of the applied electric field or current. In this sense, electroporation is a non-thermal event.

A transient high permeability state has developed. The elevated permeability persists long after **U(t)** has decayed to zero, often for seconds or longer, with the time dependence apparently depending on cell type and on temperature. This high permeability appears to allow transport of all types of ions and molecules, ranging from ubiquitous small ions to macromolecules. The maximum size of transported entities is not yet known, nor is the time range at which high permeability for the larger molecules vanishes. It also is not at all clear whether or not pores alone can account for the transport of small particles, but it is likely that one or a small number of metastable pores can account for considerable macromolecular transport by hindered diffusion.[41,46]

Membrane recovery begins to occur as **U(t)** first begins to decrease, for then the expanding pressure associated with $\mathbf{E_m^2}$ . Initial recovery is believed to be rapid[2] but after essentially complete discharge of the membrane by REB some pores may shrink and/or disappear much more slowly. As pores shrink, the cell membrane excludes smaller and smaller molecules and ions. Persistent pores, even very small pores which can pass small ions, may cause in the long term, cell stress and, eventually, cell lysis.[24,27,43]

An electroporated cell usually becomes highly stressed, through the loss of vital intracellular compounds, and also through an influx of extracellular compounds. This intracellular and extracellular exchange continues, with a time varying molecular cutoff interposed, until membrane recovery is complete. For this reason the composition of the suspending medium is believed to be important, with minimization of intracellular and extracellular differences relevant. As a result of chemical imbalances, a variety of cellular functions can become temporarily or permanently disrupted. Electroporation induced cell stress is presently poorly understood, but is crucial to understanding the conditions for cell death and cell survival.

Cells often recover completely, so that survival can occur in a significant percentage of electroporated cells. Other cells die, apparently because of delayed effects such as osmotic lysis,[24] or the unrelieved demands on metabolic systems because of membrane shunting by persistent pores. Surviving cells can retain the electroporatively introduced molecules if the molecules do not permeate the recovered membrane and are not degraded intracellularly.[23,29,48] Many of the electroporation studies have been carried out on erythrocytes or their ghosts,[19,23,49] and it should be noted that RBC ghosts, when osmotically stressed, exhibit unusually resilient recovery. Many other studies have used artificial planar bilayer membranes[4,6,8,12,13,14,22] which contain no membrane proteins and are not interacting with other cell constituents. For this reason, inferences made from RBC ghost and artificial planar bilayer membrane experiments may not be applicable to cells generally. Finally, although application of electroporation to gene transfer has involved many types of cells and protoplasts[30,31,47] these experiments yield only the biological endpoint of transformation, and therefore reveal little about the early electroporation events (eg, amount of molecular uptake, kinetics of membrane recovery). Overall, therefore, understanding of electroporation of cells is still quite incomplete.

## ELECTRICAL BEHAVIOR ASSOCIATED WITH MEMBRANE RUPTURE

Electroporation can cause non-thermal rupture, (ie, prompt destruction, of artificial planar bilayer membranes through runaway expansion of supracritical pores).[1,2] A partial rupturing of cell membranes may occur in a modified fashion. Although the closed membrane of a cell or vesicle is not globally vulnerable to the unbounded expansion of pores,[40] if portions of the cell membrane are attached to essentially permanent (on the time scale of interest) cell structures, then a portion of the cell membrane may behave similarly. That is, a portion of a cell membrane may be destroyed. This rupture can lead to an essentially permanent loss of a significant fraction of the cell's membrane. Alternatively, creation of one or a small number of large pores may result in pore trapping, with many of the same destructive consequences for the cell. In planar bilayer membranes the rupture process is prompt (eg, occurring within about $10^{-3}$ seconds), and has previously been termed "irreversible mechanical breakdown". Rupture occurs in planar bilayer membranes exposed to a transmembrane potential, **U(t)**, if **U** is approximately in the range 200 mV<**U(t)**<600 mV for times of about >$10^{-4}$ sec.[1,12,13,14,16,22 34,38,44] Prompt rupture of planar membranes involves electrical discharges with decay times significantly longer than those of REB.

This prompt rupture is termed non-thermal because the bulk temperature rise is generally small. In the case of planar bilayer membrane experiments, the lack of heating results from the experimental configuration: the high resistance membrane exists between the stimulating electrodes, which prevents significant current from flowing, thereby limiting ohmic dissipation in the medium. In this sense, rupture is not caused by heating. Instead, the existence of $U>200$ mV for times of order of $10^{-4}$ or longer leads to pores which expand beyond a critical size. The supracritical pores then expand rapidly, without limit, until a mechanical boundary is encountered, and this effectively destroys the membrane.[1,2,34,38,44] Another significant attribute of prompt rupture is that the membrane capacitance, C, changes insignificantly, which rules out significant electrocompression of the entire membrane[16,17,18] as the mechanism of rupture.[12,14] Another striking attribute of electroporation is that planar membranes often avoid rupture when exposed to the shorter, but significantly larger transmembrane potentials which cause REB.

A characteristic electrical signature of rupture is a sigmoid curve of **U(t)**, in which the destruction phase of the membrane is revealed by **U(t)** smoothly decreasing over an interval of many microseconds, which is much longer than the very rapid and history dependent decay of REB. Although early electroporation theories[1,12,14,34,38,44] correctly predicted the approximate magnitude of the critical transmembrane potential, $U_c$, associated with rupture, they did not describe the kinetics of rupture in a way which gave a description of **U(t)** which could be directly compared to experiments. In contrast, a recent version of a transient pore theory successfully describes **U(t)** during the rupture or mechanical breakdown process.[2]

## ELECTRICAL BEHAVIOR ASSOCIATED WITH REVERSIBLE ELECTRICAL BREAKDOWN

Given the existence of rupture at moderate elevated U, the existence of REB is tantalizing, as REB occurs at higher U and rupture is usually avoided. Further, the behavior of **U(t)** during reversible electrical breakdown (REB) is even more dramatic than that during rupture. Essentially by definition, REB is a rapid membrane discharge, followed by complete recovery.[4,5,6,7,8,13,37] The ability of a membrane to support REB can be qualitatively understood to be the result of larger numbers of pores appearing, with the result that their ionic conductance allows the membrane to discharge (relieving the electrical pressure) before even one pore can expand to supracritical size (which would cause rupture). A quantitative description is consistent with this view, and rests on a physical model which explicitly: 1) treats a rapidly changing population of pores with a distribution of sizes; 2) includes the local transmembrane potential decreases near conducting pores; and 3) incorporates a circuit equation to describe both the charging and discharging of the membrane.[2,8,32] Although our understanding of electroporation of artificial planar bilayer membranes is still incomplete, the success of this theory in describing several different electrical consequences of electroporation suggests that many features of the model are essentially correct.

An incomplete discharge (incomplete REB) can also occur, if the increase in **U(t)** does not result in quite enough pores (size and number) to completely discharge the membrane. This has been observed experimentally in the same membrane preparations which exhibit complete REB,[4] and is described quantitatively by a theory which also correctly describes rupture and REB occurrence.[2]

REB has several characteristic features: 1) it appears to immediately precede the longer lasting high permeability state observed in cell membranes; 2) during REB the membrane conductance, **G(t)**, of planar membranes rapidly increases to large values, and then recovers; 3) there are characteristic combinations of magnitude and time which result in REB: typical square pulse characteristics which result in REB are a pulse width, $\Delta t$, in the range $10^{-7}<\Delta t<10^{-4}$ second, and transmembrane potential magnitudes in the range $500<U<1{,}500$ mV; and 4) experiments using these short pulses show that the recovery process occurs within microseconds.[8]

If a series of short pulse experiments is carried out, REB is found to occur with an increasingly rapid discharge as the pulse amplitude, $\Delta U_i$, is increased for fixed pulse width, $\Delta t$, or as $\Delta t$ is increased at constant $\Delta U_i$. An approximate threshold potential for REB, $U_{0,c}$ is identified as the maximum value of $U_o$, which decreases as $\Delta t$ increases.[4] Finally, another significant attribute of electroporation in artificial planar bilayer membranes is the observation that rupture, REB or even incomplete REB followed by rupture can all be caused to occur by varying only the magnitude and duration of the applied pulse.[4] This interrelated behavior places significant constraints on the mechanism of rupture, incomplete REB and REB, and must be described by any unified theory of planar membrane electroporation. Although the theory is still incomplete, a recent transient aqueous pore theory has been able to accomplish this.[2]

## ELECTROPORATION IS NOT VIOLENT AT THE MOLECULAR LEVEL

The energetics of electroporation is particularly interesting, because electroporation is biochemically mild. Thus, although REB has sometimes been termed "dielectric breakdown", this is a misleading term because of the previously well established usage of dielectric breakdown for ion-pair avalanche phenomena, a very different phenomenon which involves significantly larger energy, viz. about 10 electron volts ($1.6 \times 10^{-18}$ joule) per monovalent ion.[15] In fact the transmembrane potential associated with REB (500 to 1,500 mV) corresponds to only 0.5 to 1.5 electron volts per monovalent ion, about an order of magnitude less energy, and this smaller energy is insufficient for ion-pair formation. In contrast, pore formation and expansion is able to occur at these lower energies, and provides an explanation of both rupture and REB.

## ELECTRICAL BEHAVIOR: COMPARISON OF EXPERIMENT AND THEORY

Although the molecular transport associated with electroporation is of great importance, in many circumstances the electrical behavior is more readily measured. For this reason, it is worthwhile to present several illustrations of electrical behavior. A recent theoretical model of electroporation of artificial planar bilayer membranes provides a unified, quantitative description of several aspects of electrical behavior following electroporation, and also provides some insight into pore behavior which cannot be directly measured.[2] The results shown in Figs. 2 - 5 are for conditions that generally cause REB or rupture, and are in good agreement with experimental behavior.[4] This theory is based on a number of simplifying hypothesis and approximations in the physical description of pores. One weakness is that a detailed description of the dynamic process of pore formation and pore disappearance does not yet exist. Instead, pores are assumed to be created and destroyed by a process dependent on $\mathbf{U(t)}$, with an assumed form for creation and destruction. Once formed, however, the response of the pores to both thermal fluctuations, mechanical forces and electric fields is described by a first principles physical theory.

## ELECTROPORATION INVOLVES PORES OF MANY SIZES

An important implication results from the success of this theoretical model: a large number of pores with a significant distribution in sizes is required. Different subpopulations of the total pore population are responsible for different aspects of the electrical behavior. Briefly, a small number (one or more) of large pores are responsible for rupture, while a much larger number of smaller pores are responsible for REB. The complexity of the pore population is illustrated in Fig. 4, where the number of pores, $\mathbf{n(r,t)\Delta r}$ with radii between $\mathbf{r}$ and $\mathbf{r+\Delta r}$ and is shown as a function of time during a REB curve. For the theoretical result of Fig. 4, initially fewer than ten pores are present for the macroscopic planar bilayer membrane modeled. As time progresses, the rapidly increasing $\mathbf{U(t)}$ results in both more and larger pores, and then a return as the pore population collapses upon discharge of the membrane. In a cell membrane,

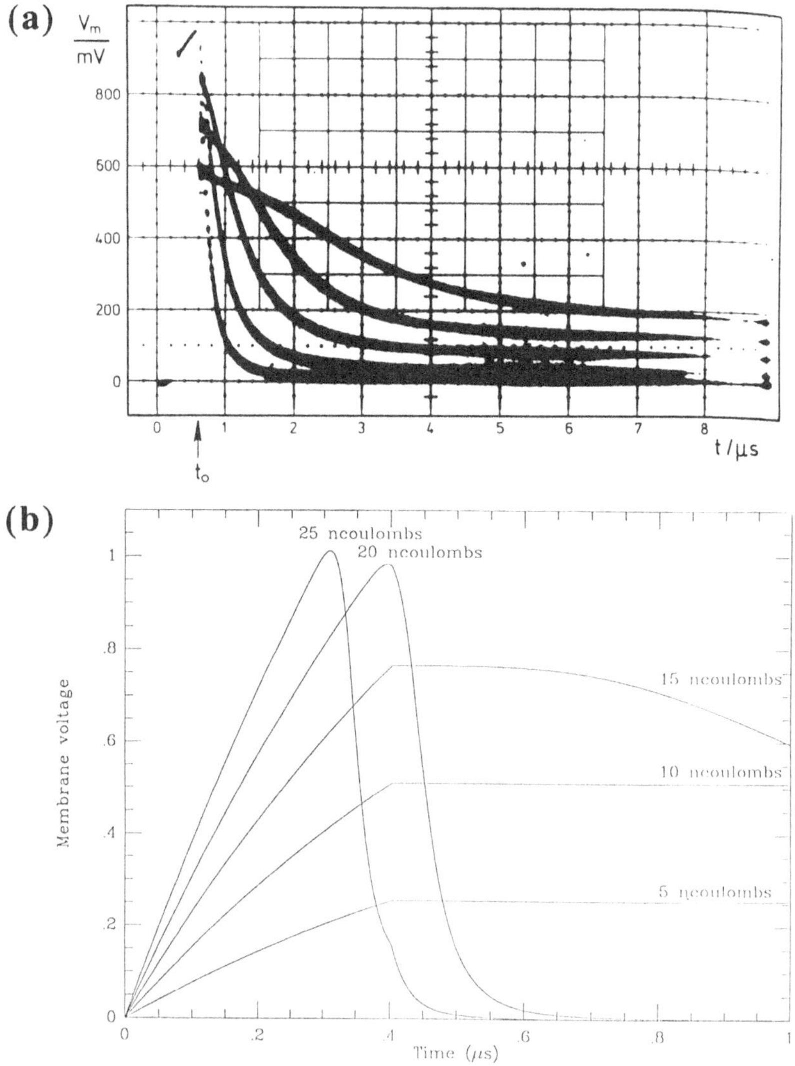


**Figure 2.** Comparison of experimental and theoretical behavior for REB caused by charge injection. (a) Experimental results showing **U(t)** after the 0.4 µsecond pulse terminates.[4] (b) Theoretical behavior of the transmembrane potential **U(t)** during REB caused by electroporation as the result of a charge injection protocol that introduced $\mathbf{Q_{inject}}$ **= 20 ncoulomb** during a 0.4 µsecond square pulse.[2] Five values of $\mathbf{Q_{inject}}$ were used (5 ncoulomb to 25 ncoulomb), with the 20 ncoulomb corresponding approximately to a threshold for REB. This result agrees reasonably with experimental results.[4]

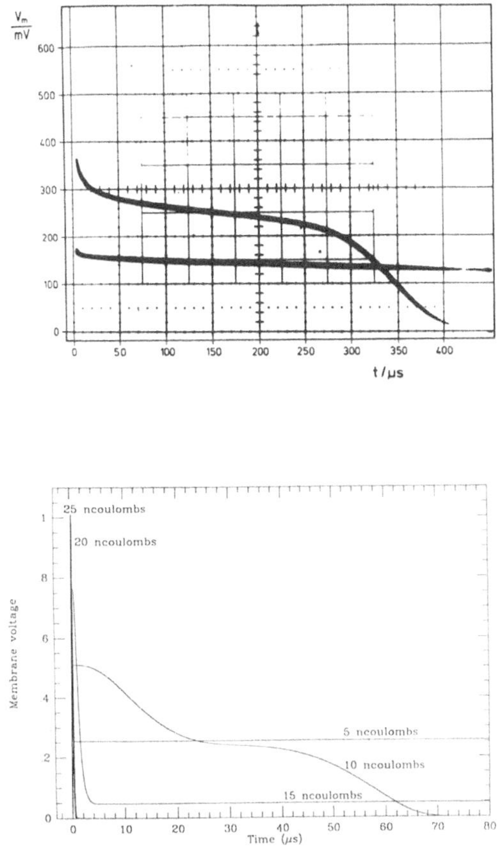


**Figure 3.** Comparison of experimental and theoretical **U(t)** for rupture for a moderate 0.4 µsecond pulse. (a) Experimental behavior associated with rupture, with the characteristic sigmoidal decay curve for **U(t)**.[4] (b) Theoretical behavior of the transmembrane potential **U(t)** during rupture, a mechanical destruction of a membrane caused by electroporation.[2] The membrane is that of Fig. 2, but with a longer time scale. Although the time scale is different, the theoretical behavior is otherwise in good agreement with the experiments.[4] Four distinguishable types of behavior are evident in Figs. 2 and 3: (1) For small injected charge (5 ncoulomb) the membrane charges to a moderately low voltage (here 230 mV), which is retained for a long time on the scale of the experiment. This essentially passive membrane charging results because few pores exist which have grown large enough to conduct significantly, and the membrane resistance therefore remains too large to rapidly discharge the membrane. In addition, there are no large pores, so membrane rupture does not occur. (2) For larger $\mathbf{Q_{inject}}$ (10 ncoulomb) the membrane develops more larger pores, and an initial slow discharge begins, levels off after about 20 µseconds, and then begins again as one or more pores have surpassed the critical size, and have begun to expand until the membrane ruptures. (3) For still larger $\mathbf{Q_{inject}}$ (15 ncoulomb) an incomplete REB is found, in which the membrane initially has sufficient pores (and corresponding large conductance) to discharge rapidly, but the pores' size and number is marginal, so that upon partial discharge enough pores shrink and/or disappear that the conductance diminishes, discharge halts, and the membrane is left with **U>0** (in this case about 230 mV). (4) For even larger $\mathbf{Q_{inject}}$ (20 and 30 ncoulomb) complete REB occurs, as sufficient pores exist transiently to completely discharge the transmembrane potential to zero, even though all pores shrink and some pores disappear during the discharge. Overall, this description of four distinguishable membrane fates is in good agreement with experimental results.[4]

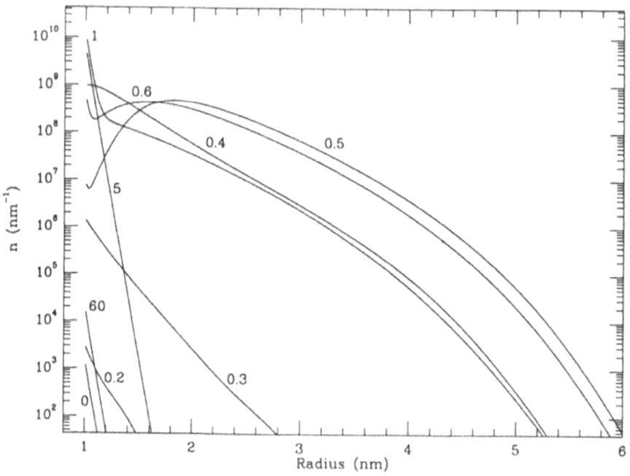


**Figure 4.** Plots of the computed pore population distribution at different times for REB (the 20 ncoulomb case of Fig. 2b) in an artificial planar bilayer membrane.[2] A wide range of pore sizes is involved. This is displayed by computing the number of pores, **n(r,t)Δr** with radii between **r** and **r+ Δr** is shown as a function of **r** for different times. Initially only a small number of pores are present (the short slanting curve labeled with "0" at the lower left corner). Then, as **U(t)** rises because of the 0.4 μsecond pulse (not instantaneously experienced by the membrane), pores increase in number and size. The pore population thus first expands (curves labeled 0.2, 0.3, 0.4. and 0.5 μseconds), and then collapses as **U(t)** decays (curves labeled 0.6, 5 and 60 μseconds). A large number of pores with diameters exceeding 6 nm are transiently present. If a cell membrane can trap, or otherwise create metastability, a single large pore, a high permeability state can exist.

any mechanism that captures one or a small number of the larger pores could provide a persistent high permeability that allows transport of fairly large molecules. Figure 5 illustrates the corresponding behavior in the case of membrane rupture for the same membrane; only the applied pulse is different.

## MOLECULAR TRANSPORT ASSOCIATED WITH ELECTROPORATION

Transport of molecules, not electrical behavior, has generated most of the recent interest in electroporation. Of particular interest is the empirical evidence that almost any macromolecule can be transported across cell membranes.[30,31,47] The introduction of genetic material into cells, but the large transient molecular fluxes are likely to have much wider applications. For example, it has recently been shown that very large numbers of molecules can be introduced into cells. In the case of the yeast S.Cerevisiae , intact cells can take up approximately $10^5$ macromolecules (FITC-dextran, 70,000 dalton at an

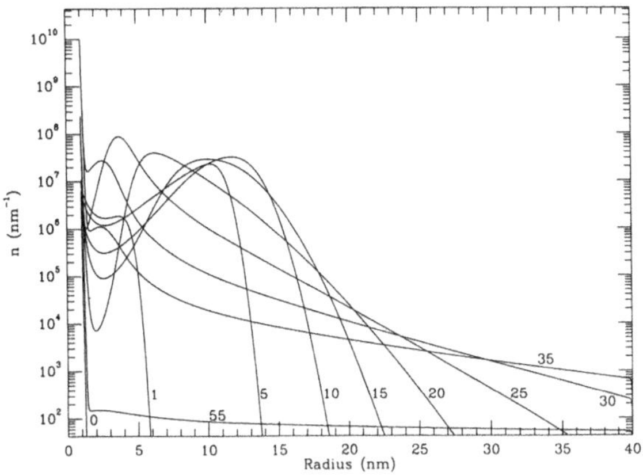


**Figure 5.** Corresponding plots of **n(r,t)Δr** for the case of rupture, viz. the curve of Fig. 2 with $Q_{inject}$ = 10 ncoulomb.[2] As in Fig. 4, the membrane initially (prior to a moderate 0.4 μsecond pulse) contains a small number of pores. The times labeling the curves are in μseconds from the time the pulse begins. As time progresses, there is a buildup of larger pores, until one or more pores becomes supracritical, and expands to rupture the membrane.

extracellular concentration of 500 μM), or approximately $10^8$ small molecules (propidium iodide, 660 dalton; extracellular concentration of 80 μM).[3] The transport of very large numbers of small molecules, large numbers of macromolecules, and at least some of DNA, RNA and even small particles, is impressive. Presently, however, there is very little quantitative understanding of molecular transport due to electroporation. Qualitatively, there is evidence that a transient high permeability state is believed to occur immediately following REB.[23,28,36,48] The nature of the openings (pores, or coalesced pores which may form cracks[39]) remains to be elucidated.

During the relatively short interval when **U(t)** is large, and many rapidly changing pores are present, significant electric fields exist across the membrane. After taking into account the spreading resistance, which results in significant potential drops within the electrolyte outside but near the pore's mouths, a field still exists within the pores. Electrokinetic transport of molecules is therefore possible[36] and is suspected of occurring in addition to hindered diffusion.[41,46]

After **U(t)** has decayed, significant molecular transport can still occur. Calculation shows that hindered diffusion is sufficient to transport large numbers of molecules across a cell membrane, even if only one large pore (eg, **r** ~ **5nm**) has a long lifetime. For example, with an initial concentration difference of

$\Delta C = C_{ext} - C_{int} = C_{ext} = 100\ \mu M$, this size pore can transport of order $10^6$ macromolecules of diffusion constant of $D_s \sim 10^{-7}\ cm^2/sec$ across a 5 nm membrane in 100 sec. Although qualitatively plausible, there is presently no detailed understanding of how the long lifetime high permeability state actually occurs, and what conditions affect its disappearance.

It is presently unknown how large metastable pores are formed, and what governs their rate of decay. A number of candidate interactions between pores and different aspects of a cell membrane have been identified, and can be considered for their significance in explaining persistent large pores. An example is a "foot-in-the-door" mechanism, in which the presence of a partially inserted macromolecule repels the pore edge, preventing pore shrinkage until the macromolecule has diffused back out or across the membrane (Barnett and Weaver, unpublished results). Such hypotheses remain to be incorporated into theoretical models in order to determine thereby whether the dramatic behavior of a persistent but metastable high permeability can be explained by such a pore mechanism.

## MEMBRANE RECOVERY AFTER ELECTROPORATION

Membrane recovery can be described by a progressively smaller molecular weight cutoff as pores shrink and/or disappear. Electrical measurements can reveal changes due to the conduction of small ions, while molecular transport measurements can probe a larger size range of pores. The relative ease of electrical measurements means that recovery probed by small ions can be readily carried out on a continuous basis, but the more difficult molecular transport measurements have not been usually reported. Early studies reported that in an artificial planar bilayer membrane the conductance, $G(t)$, decays approximately exponentially with a time constant, $\tau_{RC}$, of several microseconds,[8] while more recent studies using a different experiment arrangement report considerably longer recovery times, of the order of seconds.[13] Further, in terms of overall reversibility, it has been shown that an oxidized cholesterol planar membrane can undergo REB and recovery many times.[4] Similar behavior has been demonstrated in a model tissue.[33] Complete membrane recovery must also occur in many electroporated cells, as application of electroporation to gene transfer first causes uptake of DNA, and the subsequent obtaining of viable, transformed cells.[30,31,47] In the case of artificial planar bilayer membranes it is likely that membrane recovery takes place in two stages: 1) a rapid shrinkage of most pores as is in its decreasing phase, and 2) a reduction in pore number, as small pores are somehow destroyed.[2] Any metastable pores presumably present additional complication, as it is likely that both the shrinkage and disappearance stages will be different than those of transient aqueous pores in an artificial membrane.

## RECOVERY FROM CELL STRESS: DEATH OR SURVIVAL

The temporary disruption of the cell membrane's ability to restrict transport of all but selected molecules creates the possibility of cell stress. A number of possible mechanisms have been identified for example:

1. Loss of essential intracellular molecules
2. Admission of molecules normally excluded
3. Loading/shunting of transmembrane pumps
4. Denaturation of intracellular proteins by altered intracellular conditions; and
5. Non-recovery of a portion of the membrane (equivalent to rupture of that portion).

## DAMAGING TISSUE ELECTROPORATION

The possibility of tissue electroporation is important to evaluating medical technology that causes electric fields and current in tissue. Only a few studies to date have explicitly considered tissue electroporation, (ie, application of fields or currents which cause electroporation in some or all of the cells of the exposed tissue). As noted earlier, the sparsity of direct studies on tissue means that we must rely on studies of electroporation in isolated cells, and then make estimates. As with the more thoroughly studied isolated cell preparations, both reversible and irreversible effects can be anticipated. From the perspective of medical risks and benefits, the degree of tissue damage, rather than reversibility or irreversibility of electroporation phenomena themselves, appears most relevant.[42]

With this in mind, the studies on skeletal muscle are particularly interesting, as they provide evidence that some of the tissue injury associated with electrocution is non-thermal, and is due to electroporation.[10,25,26] Specifically, these studies show that skeletal muscle cells are lysed under conditions for which the temperature rise is negligible (less than $1^{o}C$). As might be expected from isolated cell electroporation experiments, field strength, pulse width and number of pulses were important. Changes in the membranes of the cells comprising the tissue were determined by electrical impedance measurements, with a large, permanent decrease after 100 pulses of duration 10 mseconds and separation 10 seconds for an applied field of 120 volt/cm. Overall, these studies provide strong evidence that electroporation and/or heating are important damage mechanisms. The question as to whether or not a very large number of small pulses can cause electroporation in tissue, and whether or not damage ensues, has not yet been addressed.

## NON-DAMAGING TISSUE ELECTROPORATION

Another study demonstrates that completely reversible tissue electroporation behavior can occur, without any evidence of damage.[33] In this study the evidence for the occurrence of electroporation is also compelling. Isolated, viable frog skin was used in order to provide a tissue preparation that had well defined electrical conditions, and that allowed direct electrical measurements that could be interpreted at the cellular level. The frog skin preparation presents a barrier to transport which is essentially that of a monolayer of cells which are connected laterally by tight junctions.[9] Thus, if this tissue is placed in a conventional tissue chamber, with electrodes on both sides, the resulting configuration is essentially a macroscopic monolayer of cells, and therefore similar to an artificial planar bilayer experiment. This allowed direct measurements of the transtissue potential and transtissue resistance (due mainly to one layer of cells). Results similar to REB in an artificial bilayer membrane were obtained, but with somewhat smoothed curves, which is not surprising because of order $10^6$ cells were involved. Specifically, the characteristic behavior of increasingly rapid decays, and of curve crossing was observed as: 1) the pulse amplitude was increased for fixed pulse width; and 2) the pulse width was increased for fixed pulse amplitude. This behavior is the characteristic electrical signature of electroporation, as has been previously observed in artificial planar bilayer[47,49] and in isolated cells[7] under charge injection conditions.[2]

Reversibility was also directly observed, by monitoring the transtissue resistance after a pulse on a time scale of several minutes. The resistance recovered most of its original value within 20 seconds, and returned completely within two to three minutes (25 ± 2°C). Further, there was no measurable evidence of tissue damage, as could be estimated from either transtissue resistance or open circuit potential measurements. By considering the measurement error and the resistance change that should have been caused by the complete lysis of one cell (creating a hole in the tissue), it was estimated that a single pulse causing electroporation resulted in fewer than four cells being lysed. The corresponding cell lysis rate was therefore less than about $10^{-6}$ per cell. Finally, because some preparations were subjected to twenty or more pulses without finding measurable damage, it was estimated that the cell lysis rate per pulse was even lower, having an upper bound of about $10^{-7}$ per cell per pulse.

The potential importancc of tissue electroporation to evaluation of medical technology suggests that many additional studies should be carried out. At the present time, there is strong evidence that, just as in the isolated cell studies, both reversible and irreversible phenomena can occur, and that tissue damage can be undetectable or pronounced.[48] However, the electrical conditions which separate non-damaging from damaging electroporation behavior in a particular tissue have not yet been established.

## EVALUATING TISSUE ELECTROPORATION INCIDENTAL TO BIOMEDICAL PROCEDURES

In evaluating medical technology not directed at causing electroporation, it is essential to estimate correctly the transmembrane potential of the cells of the tissue. Generally this will involve estimating local electric field, $\mathbf{E_e}$, that is experienced by the cells of the tissue, and then carrying out computations that take into account the geometry of the cells within the tissue. As part of this procedure, it may often be difficult to estimate $\mathbf{E_e}$ directly. In this case, the local current density, $\vec{J}_e$, and the corresponding electrical conductivity, $\sigma_e$ must be estimated.

Thus, for example, if it is of interest to evaluate a surgical procedure based on electrical cutting of tissue, the current density at different locations within the tissue should be measured or estimated, as should the duration of the electrical exposure. By combining these parameters with knowledge of the geometry (size and shape) of the cells in the tissue, is should be generally possible to estimate whether or not electroporation occurs in various parts of the tissue. It generally will require another step, however, to determine whether any occurrence of electroporation causes acceptable levels of tissue damage. Generally this can be expected to depend on the type of tissue and the composition of the extracellular medium at the time of electrical exposure. To date, essentially no such studies exist.

## SUMMARY

Electroporation occurs universally in cell membranes if the transmembrane potential reaches values greater than about 200mV, but also depends on the duration of this increase. In order to estimate whether or not electroporation occurs in a particular tissue as the result of a medical procedure or device, it is essential to estimate the extent and duration of changes in the transmembrane potential of the cells of the tissue. Whether or not significant tissue damage occurs requires further information, as electroporation itself can be reversible or irreversible. Cellular damage can be prompt, or delayed, and can depend significantly on the composition of the extracellular medium relative to the intracellular medium. Electroporation is incompletely understood empirically and theoretically, and what knowledge we have comes mostly from studies on artificial planar bilayer membranes or on isolated cells. Significantly more studies on mechanisms of cell damage following electroporation are needed, as are investigations of tissue electroporation.

## ACKNOWLEDGMENTS

The author thanks Tian Y. Tsong, Kevin T. Powell, Raphael C. Lee, Gail I. Harrison, Stephen H. Grund, Jonathan G. Bliss, and Alan Barnett for stimulating discussions. Supported by the Office of Naval Research through Contract N00014-87-K-0479.

## References

1. Abidor IG, Arakelyan VB, Chernomordik LV, Chizmadzhev YA, Pastushenko VF, and Tarasevich MR. Electric breakdown of bilayer membranes: I. The main experimental facts and their qualitative discussion. Bioelectrochem Bioenerget 1979; 6:37-52.

2. Barnett AS, Weaver JC. Electroporation: Unified, quantitative theory of reversible electrical breakdown and rupture. (under editorial review).

3. Bartoletti DC, Harrison GI, Weaver JC. The number of molecules taken up by electroporated cells: Quantitative determination. (under editorial review).

4. Benz R, Beckers F, Zimmermann U. Reversible electrical breakdown of lipid bilayer membranes: Charge-pulse relaxation study. J Membrane Bio 1979; 48:181-204.

5. Benz R, Conti F. Reversible electrical breakdown of squid giant axon membrane. Biochim Biophys Acta 1981; 645:115-123.

6. Benz R, Zimmermann U. Pulse-length dependence of the electrical breakdown in lipid bilayer membranes. Biochim Biophys Acta 1980; 597:637-642.

7. Benz R, Zimmermann U. Relaxation studies on cell membranes and lipid bilayers in the high electric field range. Bioelectrochem and Bioenerg 1980; 7:723-729.

8. Benz R, Zimmermann U. The resealing process of lipid bilayers after reversible electrical breakdown. Biochim Biophys Acta 1981; 640:169-178.

9. Berridge MJ, Oschman JL. Transporting Epithelia. Academic Press, New York, 1972.

10. Bhatt DL, Gaylor DC, Lee RC. Rhabdomyolysis due to pulsed electric fields. (under editorial review).

11. Bliss JG, Harrison GI, Mourant JR, Powell KT, Weaver JC. Electroporation: The distribution of macromolecular uptake and shape changes in red blood cells following a single 50 second square wave pulse. Bioelectrochem Bioenerget 1988; 19:57-71.

12. Chernomordik LV, Abidor IG. The voltage-induced local defects in unmodified BLM. Bioelectrochem Bioenerget 1980; 7:617-623.

13. Chernomordik LV, Sukharev SI, Popov SV, Pastushenko VF, Sokirko AV, Abidor IG, Chizmadzhev YA. The electrical breakdown of cell and lipid membranes: The similarity of phenomenologies. Biochim Biophys Acta 1987; 902:360-373.

14. Chizmadzhev YA, Abidor IG. Bilayer lipid membranes in strong electric fields. Bioelectrochem Bioenerget 1980; 7:83-100.

15. Condon EU, Odishaw H. Handbook of Physics, McGraw-Hill, New York, 1958.

16. Crowley JM. Electrical breakdown of bimolecular lipid membranes as an electromechanical instability. Biophys J 1973; 13:711-724.

17. Dimitrov DS. Electric field-induced breakdown of lipid bilayers and cell membranes: A thin viscoelastic film model. J Membrane Biol 1984; 78:53-60.

18. Dimitrov DS, Jain RK. Membrane stability. Biochim Biophys Acta 1984; 779:437-468.

19. El-Mashak EM, Tsong TY. Ion selectivity of temperature-induced and electric field induced pores in dipalmitoylphosphatidylcholine vesicles. Biochem 1985; 24:2884-2888.

20. Foster KR, Schwann HP. Dielectric properties of tissues. In: CRC Handbook of Biological Effects of Electromagnetic Fields (Eds. C Polk, E Postow). CRC Press, Boca Raton, FL. 1986; 27-96.

21. Geddes LA. Electrodes and the Measurement of Bioelectric Events. Wiley, New York, 1972.

22. Glaser RW, Leikin SL, Chernomordik LV, Pastushenko VF, Sokirko AI. Reversible electrical breakdown of lipid bilayers: Formation and evolution of pores. Biochim Biophys Acta 1988; 940:275-287.

23. Kinosita K, Tsong TY. Formation and resealing of pores of controlled sizes in human erythrocyte membrane. Nature 1977; 268:438-441.

24. Kinosita K, Tsong TY. Hemolysis of human erythrocytes by transient electric field. PNAS 1977; 74:1923-1927.

25. Lee RC, Gaylor DC, Bhatt D, Israel DA. Role of cell membrane rupture in the pathogenesis of electrical trauma. J Surg Res 1988; 47:709-719.

26. Lee RC, Kolodney MS. Electrical injury mechanisms: Electrical breakdown of cell membranes. Plast Reconstr Surg 1987; 80:672-679.

27. Michel MR, Elgizoli M, Koblet H, Kempf C. Diffusion loading conditions determine recovery of protein synthesis in electroporated P3X63 Ag8 cells. Experientia 1988; 44:199-203.

28. Neumann E, Rosenheck K. Permeability changes induced by electric impulses in vesicular membranes. J Membrane Biol 1972; 10:279-290.

29. Neumann E, Schaefer-Ridder M, Wang Y, Hofschneider PH. Gene transfer into mouse lymphoma cells by electroporation in high electric fields. EMBO J 1982; 1:841-845.

30. Neumann E, Sowers A, Jordan C (eds). Electroporation and Electrofusion in Cell Biology, Plenum, New York, 1989.

31. Potter H. Electroporation in biology: Methods, applications, and instrumentation. Analyt Biochem 1988; 174:361-373.

32. Powell KT, Derrick EG, Weaver JC. Quantitative theory of reversible electrical breakdown. Bioelectrochem Bioelectroenerg 1986; 15:243-255.

33. Powell KT, Morgenthaler AW, Weaver JC. Tissue electroporation: Observation of reversible electrical breakdown in viable frog skin. Biophys J (in press).

34. Powell KT, Weaver JC. Transient aqueous pores in bilayer membranes: A statistical theory. Bioelectrochem Bioelectroenerg 1986; 15:211-227.

35. Sale AJH, Hamilton WA. Effects of high electric fields on microorganisms: I. Killing of bacteria and yeasts. Biochim Biophys Acta 1967; 148:781-788.

36. Sowers AE, Lieber MR. Electropore diameters, lifetimes, numbers, and locations in individual erythrocyte ghosts. FEBS Lett 1986; 205:179-184.

37. Stampfi, R. Reversible electrical breakdown of the excitable membrane of Ranvier Node. An Acad Brasil Ciens 1958; 30:57-63.

38. Sugar IP. The effects of external fields on the structure of lipid bilayers. J Physiol Paris 1981; 77:1035-1040.

39. Sugar IP, Forster W, Neumann E. Model of cell electrofusion: Membrane electroporation, pore coalescence and percolation. Biophys Chem 1987; 26:321-325.

40. Sugar IP, Neumann E. Stochastic model for electric field-induced membrane pores: Electroporation. Biophys 1984; 19:211-225.

41. Wang H, Skalak R. Viscous flow in cylindrical tube containing line of spherical particles. J Fluid Mech 1969; 83:75-96.

42. Weaver JC. Cell membrane rupture by strong electric fields: Prompt and delayed processes electrical trauma. In: Electrical Trauma (Eds. RC Lee, JFZ Burke, EG Cravalho). Cambridge University Press, (in preparation).

43. Weaver JC, Harrison GI, Bliss JG, Mourant JR, Powell KT. Electroporation: High frequency of occurrence of the transient high permeability state in red blood cells and intact yeast. FEBS Lett 1988; 229:30-34.

44. Weaver JC, Mintzer RA. Decreased bilayer stability due to transmembrane potentials. Phys Lett 1981; 86A:57-59.

45. Weaver JC, Powell KT. Theory of electroporation. In: Electroporation and Electrofusion in Cell Biology (Eds. E Neumann, A Sowers, C Jordan). Plenum, New York. 1989; 111-126.

46. Weaver JC, Powell KT, Mintzer RA, Sloan SR, Ling H. Diffusive permeability of bilayer membranes: The contribution of transient aqueous pores. Bioelectrochem Bioelectroenerg 1984; 12:405-412.

47. Zimmermann U. Electrical breakdown, electropermeabilization and electrofusion. Rev Physiol Biochem Pharmacol 1986; 105:175-256.

48. Zimmermann U, Riemann F, Pilwat G. Enzyme loading of electrically homogeneous human red blood cell ghosts prepared by dielectric breakdown. Biochim Biophys Acta 1976; 436:460-474.

49. Zimmermann U, Schultz J, Pilwat G. Transcellular ion flow in Escherichia coli B and electrical sizing of bacterias. Biophys J 1973; 13:1005-1013.

# SECTION II

# PRECLINICAL STUDIES (IN VIVO, IN VITRO)

# METABOLIC SPECTROSCOPY FOR MONITORING ELECTROMAGNETIC MEDICAL TECHNIQUES

Clyde H. Barlow and Jeffrey J. Kelly

## ABSTRACT

Electromagnetic diagnostic and treatment devices exert direct and indirect effects on tissue metabolism. Electromagnetic devices produce spatially inhomogeneous deposition of energy in tissues and, therefore, affect tissue metabolism heterogeneously. Optical spectroscopy provides unique approaches to monitoring tissue metabolic status during medical treatment. Methods are being developed that provide both area and depth resolved metabolic measurements. This emerging technology provides high contrast images of tissue metabolism, perfusion and oxygenation for use in monitoring effects of electromagnetic radiation on tissues.

## INTRODUCTION

Regions of the electromagnetic spectrum being used for diagnosis and therapeutic purposes include microwaves, radiowaves, infrared and visible light and ultrasound. In addition magnetic and electrical field therapies are being studied. These procedures can affect properties such as the temperature, ion balance, enzyme activity and vasoconstriction of tissues. Changes in these properties produce changes in tissue metabolic activity. Increased temperature, in particular, results in vasodilation, increased rates of tissue metabolism and increased tissue oxygen consumption. These metabolic changes can be followed through changes in endogenous tissue fluorescence and absorption or phosphorescence and fluorescence of exogenous dyes. Collectively these spectroscopic techniques provide tissue metabolism, oxygenation and blood flow information.

Monitoring tissues spectroscopically has been advanced by improvements in light gathering and data processing equipment and coincident decline in their costs. In particular, the advent of silicon diode arrays, charge coupled devices (CCD), high-speed microcomputers, and rapid analog to digital conversion devices have contributed to the development of inexpensive, sensitive instruments that optically measure changes in tissue metabolism. These developments have made feasible rapid acquisition, evaluation and presentation of spectroscopic data. Conversion of spectroscopic data into high resolution images produces displays of metabolic parameters over large regions of tissue.

In this paper we present background and application of two of these spectroscopic techniques and examine the conditions that recommend their use for tissue monitoring. These techniques employ endogenous NADH fluorescence as a measure of tissue bioenergetics and exogenous palladium porphyrin phosphorescence as a measure of tissue oxygen concentration. Together with fluorescence angiography to evaluate blood flow, these techniques provide powerful tools for monitoring the effects of electromagnetic medical techniques on tissue.

## BACKGROUND

Spectroscopic methods provide noninvasive means for measuring changes in bioenergetics that do not measurably affect tissue metabolism. They lend themselves to the development of instruments that measure metabolic responses to physiological changes over a large portion of the surface of the tissue being examined. Several noninvasive methods have been developed to follow mitochondrial metabolism in functioning tissues. These methods include fluorescence of reduced nicotinamide adenine dinucleotide (NADH) and oxidized flavoprotein, visible/near-infrared reflectance and transmittance spectroscopy of mitochondrial cytochromes, and phosphorus nuclear magnetic resonance spectroscopy of phosphorus metabolites. We have developed a video fluorescence system that records ultraviolet reflectance images and NADH fluorescence images from the epicardial surface of perfused rat hearts.[13] We chose to image NADH fluorescence as it is sensitive to changes in rate of mitochondrial electron transport and provides a direct measure of the limitation of mitochondrial activity by oxygen supply.

The mitochondrial $NAD^+$/NADH couple is located at the high potential end of the mitochondrial electron transport chain. Changes in the NADH concentration can be followed fluorometrically, allowing the NADH fluorescence to serve as an intrinsic probe of changes in mitochondrial bioenergetic status. Chance and Williams [4] associated changes in $NAD^+$ reduction, along with their effects on mitochondrial electron transport components, with mitochondrial metabolic state. In 1959 Chance and Jöbsis [5] measured NADH fluorescence changes of skeletal muscle in response to muscle stimulation.

NADH fluorescence is measured by excitation near its absorption maximum of 340 nm and emission at 450 nm. $NAD^+$ exhibits no fluorescence at these wavelengths. Both cytoplasmic and mitochondrial NAD(P)H contribute to the fluorescence signal. However studies in hearts [10, 17, 25] indicate that the quantum yield of mitochondrial NADH in heart is about 8-fold higher than NADH in cytoplasm, and more than 80% of the fluorescence increase during anoxia is due to mitochondrial NADH. Mitochondrial NADH fluorescence becomes half maximal at a $pO_2$ of 0.05 torr [7] positioning NADH as an excellent measure of the bioenergetic state of the heart. At this $pO_2$ the rate of mitochondrial ATP synthesis is half maxima.[17] Thus, NADH fluorescence provides a sensitive measure of oxygen supply/demand relationships in biological tissue. The key role of mitochondrial NADH metabolism has made NADH fluorescence measurement instrumental in understanding the effects of $pO_2$, flow, $Ca^{2+}$, pH, workload and substrates in heart bioenergetics.[2, 12, 26, 28, 29]

Approaches used to detect changes in tissue NADH concentrations have typically utilized NADH excitation of a single point on the tissue from appropriately filtered mercury arc, xenon arc, or tungsten-halogen lamps and detection of the emitted fluorescence with an appropriately filtered photomultiplier. However these standard NADH fluorescence measurements provide information about mitochondrial metabolism for only a single illuminated area. NADH fluorescence imaging of tissue provides an important new dimension in tissue physiology and biochemistry. In 1976 Barlow and Chance [2] described a fluorescence photographic approach to obtain images of epicardial NADH distribution from the surface of perfused rat hearts. Ultraviolet excitation from a filtered xenon flash unit eliminated motion artifacts of the beating heart.

NADH fluorescence photographs from the epicardial surface of control hearts revealed uniformly low NADH fluorescence, indicative of well-oxygenated mitochondria. Interruption of normal oxygen delivery to the hearts resulted in increases of NADH fluorescence as expected, but with an unexpected distribution. A well-defined area of uniformly high NADH fluorescence separated from low fluorescence normoxic zones was observed.

Fluorescence angiography was used to evaluate relationships between coronary circulation and tissue metabolism. This revealed a zone of nonperfused but metabolically normal tissue separating perfused from ischemic epicardium. The width of the zone between normoxic and ischemic tissue was estimated by us to be 100-200 μm.[3, 9] Ultramicroregional chemical analysis [9] confirmed a close correlation between tissue concentrations of NADH and fluorescence intensity in freeze-trapped ischemic rabbit hearts. NADH fluorescence photographs reveal the presence of microregional metabolic heterogeneity [8] recorded during low flow ischemia and high flow hypoxia in perfused rat and rabbit hearts.[2, 26-29] Images of NADH fluorescence therefore provide excellent means for studying metabolic states of tissues.

The third factor necessary to understand properly tissue metabolic status is tissue oxygen concentration. Tissue oxygen levels may be calculated from collisional quenching of molecular fluorescence or phosphorescence by molecular triplet oxygen. This technique has the potential to overcome disadvantages of oxygen electrode measurements such as tissue damage, alterations of oxygen tension by oxygen consumption, and single point measurement. The Stern-Volmer[21] relationship, as modified for fluorescence and phosphorescence lifetimes, describes oxygen quenching in diffusion-limited systems:

$$t_O/t = 1 + t_O k_q [O_2],$$

where $t_O$ and t are lifetimes in the absence and presence of $[O_2]$ respectively, and $k_q$ is the second-order quenching rate constant.

Vaughn and Weber [24] demonstrated in 1970 that oxygen concentrations in physiological ranges quenched pyrene-3-butyric acid (PBA) fluorescence. Knopp and Longmuir [14, 15] and Jöbsis with Mitnick and Snow [16, 19] subsequently demonstrated the utility of PBA fluorescence quenching in tissues. PBA has not been widely used as an oxygen indicator, however, in major part due to its short fluorescence lifetime, 140 ns, in the absence of oxygen. This has two consequences. First, the short lifetime results in only 17% quenching of PBA fluorescence at an oxygen tension of 100 torr, limiting the sensitivity of the dye in the physiological range. Secondly, fluorescence intensity is a function of both dye concentration and oxygen concentration, while fluorescence lifetime is a function of oxygen concentration alone, assuming constant diffusion environment. Nanosecond lifetime measurements necessary for quantitative studies require specialized, expensive instrumentation which limits access of this methodology.

Recently, Vanderkooi and Wilson made a major breakthrough that promises to allow practical optical monitoring of oxygen concentration in tissues. [18, 23, 30] Their approach is based upon oxygen quenching of phosphorescence. Closed-shell palladium porphyrins exhibit high phosphorescence yields and are essentially free from short-lived fluorescence. Pd coproporphin and Pd meso-tetra(p-sulfanatophenyl) porphin exhibit long-lived fluorescence of 1.2 and 1.0 msec respectively in deoxygenated phosphate buffer containing 2% albumin. With second-order oxygen collisional quenching rate constants of 8.3 x 108 and 4.1 x 108 $M^{-1}$ $sec^{-1}$, 50% quenching of these Pd porphyrins will occur at oxygen concentrations of 8.3 μM and 2.4 μM respectively. The useful range extends from air saturation down to the nanomolar range.[23] Vanderkooi and

Wilson have studied Pd coproporphin with enzyme systems,[23] isolated mitochondria ,[30] and perfused rat liver [18] and have found no cytotoxicity at the 1 μM concentrations used, no self quenching in albumin solutions, no quenching by physiological concentrations of paramagnetic species, and no tissue damage.

Thus tissue oxygen concentration measurements by phosphorescence decay appear practical. High signal to noise ratios available from submicromolar concentrations of Pd porphyrins using flash phosphorescence video detection, make practical two and three-dimensional imaging of epicardial oxygen concentration by gated-imaging of palladium porphyrin phosphorescence decay. In the following experimental section, a videofluorometer system is described that combines capabilities for imaging both NADH fluorescence and Pd porphyrin phosphorescence from the same tissue. The device combines a computer-controlled fluorescence excitation system with digital image acquisition to quantify tissue bioenergetics in both spatial and time domains. The digital nature of the acquired image allows detailed calculation and analysis of physiological features and their time dependence. Through the combination of information from fluorescence, phosphorescence and reflectance measurements, compensation for changes in tissue absorption, blood volume and light scattering can be made and corrected images can be displayed. This instrument is a powerful new tool for the study of spatial and temporal relationships between mitochondrial metabolism and tissue oxygen distribution as a means of monitoring the effects of electromagnetic medical techniques.

## EXPERIMENTAL

### A. Videofluorometer System

The videofluorometer system consists of a microcomputer with I/O capabilities, separate video monitor, fluorescence excitation flash apparatus and camera and image collection hardware. The design and performance specifications of the instrument have been described by Kelly et al.[13] Figure 1 details the major units of the system and the control lines involved in synchronizing the system components. An AST Research Inc. Model 80 microcomputer with 70 Mbyte Miniscribe hard disk, Data Translation, Inc. DT2851 high resolution frame grabber, Data Translation, Inc. DT2858 auxiliary frame processor and MetraByte Corp. Dash-16 I/O board were used for data and image acquisition and control. Computer control of xenon flash triggering, imaging capture and image display and manipulation was accomplished by programs written in Microsoft Fortran.

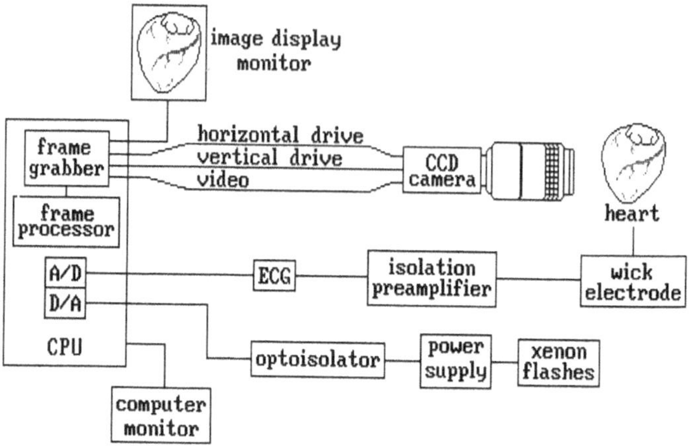


**Figure 1.** Videofluorometer design showing interconnections between major components of the instrument.

For studies of perfused rat hearts, the electrocardiogram (ECG) was collected from wick electrodes placed against the ventricular epicardium of the beating heart. It was amplified by a Quinton Instruments preamplifier and displayed continuously on a Quinton Instruments Model 621B ECG monitoring system. The preamplified ECG signal was input to an A/D port of the MetraByte board. This signal was displayed on the computer CRT to determine the trigger level for the QRS spike during ventricular depolarization. A selected ECG amplitude, corresponding to 70-80% of the spike, initiated the trigger sequence. Image acquisition was timed to occur during the diastolic phase of the cardiac cycle (to minimize heart movement) at an appropriate delay after the QRS spike. The length of this delay was calibrated by using the videofluorometer to image the ECG trace on an oscilloscope, thus verifying the timing of image collection.

In order to determine the best delay time for the minimization of heart movement, difference images were generated by subtracting images obtained at the same delay time. Ideally these difference images would be completely dark, indicating no difference in heart position. The sums of the pixel intensities from a number of the difference images at a fixed delay time were compared to determine the extent of heart motion. After performing this evaluation at different delay times, it was concluded that delays of 50 to 70% of the cardiac cycle after a QRS peak gave the most reproducible images. Image correction for the 30 ms camera retrace variation were performed by software translations in x and y. These corrections were found to be quite adequate for image comparison within selected windows.

Fluorescence, reflectance and phosphorescence images were obtained using an XC-77 Sony CCD videocamera (760x480 pixel array) with a C-mount adapter to a 55mm *f*2.8 Micro-Nikkor lens set at *f*4.0. Two xenon flash tubes (EG&G Corp. model #FX47-C3), mounted on either side of the videocamera and excited with a Novatron-800 800J power supply, provided ultraviolet light excitation.

In NADH fluorescence mode, Corning 7-60 filters covered the flash tubes to provide 330-380 nm excitation of tissue NADH. Hoya CM500 and L42 filters were placed before the camera lenses to allow transmission of the 400-600 nm region, permitting the 450 nm (maximum) NADH fluorescence to be transmitted but blocking ultraviolet excitation light. For uv reflectance images, neutral density filters, added to the Corning 7-60 filters in front of the flash tubes, reduced the illumination intensity by 102 and a Corning 7-60 replaced the filters in front of the camera to transmit only uv radiation. In palladium porphyrin mode, the flash tubes were covered with Corning 4-96 filters which provided excitation for the porphyrin's 408 nm (maximum) Soret band and the camera lens covered with Corning 2-58 630 nm cutoff filter for porphyrin phosphorescence at 700 nm (maximum). Figure 2 provides a diagram of the optics of the videofluorometer.

B. Physiological Procedures

Hearts were rapidly excised from anesthetized (sodium pentobarbital, 60 mg/kg or urethane, 1.5 g/kg) 250-400g Sprague Dawley rats, the aorta

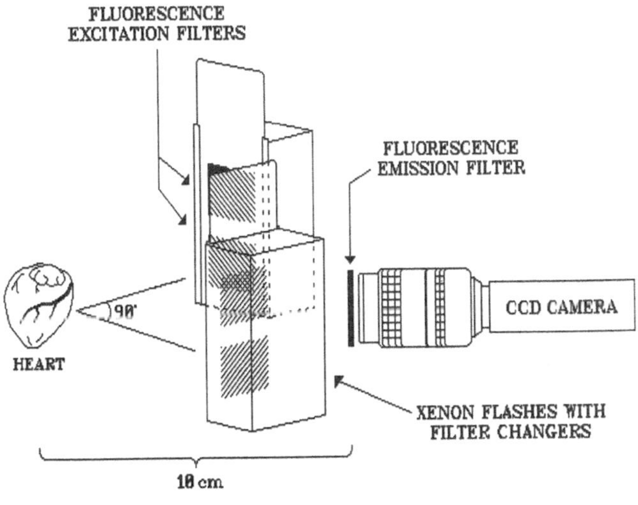


**Figure 2.** Relative position of optic components of videofluorometer.

cannulated and retrograde perfusion initialized. Perfusion was performed on a modified Langendorff perfusion apparatus with a Krebs-Ringer bicarbonate buffer containing 2.5 mM $Ca^{2+}$ and 5 mM glucose at 37°C. The perfusate was equilibrated with 95% oxygen, 5% carbon dioxide and filtered through two 0.45 µm Millipore filters. The perfusion apparatus was equipped with an overflow reservoir maintaining aortic perfusion pressure at 100 cm $H_2O$. Ischemia was induced globally by lowering the perfusion pressure to 40 cm $H_2O$ which produced heterogeneous distribution of increased NADH fluorescence, or locally by ligation of the left main coronary artery on the free wall of the left ventricle. The latter resulted in a contiguous ischemic area of high NADH fluorescence.

## RESULTS

### A. NADH Fluorescence Images

The videofluorometer produces images of buffer-perfused hearts with outstanding contrast and resolution. The videocamera and the macro lens give 512x480 pixel (width x height) images of perfused hearts that vary from about 15x12.5 mm to 30x25 mm in tissue area. This corresponds to pixel dimensions in the range 30x25 µm to 60x50 µm for the heart images.

Figure 3 illustrates the fluorescence images obtained with the videofluorometer. Figure 3a is an NADH fluorescence image of a perfused rat heart under control conditions (100 cm $H_2O$) of adequate flow and oxygen supply. The view is of the epicardium of the left ventricle with left atrium and fatty tissue at the top of the image. The ventricle exhibits low NADH fluorescence of the highly oxidized state of the $NAD^+$/NADH couple. Clearly evident are the slightly darker surface veins that drain the ventricle tissues.

Figure 3b is an image of the same heart following coronary artery ligation. A major portion of the tissue is clearly delineated by bright NADH fluorescence from the induced ischemic area. At the left edge of the ischemic area are several dark fingers of normoxic tissue where small surface veins continue to drain volumes of the heart still being perfused. This surface flow of oxygenated perfusate causes the veins in the ischemic region to be somewhat darker than the surrounding tissue,[22] but NADH fluorescence from the underlying ischemic tissue is still visible. A sharp transition between normoxic and ischemic tissue can be seen at the boundary of the ischemic area indicating a relatively sharp border zone between tissue with high NADH and low NADH content.

The time course of the development of ischemia as determined by increasing NADH fluorescence following artery ligation was measured to

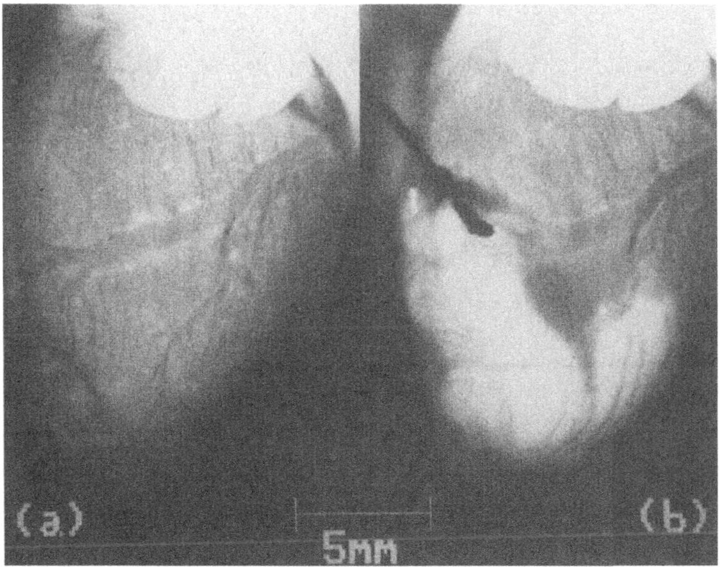


**Figure 3.** NADH fluorescence images of buffer-perfused rat heart under control conditions (left) and three minutes after coronary ligation (right).

demonstrate the ability of the videofluorometer to monitor changes in metabolic state with time. A full image of the heart was obtained at 2 seconds after ligation and a 512x40 pixel portion of the image stored in the second buffer of the frame grabber. Twelve such windows were taken over the next 150 s and transferred to different portions of the second frame buffer. The images included a region of increasing NADH fluorescence (ischemic area), a normoxic region of constant low fluorescence and, along one edge, a stripe of 3-M Scotch black Photographic Tape #235 which provided a fluorescence intensity reference to monitor variations in flash intensity. The relative NADH fluorescence was quantified by summing the pixel intensities in a window from the ischemic area and taking the ratio with the summation of a constant window from the reference tape.

Figure 4 presents kinetics data of the onset of local ischemia due to coronary artery ligation in four different hearts as determined by the quantification method above. The zero time value was taken from the normoxic heart before ligation using the same reference and heart image window areas. The experiments yielded a total increase in NADH fluorescence of 70-80%. The information presented in Fig.4 is the summation of the NADH fluorescence over an 3.0x1.2 mm area of the heart surface. The data support a biphasic formation of NADH following ligation. This is more readily seen in Fig. 5, which

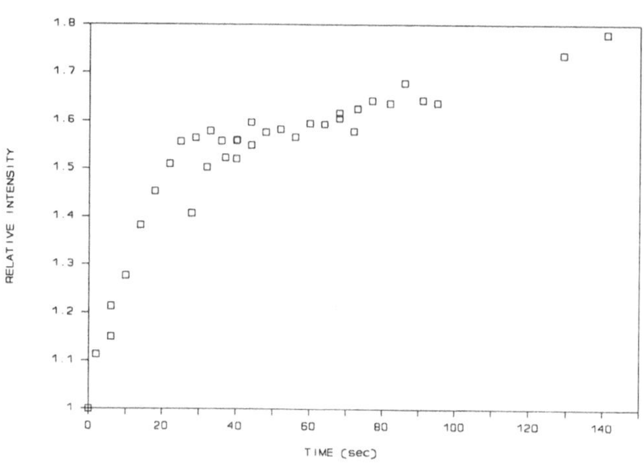


**Figure 4.** Time course of relative NADH fluorescence from three different hearts during the onset of ischemia following coronary artery ligation.

represents the data of Fig. 4 as a natural logarithm plot of the final NADH fluorescence intensity, NADH(t∞), as estimated from heart images taken 10 minutes after ligation, divided by the fluorescence intensity at time t. The two straight lines were determined by linear regression and have regression correlation coefficients of 0.978 for the fast component and 0.900 for the slow. The ratio of fast/slow component, estimated from the data to be 7.7, strongly suggests that the two signals are from the reduction of mitochondrial (fast) and cytoplasmic (slow) $NAD^+$ respectively. This ratio agrees well with the observations of Chance [6] and Chapman.[10] In addition to data presentation as in Figs 4 and 5, the videofluorometer actually yields positional information that is averaged in these kinetic plots. Indeed, three dimensional plots displaying fluorescence intensity profiles across any line through the heart can be generated with time as the third axis.

Global low-flow ischemia was induced by lowering the heart perfusion pressure. This alteration in tissue environment provides an excellent test of the videofluorometer's ability to monitor changes in conditions affecting tissue metabolism. Figure 6 presents NADH videofluorometer images of low-flow ischemia at 40 and 20 cm of $H_2O$ pressure respectively. The fluorometer provides images with good contrast for the analysis of the increase in NADH fluorescence under these different physiological conditions. Figure 7 presents the relative NADH fluorescence, produced by windowing the same region of epicardial surface, as a function of perfusion pressure. These results demonstrate the ability of the instrument to not only detect and displayed regional alterations

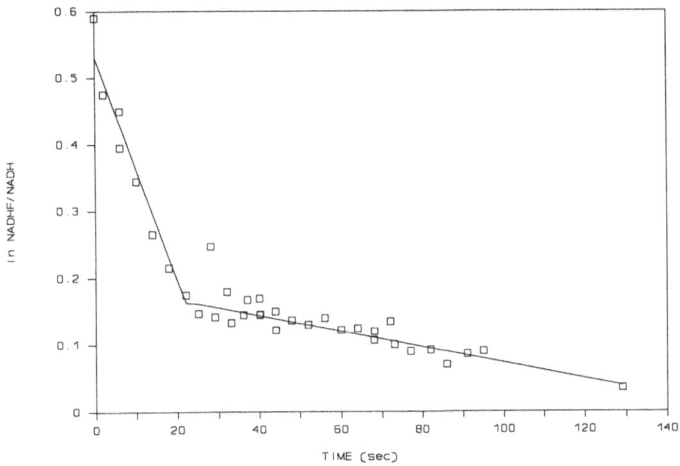


**Figure 5.** Natural logarithm plot of ratio of the final NADH fluorescence intensity to the fluorescence level at a given time, NADH(t∞)/NADH(t).

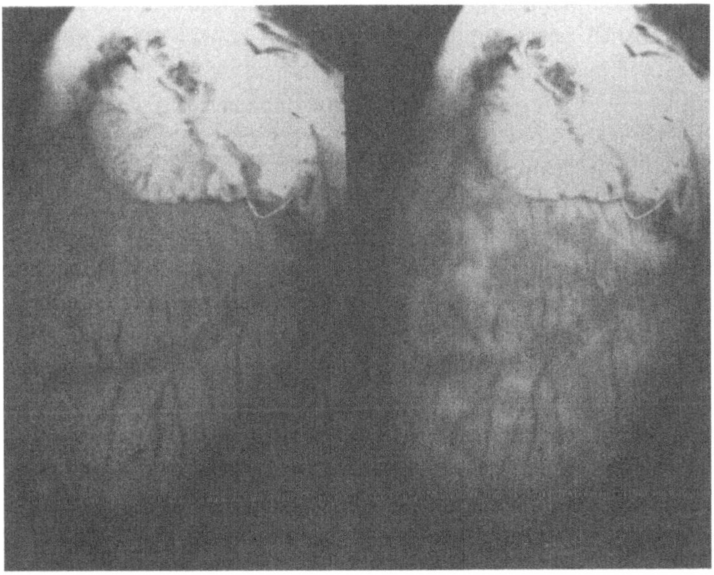

**Figure 6.** NADH fluorescence images of buffer perfused rat heart at 40 cm $H_2O$ (left) and 20 cm $H_2O$ (right).

in tissue metabolism induced by external causes, but also to quantify the metabolic response.

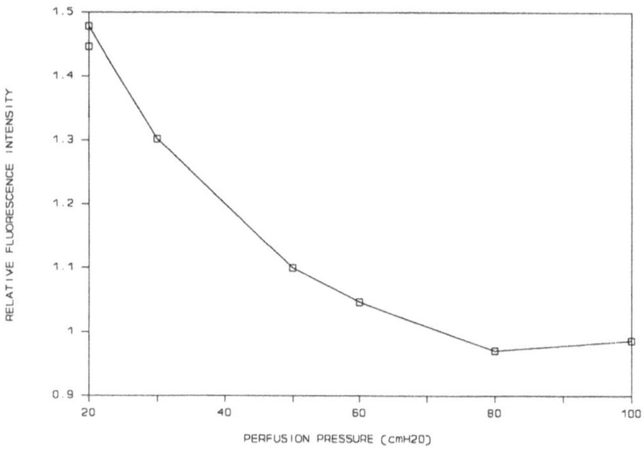


**Figure 7.** Relative NADH fluorescence as a function of perfusion pressure of a buffer perfused rat heart during global low-flow ischemia.

## B. Palladium Porphyrin Phosphorescence Images

Minor modification of the videofluorometer allow imaging of porphyrin phosphorescence and computation of oxygen concentration images. With NADH providing a sensitive monitor of mitochondrial bioenergetics and Pd porphyrin providing a sensitive measure of epicardial oxygen concentration, these two techniques open unique new views into the dynamics of bioenergetic processes of biological tissue.

Palladium (IV) meso-tetra(p-sulfanatophenyl)porphin (TSPP), a highly charged, water soluble, phosphorescent porphyrin was synthesized by cyclization of pyrrole and benzaldehyde by the method of Adler et al., [1] sulfonation with concentrated sulfuric acid according to Srivastava et al.,[20] followed by palladium insertion using the imidazole melt method of Gouterman et al.[11] Structure and purity were assessed by chromatographic and spectroscopic methods by us and by Dr. Vanderkooi at the University of Pennsylvania. The absorption spectrum of the synthesized Pd porphyrin (obtained on a Cary 17 spectrophotometer) and phosphorescence emission spectrum (obtained on a Perkin-Elmer LS-5 spectrofluorometer) are presented in Figs 8 and 9 respectively.

Imaging system optical filters passed excitation of the 408 nm absorption maximum of TSPP and emission filter transmitted only light with wavelength longer than 620 nm to pass the 700 nm emission of TSPP phosphorescence. The near infrared blocking filter standard with the CCD camera was removed. Rat hearts were perfused as before but with inclusion of 2% albumin and

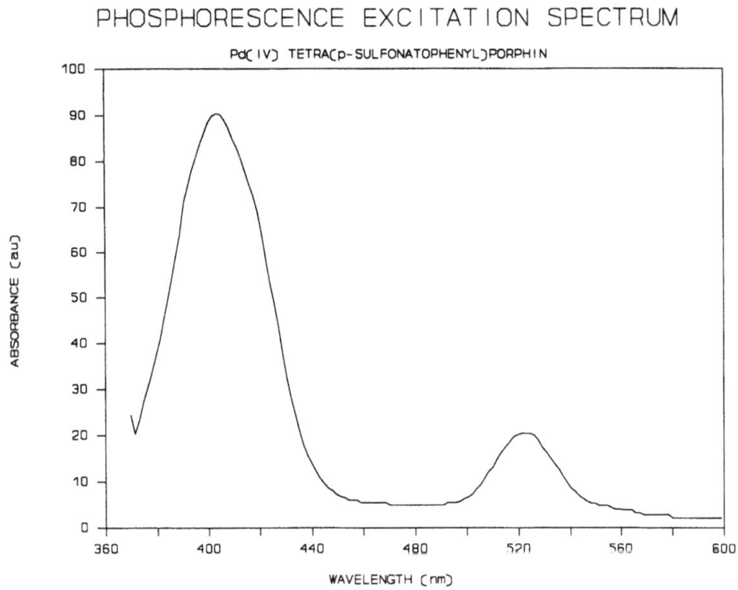


**Figure 8.** Phosphorescence excitation spectrum of Palladium (IV) meso-tetra(p-sulfanatophenyl)porphin in deoxygenated water.

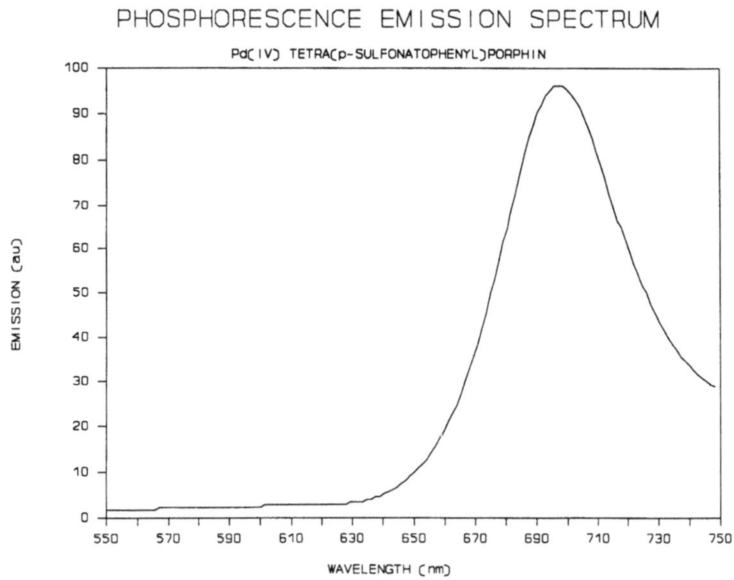


**Figure 9.** Phosphorescence emission spectrum of palladium (IV) meso-tetra(p-sulfanatophenyl)porphin in deoxygenated water.

1 μM TSPP. With the current videofluorometer an *f* stop of 5.6 was used to prevent saturation of the CCD detector from the intense TSPP phosphorescence. With *f*1.4 lenses commonly available, we could have decreased the Pd TSPP concentration 16 fold and obtained images of similar intensity but with shallower depth of field. No detectable signal was observable when hearts not containing TSPP were imaged in the same manner at *f*2.8, the minimum *f* stop of the Micro Nikkor 50 mm lens.

A series of Pd porphyrin phosphorescence images of a 6.6x3.5 mm window of a rat heart perfused with 1.1 μM Pd porphyrin in the perfusate were obtained during a pressure titration from 100 to 0 cm $H_2O$ pressure. The images showed increased phosphorescence intensity as the pressure was lowered and demonstrated the ability of the videofluorometer to detect Pd porphyrin from micromolar perfusate. Clearly Pd porphyrin phosphorescence can be imaged in vivo and responses to the decreased oxygen concentration that develops at lower perfusion pressure. Summing window intensities gives an indication of the relative phosphorescence at different perfusion pressures. Figure 10 presents such summations from six images during the pressure titration and demonstrates capability of the current video system to monitor intensities of Pd porphyrin phosphorescence.

Since quenching of phosphorescence intensity is proportional to quenching of phosphorescence lifetime, one can rewrite the Stern-Volmer equation, as

$$t_O k_q [O_2] = I_O/I - 1.$$

By using the summed-window phosphorescence intensity of the zero flow heart as $I_O$, and assuming $[O_2] = 0$ at this pressure, the above relationship can be used to calculate a quantity proportional to average $O_2$ concentration in each window (but without an absolute reference value). Figure 11 is a plot of this relative tissue $[O_2]$ as a function of perfusion pressure. Figure 11, relative $[O_2]$ vs perfusion pressure, and Fig. 7, relative NADH fluorescence vs perfusion pressure, illustrate the relationship between epicardial changes in $[O_2]$ and NADH. These results are for data averaged over several square mm of epicardium during a process we know to form heterogeneous metabolic patterns.

To test this complimentation of the two metabolic imaging techniques, NADH fluorescence and Pd porphyrin phosphorescence images were obtained from the same rat heart both before and after coronary artery ligation. Figure 12A and 12B present the NADH fluorescence images before and after ligation, while 12C and 12D are the corresponding Pd porphyrin phosphorescence images.

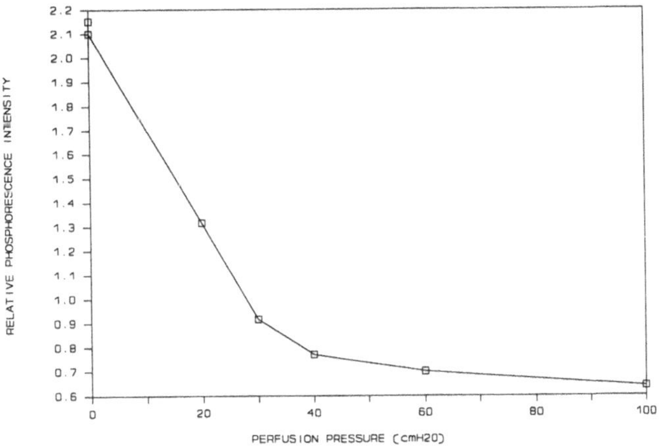


**Figure 10.** Relative Pd porphyrin phosphorescence intensities during global low-flow ischemia as a function of perfusion pressure.

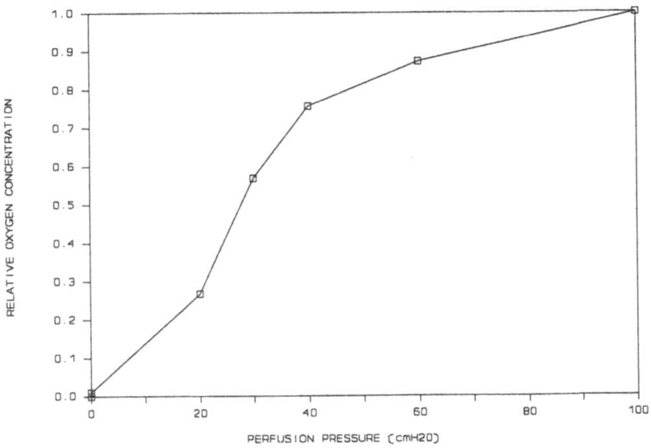


**Figure 11.** Relative oxygen concentration as a function of perfusion pressure.

The ischemic region of the heart epicardium is highlighted by increased pixel intensity of both NADH fluorescence in Fig. 12B and Pd porphyrin phosphorescence in 12D. In each pixel in the Pd porphyrin phosphorescence images can be scaled by the Stern-Volmer relationship described in the last paragraph, to produce relative oxygen concentration images (Fig. 13A and B). In these images high intensity now corresponds to high relative oxygen concentrations and low oxygen concentration in the ischemic area is clearly indicated by low intensity.

Figures 12 and 13 show the utility of the videofluorometer to provide detailed images for detecting changes in tissue metabolism and oxygenation. Such images allow the quantitative evaluation of these tissue properties in both time and space. The digital nature of the acquired images allows detailed analysis of spatial variations in NADH and oxygen concentrations. Figure 14 indicates a line across the ligated heart's surface from the ischemic to the normoxic zones of the heart.

Images of oxygen and NADH concentrations in rat heart epicardium are qualitatively similar but quantitatively different. Perfused rat hearts have arterial oxygen tensions of approximately 650 torr, which decrease to essentially 0 torr in ischemic areas. Mitochondria reside at the bottom of self generated oxygen wells and have high oxygen affinity. They therefore respond maximally to anoxia over a relatively narrow oxygen concentration range. Figure 15 is a graph of relative NADH fluorescence from Fig. 12B and relative oxygen concentration from 13B as a function of distance along the line in Fig. 14. The changes in NADH emission and oxygen concentration across an ischemic boundary are clearly depicted.

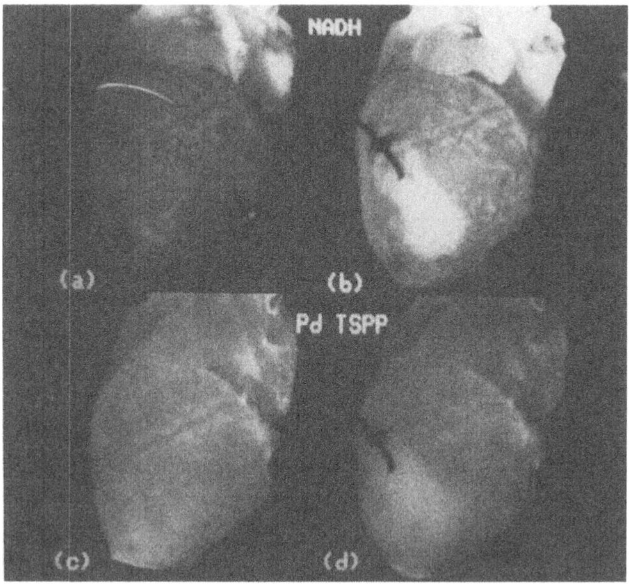


**Figure 12.** Images of perfused rat heart before, 12a and 12c, and 1 minute after, 12b and 12d, coronary artery ligation. 12a and 12b are NADH fluorescence. 12c and 12d are Pd TSPP phosphorescence.

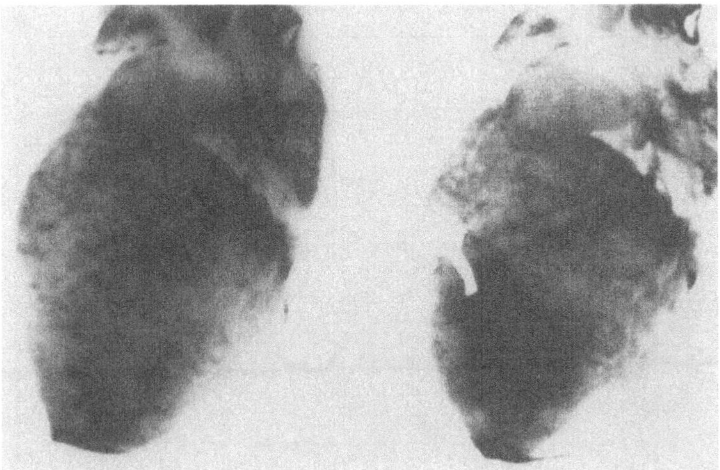

**Figure 13.** Relative oxygen concentration image of perfused rat heart computed from images 12c and 12d using Stern-Volmer relationship for phosphorescence quenching. The dark ischemic area was defined as zero oxygen and the linear gray scale correlation to oxygen concentration is based on the ischemic area as a reference for $t_0k_q[O_2] = I0/I - 1$.

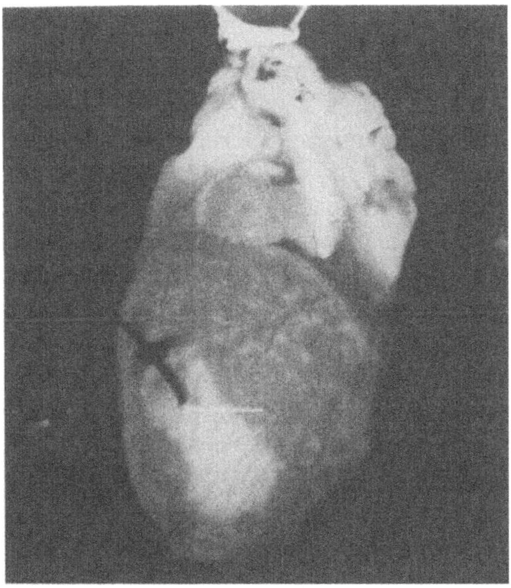

**Figure 14.** NADH fluorescence image of heart from Figure 12b with line indicating epicardial region used for graphs of NADH fluorescence intensity and oxygen concentration in Figure 15.

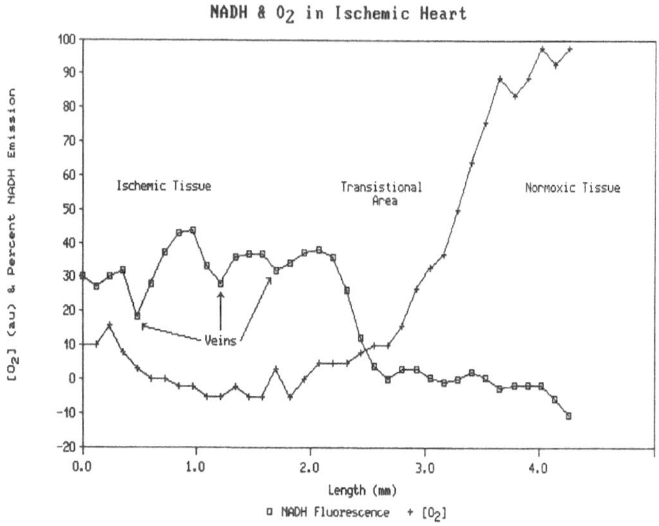


**Figure 15.** NADH fluorescence intensity and relative oxygen concentration along a line passing from ischemic to normoxic epicardium. NADH gradients are much steeper than those for oxygen, reflecting the low critical $PO_2$ for alterations in mitochondrial NADH/NAD$^+$ and the ability for mitochondria to maintain normal function at reduced oxygen concentration.

The NADH trace shows a much sharper transition compared to the broader oxygen trace. The steeper gradient in NADH vs $[O_2]$ results from NAD$^+$ in actively respiring mitochondria becoming half reduced only at very low oxygen tension (<0.1 torr). Thus the NADH fluorescence image provides a sensitive monitor of mitochondrial metabolism while the Pd porphyrin images provides a wide dynamic range of oxygen concentration in tissue. Together, digital images of oxygen and NADH allow visual and quantitative evaluation of local tissue oxygen supply/demand distribution in functioning tissue.

## CONCLUSION

Imaging of tissue metabolic status from NADH fluorescence, flow distribution from fluorescence angiography, and oxygen distribution by palladium TSPP phosphorescence provides a powerful triad of techniques sensitive to physiologic and metabolic alternations of tissues. All three provide important and unique information. NADH metabolic images clearly delineate regions where oxygen demand outstrips oxygen supply and therefore can indicate tissue areas where mitochondrial metabolism has been disrupted. Pd porphyrin phosphorescence images reflect tissue oxygen concentrations, the most critical external metabolic resource. These images monitor a greater dynamic range of tissue metabolic response than NADH. Oxygen images show variation before NADH images and can act as an early warning for regional tissue variations due to external causes. Flow distribution is important but may not reflect tissue oxygen or metabolism as in the case of arterial venous shunting. Thus each of the measurements - NADH, flow, and oxygen, are important individually, but together form an analytical picture of regional changes in tissue metabolism. Approaches such as these are critical for evaluating effects of therapeutic interventions on tissue metabolism.

## ACKNOWLEDGMENTS

The authors thank Dawn A. Rorvik, Keith N. Richmond and Thomas Reinbold for technical assistance and Dr. Jane Vanderkooi for helpful discussions that initiated some of these studies. This research was supported by National Institutes of Health Grants HL34498 and HL40775.

## References

1. Adler AD, Sklar L, Longo FR, Finarelli JD, Finarelli MG. A mechanistic study of the synthesis of meso-tetraphenylporphin. J Heterocycl Chem 1968; 5:669-678.

2. Barlow CH, Chance B. Ischemic areas in perfused rat hearts: measurement by NADH fluorescence photography. Science 1976; 193:909-910.

3. Barlow CH, Harken AH, Simson MB, Harden W, Rastegar H, Chance B. Resolution of ischemic borderzone by NADH fluorophotography in isolated perfused rabbit hearts. In: Frontiers of Biological Energetics: From Electrons to Tissues (Eds. PL Dutton, JS Leigh and A Scarpa). Academic Press, New York. 1978; 1533-1540.

4. Chance B, Williams GR. Respiratory enzymes in oxidative phosphorylation:III. the steady state. J Biol Chem 1955; 217:409-427.

5. Chance B and Jöbsis FF. Changes in fluorescence in a frog sartorius muscle following a twitch. Nature (London) 1959; 184:195-196.

6. Chance B, Schoener B, Krejci D, Russman W, Weseman W, Schnitger H, Bücher T. Kinetics of fluorescence and metabolite changes in rat liver during a cycle of ischemia. Biochem Z 1965; 341:325-333.

7. Chance B. Pyridine nucleotide as an indicator of the oxygen requirements for energy-linked functions of mitochondria. Circ Res 1976; 38:I31-I38.

8. Chance B, Barlow CH, Nakase Y, Takeda H, Mayevsky A, Fischetti R, Graham N, Sorge J. Heterogeneity of oxygen delivery in normoxic and hypoxic states: a fluorometer study. Am J Physiol 1978; 235:H809-H820.

9. Chance B, Barlow C, Haselgrove J, Nakase Y, Quistorff B, Matschinsky F, Mayevsky A. Microheterogeneities of redox states of perfused and intact organs. In: Microenvironments and Metabolic Compartmentation (Eds. PA Srere and RW Estabrook). Academic Press, New York. 1978; 131-148.

10. Chapman JB. Fluorometric studies of oxidative metabolism in isolated papillary muscle of the rabbit. J Gen Physiol 1972; 135-154.

11. Gouterman M, Schumaker CD, Srivatava TS, Yonetani T. Absorption and luminescence of yttrium and lanthanide octaethylporphin complexes. Chem Phys Let 1976; 40:456-461.

12. Harden WR, Barlow CH, Simson MB, Harken AH. Temporal relation between onset of cell anoxia and ischemia contractile failure: myocardial ischemia the isolated, perfused rabbit heart. Am J Cardiol 1979; 44:741-746.

13. Kelly JJ, Rorvik DA, Richmond KN, Barlow CH. Videofluorometer for imaging tissue metabolism. Review of Scientific Instruments 1989; submitted for publication.

14. Knopp JA, Longmuir IS. Intracellular measurement of oxygen by quenching of florescence of pyrenebutyric acid. Biochim et Biophys Acta 1972; 279:393-397.

15. Longmuir IS, Knopp JA. Measurement of tissue oxygen with a fluorescent probe. J Appl Physiol 1976; 41:598-602.

16. Mitnick MH, Jöbsis FF. Pyrenebutyric acid as an optical oxygen probe in the intact cerebral cortex. J Appl Physiol 1976; 41:593-597.

17. Nuutinen EM. Subcellular origin of the surface fluorescence of reduced nicotinamide nucleotides in the isolated rat heart. Basic Res Cardiol 1984; 79:49-58.

18. Rumsey WL, Vanderkooi JM, Wilson DF. Imaging of phosphorescence: a novel method of measuring the oxygen distribution in perfused tissue. Science 1988; 241:1649-1551.

19. Snow TR, Jöbsis FF. (absract) Characterization of the noninvasive oxygen fluorochrome pyrenebutyric acid in heart mitochondria. Physiologist 1975; 18:397.

20. Srivastava TS, Tsutsui M. Preparation and purification of tetrasodium meso-Tetra(p- sulfophenyl)porphine. An easy procedure. J Org Chem 1973; 38:2103.

21. Stern O, Volmer M. Über die abklingungszeit der fluoreszenz. Phys Z 1919; 20:183-188.

22. Taira Y, Kanaide H, Nakamura M. Coronary venous perfusion of the ischemic myocardium during acute coronary artery occlusion in isolated rat hearts. Circ Res 1985; 56:666-675.

23. Vanderkooi JM, Maniara G, Green TJ, Wilson DF. An optical method for measurement of dioxygen concentration based upon quenching of phosphorescence. J Biol Chem 1987; 262:5476-5482.

24. Vaughan WM, Weber G. Oxygen quenching of pyrene-butyric acid fluorescence in water. A dynamic probe of the microenvironment. Biochem 1970; 9:464-473.

25. Williamson JB. Glycolytic control mechanisms. II. Kinetics of intermediate changes during the aerobic-anoxic transition in perfused rat heart. J Biol Chem 1966; 241:5026-5036.

26. Williamson JR, Davis KN, Medina-Ramirez G. Quantitative analysis of heterogenous NADH fluorescence in perfused rat hearts during hypoxia and ischemia. J Mol Cell Cardiol 1982; 14:29-35.

27. Williamson JR, Kobayashi K. Use of the perfused rat heart to study cardiac metabolism: retrospective and prospective views. Basic Res Cardiol 1984; 79:283-291.

28. Williamson JR, Steenbergen C, Rich T, DeLeeuw G, Barlow CH, Chance B. The nature of ischemic injury in cardiac tissue. In: Pathophysiology and Therapeutics of Myocardial Ischemia (Eds. AM Leffer, GJ Kelliher and MJ Rovetto). Spectrum Publications Inc, New York. 1977; 193-225.

29. Williamson JR, Steenbergen C, DeLeeuw G, Barlow CH. Control of energy production in cardiac muscle: effects of ischemia in acidosis. In: Recent Advances in Studies on Cardiac Structure and Metabolism, Vol. II, Heart Function and Metabolism (Eds. K Kobayashi, T Sano and NS Dhalla). University Park Press, Baltimore, MD. 521-531.

30. Wilson DF, Rumsey WL, Green TJ, Vanderkooi JM. The oxygen dependence of mitochondrial oxidative phosphorylation measured by a new optical method of measuring oxygen concentration. J Biol Chem 1988; 262:2712-2718.

# TRANSCATHETER MICROWAVE TECHNOLOGY FOR TREATMENT OF CARDIOVASCULAR DISEASES

James C. Lin

## INTRODUCTION

Therapeutic options for managing a variety of cardiac and vascular diseases have concentrated on pharmacologic agents and surgical interventions. In recent years, there has been increasing interest in nonpharmacological therapy and minimally invasive procedures. In a small significant fraction of patients with ventricular tachycardias, available drug therapy has been found unsatisfactory because of a lack of meaningful response or unacceptable side effects. Transvenous catheter electrical ablation of the atrioventricular (AV) junction to effect complete AV block has become an increasingly accepted technique for treating patients with drug resistant supraventricular arrhythmias.[2,14]

To a greater extent, percutaneous transluminal balloon angioplasty has become a widely used method in the treatment of arteriosclerotic diseases.[3] Unlike ablation, angioplasty is a palliative treatment procedure. Its primary goal lies in dilatation and recanalization of obstructions in arteries to achieve maximum function for a prolonged period of time. Since 1987 percutaneous transluminal coronary angioplasty has become the frontline treatment of symptomatic patients with silent ischemia and abnormal exercise stress test results. But problems of abrupt closure and late restenosis of the dilated segment have prompted the investigation of thermal techniques to prevent occlusion and restenosis.[4] This paper describes the advantage of microwave heating technology for transluminal catheter treatment of tachyarrhythmias and arteriosclerotic diseases. It is noted that aside from microwaves other energy sources such as lasers, electrical, and radio frequency currents have also been used both for angioplasty and cardiac ablation. Some of the major limitations of these techniques include perforation of heart and vessel walls as well as nonfocussed damage to cardiovascular tissues.

## ADVANTAGE OF MICROWAVE TECHNOLOGY

The absorption of microwave energy in biological tissue is governed by the source frequency, antenna configuration, as well as tissue geometry and permittivity. The frequency within a decade of one gigahertz (GHz) is readily absorbed by biological tissues since only a small fraction of the incident energy is reflected at tissue-tissue boundaries.[7] For a plane wave propagating in a planar infinite tissue medium, the effective absorption depth (half power point) is about 10 and 50 mm for tissues of high and low water contents, respectively. However, the wavelength contraction that takes place naturally in tissue medium and the diverging character of radiating near field of small antennae help to reduce the effective absorption depth by factors of five to ten. Since the absorption of microwave energy is governed by the radiating behavior of the antenna, the distribution of absorbed energy can be controlled by judicious design of the antenna that will not only permit greater penetration of microwave energy into tissue but also allow discrete localization of microwave energy absorption.

The absorbed microwave energy is converted into heat by the dielectric loss mechanism which characterizes biological tissue at microwave frequencies. Hence, percutaneous transluminal microwave angioplasty may be attempted by advancing a microwave catheter antenna through the stenosis to reduce arterial narrowing by microwave induced heating of atherosclerotic plaque. Alternatively, microwave thermal angioplasty may be used in combination with balloon angioplasty: a coaxial microwave catheter antenna applies microwave energy to heat either the balloon or the plaque during conventional percutaneous angioplasty in remodeling the atherosclerotic lesion. In percutaneous transluminal catheter ablation of tissue substrate responsible for tachycardia, under the guidance of endocardiac electrogram, microwave energy is delivered through a catheter to thermally ablate the cardiac conducting tissue. Since tissue heating is the principal mechanism in producing pathological alterations following microwave irradiation of biological tissue, it is possible to titrate, in stepwise manner, the microwave energy delivered to cause various degrees of tissue modification. The range of control afforded by microwaves is a distinct advantage over other modalities both for ablation and angioplasty. Lower energy levels may help to localize and confirm the site to be ablated or to identify the degree of plaque calcification to allow adjustment of microwave power to atherosclerotic plaque to effect vascular recanalization.

## MAJOR CONPONENTS OF A CLINICAL SYSTEM

There are three major components of a clinical transcatheter microwave system; the microwave power source, the transmission cable and the catheter antenna. Microwave power sources of low cost and capable of delivering up to a few hundred watts are commercially available. Successful interfacing of microwave sources with flexible coaxial transmission lines is relatively trivial, even with the diminished size cables required for ablation and angioplasty. The most critical element in the design of the transcatheter microwave delivery system is the distal antenna. It is therefore useful to discuss the catheter antenna in terms of specific designs.

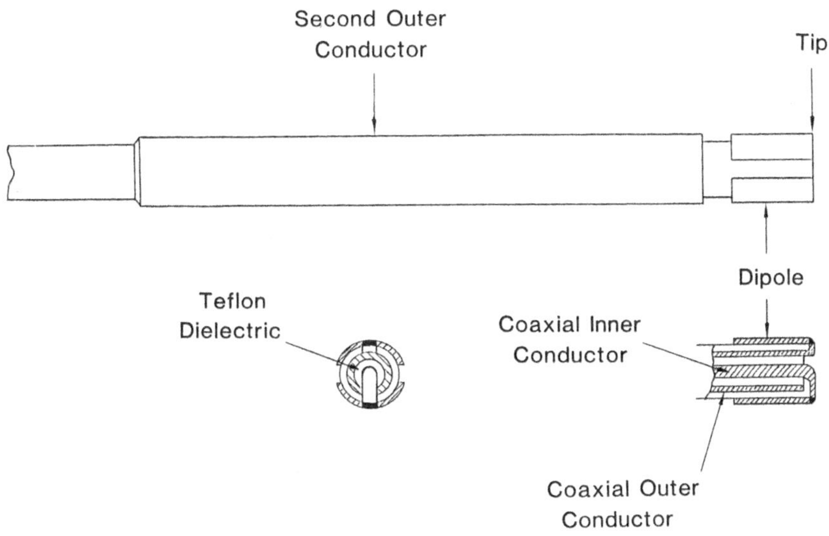


**Figure 1.** A catheter antenna with enhanced power deposition near the tip. The folded arms of the dipole are connected to the inner and outer conductors of a flexible coaxial cable. A conducting sleeve is shorted to the outer conductor of the transmission line to form a matching balun.

A. Cardiac Ablation

A transcatheter microwave antenna designed for ablation of conductive tissue with enhanced heating near the tip is shown in Fig. 1. It consists of a balun-fed, folded dipole in which the two arms of the dipole are connected to the inner and outer conductor of the coaxial cable. A conducting sleeve, one quarter wave in length, is attached to the outer conductor of the input coaxial cable to form a balun.[9] Thus the effective impedance between the balun and the outside of the transmission line at the dipole junction is made sufficiently high to permit efficient power transmission and to eliminate standing waves on the coaxial line. The antenna is constructed using silver-coated flexible coaxial cable (RG/178BU 2 mm in diameter). The sleeve is 20 mm in length and the folded dipole extends 3 mm beyond the 10 mm junction. A saline phantom consisting of a test tube filled with 50 ml of 0.9 percent NaCl was used to test the performance of the antenna in tissue. Since there is significant heat transfer in saline solution through conduction and convection, results obtained should mimic the worst performance of the antenna. Aside from a slight adjustment that may be required for matching purposes, its performance in soft tissue should be superior. The antenna was inserted along the axis of the saline-filled test tube with predetermined depth. A Vitek thermistor probe was attached to a carrier on

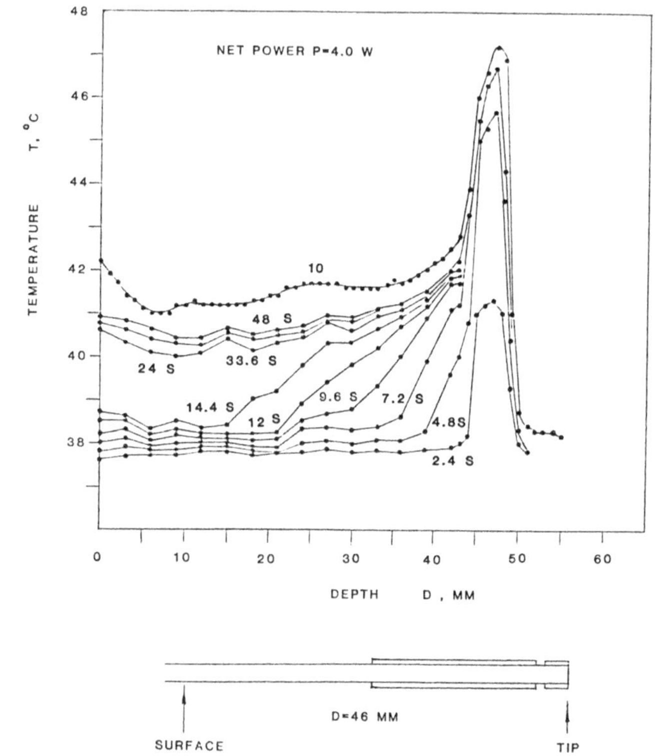


**Figure 2.** Temporal variation of temperature distribution along the folded dipole catheter antenna.

a stereotaxic manipulator to allow precise placement of the probe relative to the antenna. Microwave power was supplied through an amplifier and a generator operating at 2.45 GHz. A cool-down period of 306 min was needed between each measurement to establish temperature equilibrium.

The transient temperature distributions along the length of the antenna from the phantom surface of the antenna tip are shown in Fig. 2. It is seen that the temperature distribution at 48 s approaches that of the steady state (10 min). The curves for 2.4, 4.8 and 7.2 s clearly show the greatly increased heating near the tip. It should be noted that effects of heat conduction and convection in saline phantom become gradually significant only for heating times greater than 7 s. However, even in steady state, temperature distributions along the 46 mm length of the antenna in the saline phantom show peak heating in the tip region. While heating at a radical distance of 1 mm from the tip is 50 percent higher than the rest of the antenna; at 2 mm, tip heating is the same as elsewhere. Thus the heating pattern is highly localized. These results suggest that the antenna would be most appropriate for ablating cardiac tissue.

In preliminary tests,[1,10,11] the entire catheter antenna was insulated except for the tip and was used for recording bipolar electrograms. The 2.45 GHz magnetron source was able to deliver up to 300 watts power with output power and adjustable timing circuits. The microwave catheter antenna was inserted percutaneously through the femoral vein or the carotid artery, first, to record from the His bundle using the bipolar electrode and, then, to ablate the atrio-ventricular conduction system in dogs. Microwave energy was applied incrementally until temporary or permanent atrio-ventricular block occurred. Permanent atrio-ventricular block was achieved in all dogs in which good His bundle potential could be recorded. The incident power varied from 50 to 100 watts. With durations of microwave application from 2 to 7 sec, the temperature at the tip of the catheter antenna reached 80°C

B Microwave Angioplasty

The basic geometry of an antenna suitable for angioplasty is shown in Fig. 3(a). The active region of the antenna is comprised of sections before and after the antenna junction. The junction creates a voltage between the end of the outer

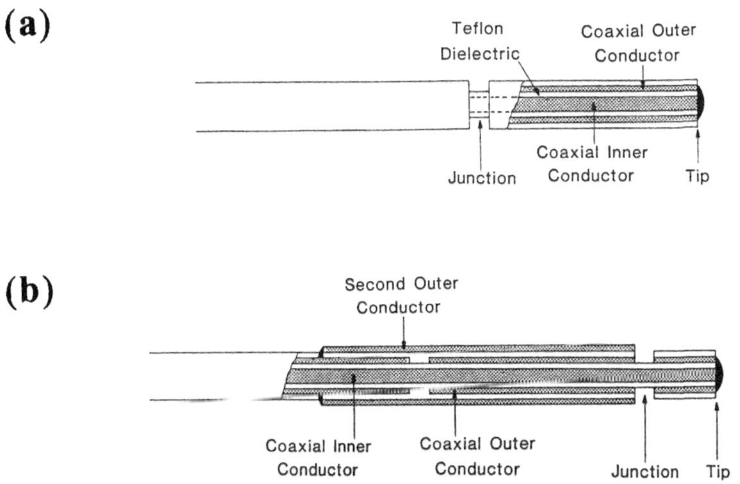


**Figure 3.** (a) A simple catheter antenna based on conventional dipole design; (b) A sleeved coaxial slot antenna with more uniform power deposition. The antenna is short circuited at the tip and at the proximal end of the second outer conductor (sleeve).

conductor and the start of extension of the inner conductor. This voltage establishes the currents that flow through the active regions of the antenna, and radiate microwave energy into the biological tissue.[8] The antenna is constructed using flexible silver-coated coaxial cable (Belden RG/178BU). The length from the center of the junction to the shorted tip of the antenna is 10.8 mm, which is equal to 1/8 wavelength in Teflon at the frequency of 2.45 Ghz. The active region of the antenna is completely insulated. When the antenna junction is placed 30 mm below the surface of the normal saline, this antenna is well matched to the coaxial cable with a power reflection coefficient less than 3.5 percent.

An antenna designed to provide more even heating is shown in Fig. 3(b). This coaxial antenna has two slots and a second outer conductor. The distance between the two slots and the center of the distal slot and the tip are 10.8 and 3.5 mm, respectively. The second outer conductor is 15 mm in length and is short-circuited at the proximal end to the outer conductor of the coaxial cable. In addition, the antenna is short-circuited at the tip.[8] The antenna is constructed using flexible silver-coated coaxial cable (RG/178BU), with the addition of a thin-walled brass outer conductor. The antenna is completely insulated using plastics. The principle of operation is easily understood by recognizing the antenna as consisting of an array of two radiating slots separated by an 1/8 wavelength over a ground plane. This configuration gives rise to a rather broad field pattern. By wrapping the ground plane around the slots in the form of a sleeve, the design forces the field pattern to shift toward the antenna tip.

The temporal dependence of temperature distribution along the basic antenna length is shown in Fig. 4. The transient behaviors are illustrated for 48 s after microwave power application. The steady-state temperature at 10 min of microwave power application is shown as the top curve. The temperature distribution at 48 s approaches that of steady state. The effect of conduction and convection in the saline phantom become gradually more significant for heating times greater that 7 s. Their influence is negligible for periods less than 5 s. The curve at 4.8 s thus represent the power deposition pattern of this insulated antenna in saline phantom. Clearly, power deposition is concentrated around the antenna junction and is minimal at the tip of the antenna.

The temperature distribution produced by a 2.45 GHz sleeved coaxial slot radiator with the parameters given above are shown in Fig. 5. The transient behavior of the temperature distribution pattern is seen in the initial 48 s of microwave power application. As mentioned above, the influence of heat conduction and convection in the saline phantom is negligible for the first 5 s. For times longer than 7 s, heat conduction and convection would tend to flatten

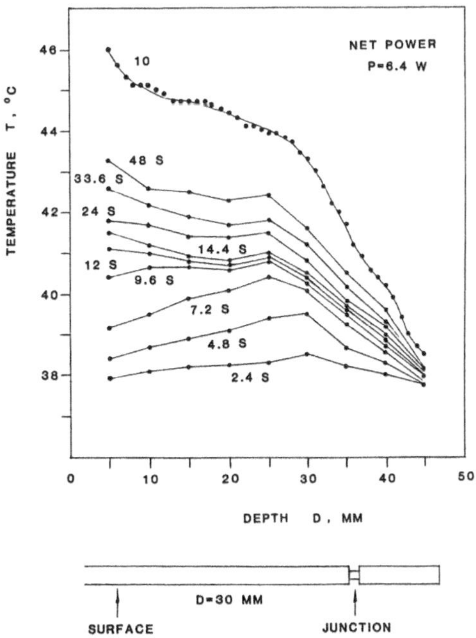


**Figure 4.** Temperature distribution along the simple conventional dipole catheter antenna as a function of time.

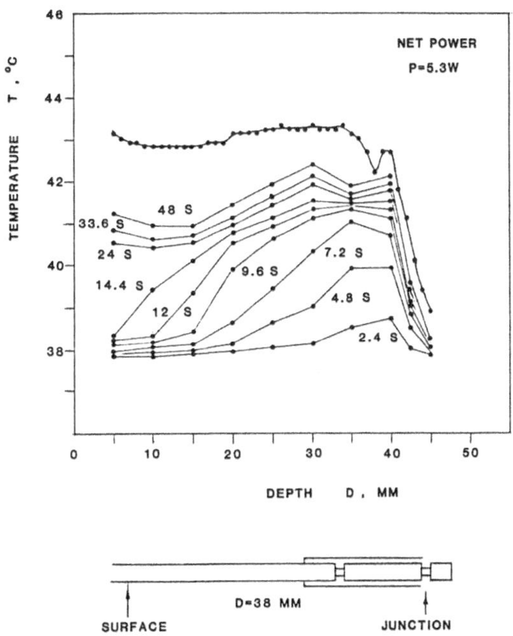


**Figure 5.** Temperature change along the sleeved coaxial slot catheter antenna as a function of time.

the temperature distribution gradually along the antenna length and allow the distribution to reach steady state in 10 min (top curve shown in Fig. 5). It is important to note that the power deposition pattern represented by the curve at 4.8 s stays near maximum to the tip of the coaxial slot radiator, in contrast to the cold zone at the tip of a simple insulated antenna.

A coaxial dipole antenna similar to that shown in Fig. 3(a) has been tested for use in a conventional balloon angioplasty catheters (personal communication).[13] The antenna was centered within the balloon and 915 MHz microwave energy was delivered through a thin, flexible coaxial cable to the antenna and radiated to the surrounding arterial wall in a rabbit model. Local tissue temperature was raised to 65°C with 30 watts of input power. Preliminary results suggest that microwave assisted balloon angioplasty can produce such tissue modifications as softening and spreading of the fatty arterial plaque.

## CONCLUSION

Cardiovascular applications of microwave technology in the past have focused on the noninvasive sensing of heart and vascular wall movements for diagnostic and monitoring purposes.[5,6,12] Recent developments have shown that substantial quantities of microwave energy could be transported via flexible coaxial cables of very small dimensions. The principal mechanism of microwave-tissue interaction revolved around the fact that such energy could produce tissue heating, this suggests potential percutaneous treatment of cardiovascular diseases amenable to heat therapy. To highlight this potential, this paper describes the application of this technology for intracardiac ablation of conductive tissue and heated balloon angioplasty. Preliminary test have demonstrated that it is feasible both to ablate pathologic cardiac tissue and to remodel arterial lesion using controlled microwave thermal therapy.

It is emphasized that more basic design and experimental work are needed before percutaneous cardiovascular microwave therapy can become a clinical reality. A major feature of this work would involve the design of catheter antennae to deliver microwave energy in appropriate fashion to the cardiovascular system. The basic experimental work needed would include: 1) energy level; 2) power intensity and exposure duration of microwave for various lesions and applications; 3) the effect on normal tissue structure adjacent to the lesion; 4) potential complications resulting from the procedure such as thrombogenesis; and, 5) methods for identifying, pinpointing and targeting the exact pathology for microwave therapy so that thermal injury is not extended beyond the lesion area. Percutaneous translumenal microwave has potential uses in peripheral vascular obstructive lesions as well as other cardiovascular disorders. I believe within a few years, it will emerge as an effective, nonsurgical, nonpharmacological modality in the treatment of cardiovascular diseases. It is hoped that this discussion will provide the stimulus for additional in vivo animal studies and the initiation of clinical investigations of this exciting new therapeutic modality.

## References

1. Beckman KJ, Lin JC, Wang Y, Illes RW, Papp MA, Hariman RJ. (abstract) Production of reversible and irreversible atrioventricular block by microwave energy. Circulation 1987; 16:1612.

2. Breithardt G, Borggrefe M, Zipes DP. Nonpharmacological therapy of tachyarrhythmias. Futura Publishing Co., Mount Kisco, New York, 1988.

3. Chokshi SK, Meyers S, Abi-Mansour P. Percutaneous transluminal coronary angioplasty: Ten years' experience. Progress in Cardiovascular Diseases, 1987; 3:147-210.

4. Isner JM, Steg PJ, Clarke RH. Current status of cardiovascular laser therapy. IEEE J of Quantum Electronics 1987; 23:1756-1771.

5. Lin JC. Microwave propagation in dielectrics with application to cardiopulmonary interrogation. In: Medical Application of Microwave Imaging (Eds. LE Larsen, Jacobi). IEEE Press, New York. 1986; 47-58.

6. Lin JC. Microwave noninvasive sensing of physiological signatures. In: Electromagnetic Interaction with Biological Systems. Plenum Publ Corp., New York. 1989; 3-25.

7. Lin JC. Electromagnetic Interaction with Biological Systems. Plenum Publ Corp, New York, 1989.

8. Lin JC, Wang Y. Interstitial microwave antenna for thermal therapy. Int J Hyperthermia 1987; 3:37-47.

9. Lin JC, Wang Y. An implantable microwave antenna for interstitial hyperthermia. Proceedings IEEE 1987; 75:1132-1133.

10. Lin JC, Hariman RJ, Beckman KJ. (abstract) Transcatheter cardiac ablation using microwave energy. Bioelectromagnetics Soc. Annual Meeting, Stamford, CT, 1988.

11. Lin JC, Beckman KJ, Hariman RJ. Microwave ablation tachycardia. Proc IEEE Engineering in Med and Biol Soc Meet. Seattle, WA, 1989.

12. Papp MA, Hughes C, Lin JC, Pouget J. Doppler Microwave: A clinical assessment of its efficacy as an arterial pulse sensing technique. Invest Radiol 1987; 22:569-573.

13. Rosen A, et al. Percutaneous transluminal microwave angioplasty catheter. Presented at IEEE MTT-S Intern Symp. Long Beach, CA, 1989.

14. Scheinman MM. Catheter Ablation of Cardiac Arrhymias. Martinus Nijhoff Publishing, Boston, MA, 1988.

# RF ENERGY FOR WARMING DIVERS' HANDS AND FEET

Richard G. Olsen

## INTRODUCTION

It has been well documented that cold water is a prime cause of impaired diver performance.[2,4] Decrements in dexterity typically occur at finger temperatures below about 15°C.[3,5] Purely passive thermal hand protection is self defeating because it necessitates the use of bulky gloves or mittens and precludes the full usefulness of the hands.

To date, active thermal protection of the extremities has involved surface-applied heat such as warm water or direct electrical current to heat a resistive wire. Surface-applied heating methods tend to be inefficient in cold water because of large conductive heat losses and because of the thermal limitations due to skin burning. Radio frequency (RF) energy has received little attention in this area even though the deep-heating properties of helical RF coils for tumor therapy and hypothermia resuscitation have been demonstrated.[6,8] A system that deposits energy within the volume of an extremity rather than at its surface might be efficient enough to maintain a diver's manual dexterity without the need for a prohibitively large and bulky electrical power source. This study was, therefore, conducted to investigate the feasibility of using a resonant RF applicator system to efficiently provide warmth to a diver's extremities. The results were sufficiently encouraging to suggest that direct RF heating of the limbs could have a wide range of applications.

## MATERIALS AND METHODS

### A. Subjects and Apparatus

Voluntary, informed consent was obtained from six healthy, male subjects, 26 to 50 years of age, weighing 71 to 107 kg.

#### 1. RF Coils

Figure 1 shows the RF hand coil assembly. It was fashioned as a series connection of finger coils from 1.8 m of solid, 18 AWG insulated copper hook-up wire. Opposite ends of the coil were returned to a central location on the

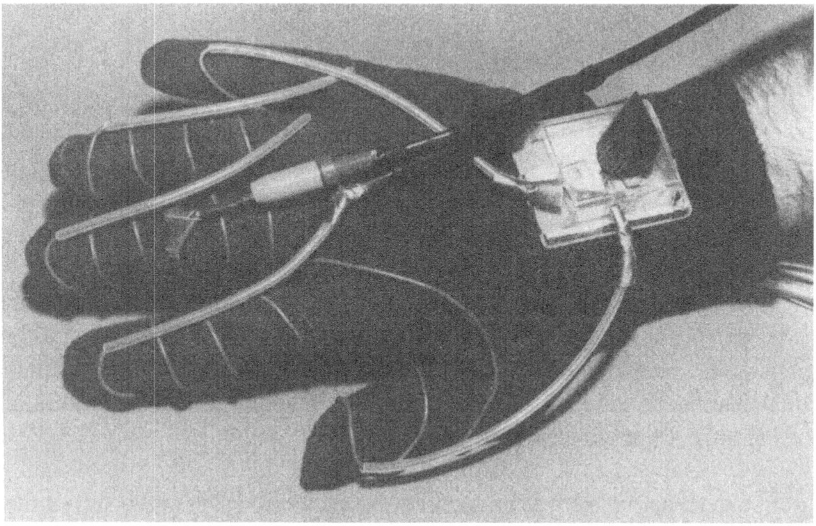

**Figure 1.** The RF hand-warming coil was a series connection of finger coils attached to a cotton work glove.

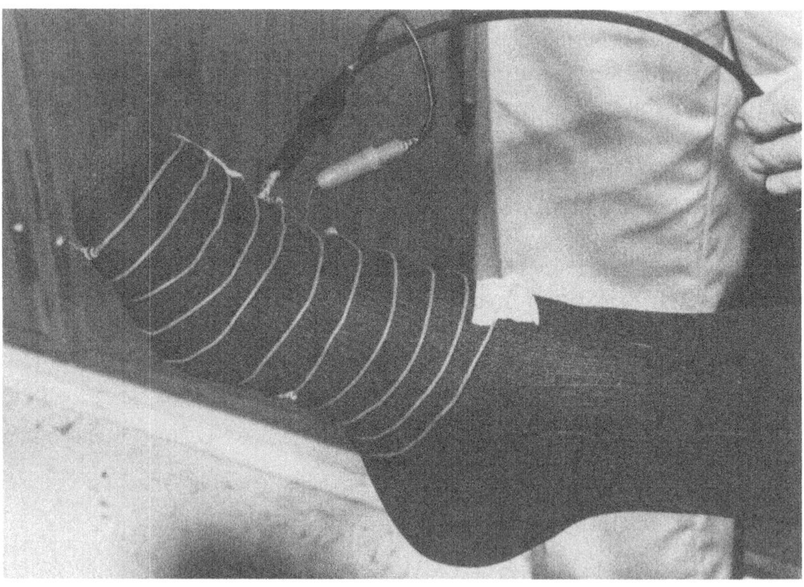

**Figure 2.** The RF foot-warming coil was attached to a sock worn by each subject.

back of the hand and connected to a 3-50 pf variable capacitor. As shown in Fig. 1 the coil was wound over a (right hand) cotton work glove to facilitate repeated use by multiple subjects. Radio frequency energy at 27.12 MHz, an industrial, scientific, and medical (ISM) channel, was fed to the central portion of the hand coil through a length of RG/58A coaxial cable and a 1-ampere, type 3AG fuse. A banana plug and jack formed the other electrical connection, and insulating tape covered all bare conductors. Adjustment to a 50-ohm impedance match at 27.12 MHz was made with the aid of a network analyzer. Small subject-to-subject offsets in resonant frequency were accommodated by adjusting the variable capacitor to permit operation entirely at 27.12 MHz.

Figure 2 shows the simpler helical foot coil assembly. The coil consisted of a 2.5 m length of the same type of wire that comprised the hand coil with the same type of RF power connection scheme. For each experiment, the foot coil was wound over a nylon sock and secured with medical adhesive tape rather than being permanently attached to a garment as was the hand coil.

2. Thermometry Systems

A 12-channel, optical fiber thermometry system (Clinitherm model T-1200) was used in the hand warming experiments. Before donning the glove-mounted hand coil assembly, 12 temperature probes were attached to the subject's right hand, two each on the palm side of each finger (ball of fingertip opposite the nail and fleshy region between the two most distal joints). Two additional probes were placed on the top and bottom of the central palmar region. Temperature probe output was printed at 30-second intervals during each experiment.

A four-channel optical fiber thermometry system (Luxtron model 750) was used in the foot warming experiments. Before the subject placed his foot in the nylon sock, a temperature probe was taped in place on the bottom of the big toe, on the bottom of the little toe, on the top of the foot under the central portion of the coil, and on the bottom of the foot at the instep. As in the hand warming experiments, temperature probe output was printed at 30-second intervals.

3. RF System

A 50-watt military radio transmitter (AN/WRC-1) was used to generate and amplify the 27.12 MHz RF signal. A Bird model 43 in-line power meter was used to monitor net RF power applied to the coils.

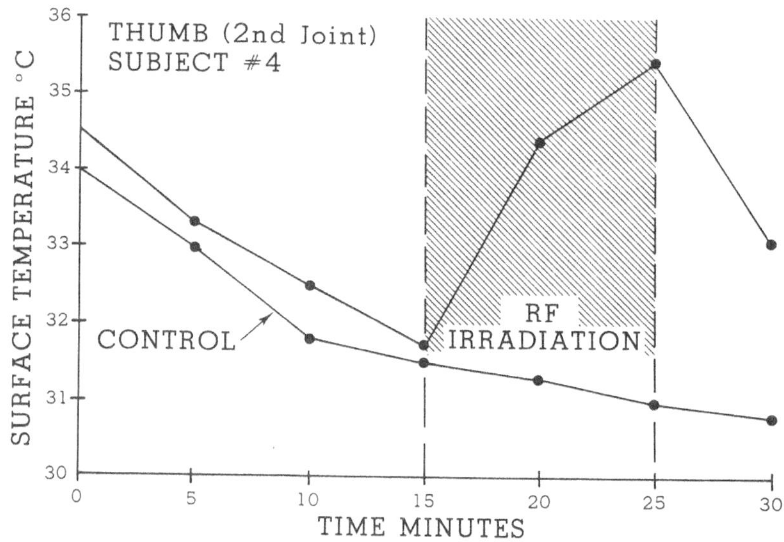


**Figure 3.** The timecourse and temperature data are shown for typical control and RF irradiation experiments.

B. Procedure

For the hand- and foot-warming experiments, both control and irradiation experiments were conducted with each subject serving as his own control. Control experiments were identical to those involving RF irradiation except that the RF generating apparatus was not energized. Hand-warming experiments used 20 W net RF power, and foot-warming experiments used 25 W. These power levels were selected as being comfortable for all subjects and less than the average, whole-body specific absorption rate (SAR) allowed by the safety guideline of the American National Standards Institute (ANSI).[1] After the coil assembly had been secured to either the right hand or the right foot, two polyethylene bags were placed over the extremity before it was immersed into a 15-gallon plastic tub containing simulated sea water, a three percent (by weight) salt solution at room temperature (24 ± 2°C). The rectangular plastic bags were sufficiently long to encase the submerged limb within a protective dam; thereby, excluding the simulated sea water from intimate contact with the coil assembly. In the foot-warming experiments, three layers of elastic bandages were wrapped over the coil assembly to simulate the insulating effect of a heavy sock and dry boot. No bandages were used over the hand coil to simulate the limited insulating effect of a lightweight work glove. After immersion, data acquisition from the thermometry system was started. A 15-minute equilibration period was then commenced followed by a 10-minute irradiation (RF experiments only) which was followed by a five-minute post irradiation period for a total immersion time of 30 minutes. Figure 3 shows this regimen for a single probe location.

### C. Data analysis

For the 10-minute period of RF irradiation, experimental and control temperatures were statistically compared at two-minute intervals beginning with the start of irradiation. For each subject and probe location, a normalized temperature increment ($\Delta T$) was obtained by subtracting the irradiation temperature from the control temperature and further subtracting any initial temperature offset such that at zero irradiation time, the normalized temperature

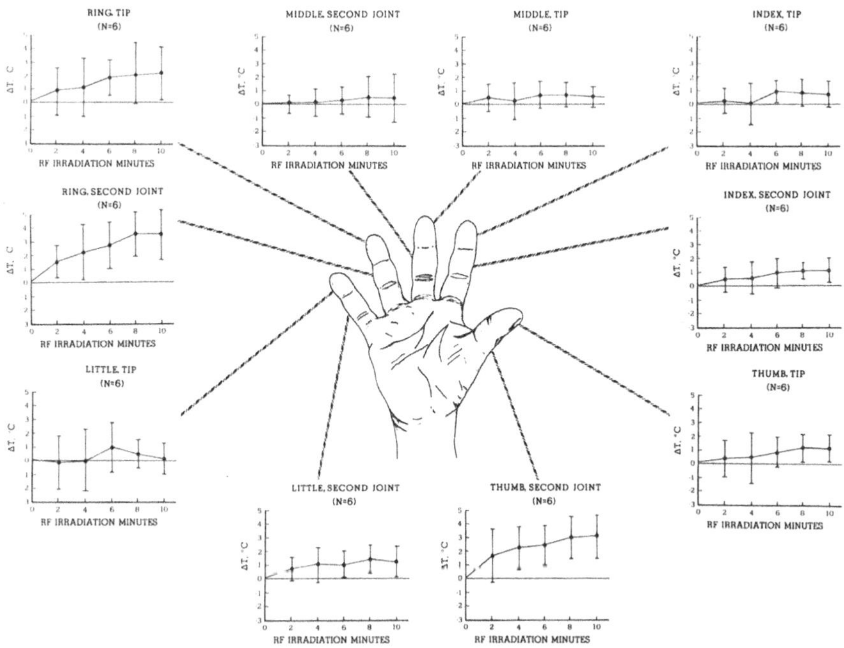


**Figure 4.** Normalized mean (± SD) temperature increment ($\Delta T$) for ten hand locations during RF irradiation. Twenty watts at 27.12 MHz was used.

increment (ΔT) was zero. The resulting distributions of normalized RF-induced thermal rises were analyzed using a repeated measures analysis of variance (ANOVA) and tested at the $p = 0.05$ level of significance for any change over time.

## RESULTS

Figure 4 shows the RF-induced temperature rises for the ten finger locations of the surface temperature probes. Nearly all temperatures showed upward trends, but only four locations (the second joint region of the thumb, index, ring, and little fingers) were significantly different from controls during the irradiation period. Of the two other recorded locations (top and bottom of the palm), one location (top) was statistically different from the control condition with a mean temperature increment of 1.2°C at the end of irradiation.

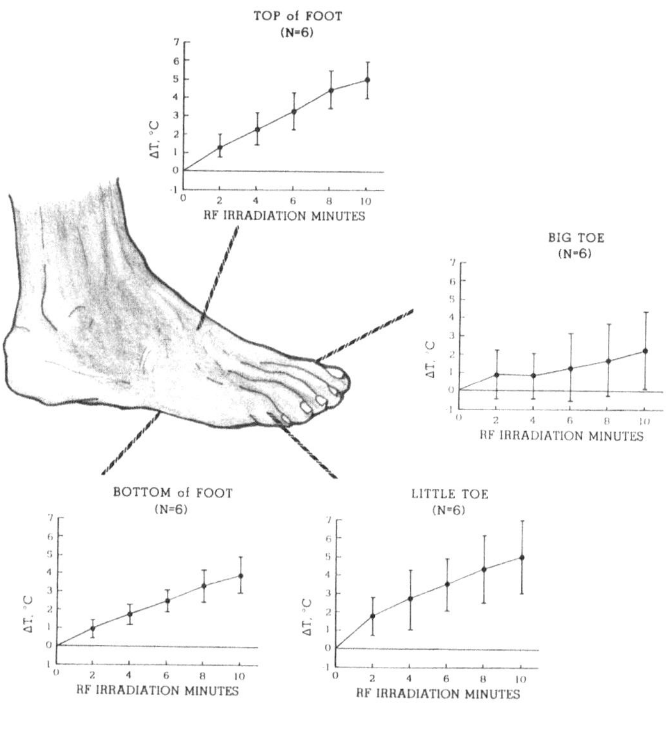


**Figure 5.** Normalized mean (± SD) temperature increment (ΔT) for four foot locations during RF irradiation. Twenty-five watts at 27.12 MHz was used.

Figure 5 shows the RF-induced temperature rises for the foot probe locations. Higher temperature increments were observed on the foot as compared to the hand, probably due to the combination of higher RF power and the insulating layers of bandage over the foot. All four of the recorded foot temperatures were significantly different (warmer) during irradiation as compared to the control condition.

## DISCUSSION

These experiments used highly simplified conditions in order to represent how RF energy might be used to provide warmth to a diver's extremities. Much development effort, however, remains to be accomplished before practical underwater RF warmers are commercially available. The surface-mounted temperature probes could not be expected to properly indicate the internal RF-induced heating patterns; nevertheless, significant heating was observed at nine of the 16 measured locations. The effect of placing thermal insulation over the coil and limb was seen in the higher temperature increments in the foot. Obviously, more hand heating would have been observed if more insulation had been employed there.

The RF-induced heating appeared to be highly efficient with little energy being deposited directly in the simulated sea water or being lost from the skin surface. In order to gain a qualitative understanding of the salient differences in RF and conventional heating, a nichrome wire hand coil closely resembling the RF hand coil was placed on the right hand of two of the subjects and energized (in free space) with direct current (DC) from a small laboratory power supply. When the DC power supplied to the resistive wire was adjusted to 15 W, no more than about three minutes of heating could be tolerated by the subjects because of uncomfortable skin temperatures. This feeling was quite different than the subtle warmth produced by the RF hand coil at this power level. Obviously, more experimentation is needed to compare RF heating with resistive wire heating on the basis of maximum tolerable skin temperature, as well as, overall energy efficiency.

Because of the simple nature of this study, we do not know what extremity temperature can be maintained for a given RF power level, glove/sock type, or water temperature, and more study is needed in this area. Questions also arise concerning the efficacy of using a much simpler RF coil system on the forearm rather than on the hand. Our preliminary efforts used a forearm coil, but the hand-warming ability of such a device appeared to be much below that of the hand coil used in the present study. Physiologists might argue, however, that whole-body, cold-water immersion experiments must be conducted in order to properly compare forearm coils with hand coils.

## CONCLUSIONS

This study presents evidence showing that the technology of RF-induced extremity warming has potential for significant beneficial impact in the area of diver dexterity and in other occupations and activities where cold-induced performance decrements occur. Expansion of this technology could eventually lead to the development of many useful devices, possibly including a highly efficient pair of RF-powered (all-day) "hunter's socks" to replace the simple battery-operated models of today.

The presence of relatively strong RF fields near the new devices will have to be carefully considered, however, and during irradiation, certain changes in the way work is accomplished will have to be made. Changes similar to these, however, have already been implemented and are common when one considers the special procedures and precautions associated with the use of a magnetic resonance imager. As our conventional energy resources continue to decline, emphasis on thermal efficiency will increase in all sectors, and the special advantage of RF-induced tissue heating, for many purposes, will become more important.

## DISCLAIMER

This research was sponsored by the Naval Medical Research and Development Command under work unit 61152N-MR0001.001-7044, accession number DN247513. The views expressed in this article are those of the author and do not reflect the official policy or position of the Department of the Navy, Department of Defense, nor the U.S. Government. Volunteer subjects were recruited, evaluated, and employed in accordance with the procedures specified in Department of Defense Directive 3216.2 and Secretary of the Navy Instruction 3900.39 series. These instructions are based upon voluntary informed consent and meet or exceed the provisions of prevailing national and international guidelines.

### References

1. American National Standards Safety Levels with Respect to Human Exposure to Radio Frequency Electromagnetic Fields, 300 kHz to 100 GHz. ANSI C95.1-1982. The Institute of Electrical and Electronic Engineers, Inc. New York.

2. Bachrach AJ, Egstrom GH. Human performance underwater. In: Diving Medicine (Ed. RH Strauss). Grune and Stratton, New York. 1976; 1831-1896.

3. Clark RE. The limiting hand skin temperature for unaffected manual performance in the cold. J Appl Physiol 1961; 45:193-194.

4. Curley MD, Bachrach AJ, Langworthy HC. Wet-suited scuba diver performance in 5 - 25 degree C water. NMRI-81-51. Naval Medical Research Institute, Bethesda, MD, 1981.

5. Dusek ER. Effect of temperature on manual performance. In: Protection and Function of the Hands in Cold Climates(Ed. FR Fisher). National Academy of Science, Washington DC. 1957; 63-76.

6. Olsen RG. Reduced temperature afterdrop in Rhesus monkeys with radio frequency rewarming. Aviat Space Environ Med 1988; 59:78-80.

7. Olsen RG, David TD. Hypothermia and electromagnetic rewarming in the Rhesus monkey. Aviat Space Environ Med 1984; 55:1111-1117.

8. Ruggera PS, Kantor G. Development of a family of RF helical coil applicators which produce transversely uniform axially distributed heating in cylindrical fat-muscle phantoms. IEEE Trans Biomed Eng 1984; BME-31:98-106.

# EMR AND THE BRAIN: A BRIEF LITERATURE REVIEW

Paul Tyler

## HISTORICAL BACKGROUND

Man has lived in the presence of electric fields for many centuries most of which predate the invention of the electrical generator. Current theory, as well as history, tells us that early man recognized that he had no control over lighting and that magnetic mineral ore, lodestone, did not appear to have any biological effect. However, contact with the fish, Torpedo, which inhabits the Mediterranean, could produce shocking results when in contact with the body. The biological effects of the Torpedo fish were described by Plato, Aristotle and many others in ancient Greek and Roman texts. The first medical use of electric fish is found in a report by S. Largus in the first century.[22] Prior to the Renaissance, electric fish were used by Arab physicians to treat sleeping disorders,[37] migraine, melancholy and epilepsy.[50] This use of electric fish represents an ancient precursor of electroshock therapy for severe depression. The ancient physicians did not recognize that these fish were producing an electrical current, rather the cure was thought to result from some excretion of a semi-material nature. It was recognized, however, that the electric fish had to be alive to produce any beneficial effects. Thus, therapy was limited to sites near the seashore where fish could be caught and maintained fresh and alive until used for treatment.

With the advent in the late 1800's and the early 20th century of the electric battery and other devices that produce electrical energy, electrotherapy became a panacea for the medical profession. Legitimate physicians as well as charlatans were using all types of electrical devices to produce cures for almost every conceivable disease known to man. Along with this extensive use of electric fields, there was an appreciation that electricity could have unwanted effects on the brain with indiscriminate use. As stated by a physician in 1897:

> "Persons of little experience should not use the current about the head as great caution should be used with electricity about the head".[23]

It was estimated that by 1894, 10,000 physicians in the United States were using electrotherapy daily in their practice.[8] During this period, in addition to legitimate use, there was a great deal of abuse and fraudulent use of electromagnetic fields. Many individuals were charging large amounts of money for useless electrotherapy. This excessive use of electrotherapy was one of the major factors that lead the Carnegie Foundation to establish a medical commission to study medical education and standards of medical practice. The final report of this commission named after the chairman of the commission, Abraham Flexner, lead to an almost instantaneous, major revision of medical education and licensure of physicians. An additional fallout from the Flexner Report was the demand that all medical therapies rest upon a sound scientific basis. Given the scientific knowledge of that time regarding electrotherapy, it became insupportable and fell into disrepute. Unfortunately, legitimate uses for electromagnetic therapy in the treatment of human disease were also lost.

## PARAMETERS INFLUENCING BIOLOGICAL EFFECTS

Past research indicates that specific biological effects depend upon the electromagnetic parameters utilized. By changing the electromagnetic parameters, biological effects may be enhanced, abated or altered.

The frequency of the radiation is one of the more critical parameters. The laboratory headed by Adey[9] was one of the first to report frequency specific biological effects. They reported that their observation of calcium efflux from the isolated forebrain of the neonatal chick was dependent upon the frequency of modulation. The maximum efflux occurred at 16 hertz modulation. The calcium efflux was not significant when the modulation was less than 6 hertz or greater than 20 hertz. Webb also reported frequency specific effects in the growth and synthesis of DNA, RNA, and protein in bacteria using different frequencies in the gigahertz range.[58,39]

The power level of the radiation is another critical parameter when evaluating biological effects. Again Adey's laboratory was among the first to report the presence or absence of biological effects as a result of altering the power levels of the radiation source. Adey was the first to apply the term "window" to the concept of frequency and power specificity. When the frequency or power were outside the window, no effects were observed.[1,26]

As would be expected, the duration of exposure also influences the biological outcome. However, it does not appear to represent a simple linear effect. There is evidence in some cases that as the duration of exposure increases the biological effect disappears. The Eastern European, particularly Soviet literature, discusses this concept as an adaptation process in which there are immediate changes following exposure. With continued exposure the effects disappear as the body adapts to the new conditions.[32]

Other factors which influence biological systems include whether the wave form is continuous or pulsed. If it is a pulsed wave, the repetition rate of the pulses is important. Even the shape of the pulse appears to influence the production of biological changes in experimental subjects under some experimental conditions.[24] Different biological effects can also be produced depending on whether the exposure is to a magnetic field, an electric field or electromagnetic fields.

The depth of penetration in classical electrophysics is also dependent upon the frequency of the radiation. In general, the human body is invisible to the very low frequencies because the body is so small in relation to the length of the electromagnetic wave (eg, a 6 foot man exposed to a 10 hertz field for which the wavelength would be approximately 25,000 miles long). As the waves get smaller, (ie, higher frequency) and approach the size of man, the absorption of the wave's energy becomes greater and greater. When the electromagnetic wave becomes smaller than man at higher and higher frequencies, basic physics predicts that the depth of absorption again becomes less and less. While this generalization may be valid in some cases, there are many research reports that indicate both extremely low frequencies (ELF) and very high frequencies (visible light) produce biological effects that cannot be accounted for by superficial absorption on the skin. For example, Mantle and Persinger[39] reported that alternations in subjective evaluations, a cognitive function, was noted with exposure to 5 hertz magnetic field devices but not to exposure to a 9 hertz field. At the high frequencies the light can penetrate into the brains of fish, amphibians, reptiles, birds, and mammals as large as sheep.[40,27]

## CENTRAL NERVOUS SYSTEM (CNS) EFFECTS: LABORATORY STUDIES

### A. Electromagnetic Effects Upon the Electroencephalogram

As with many areas of electromagnetic research on biological systems, there is a wide variety in reported effects upon the EEG. Some research papers report no effects while others report significant alternations. The majority of these differences can be attributed to variation in the exposure parameters and differences in the design of the protocols. Bartko, et al.[7] studied patients with various neurological diseases, along with normal control subjects. A magnetic field of 50 gauss, 50 hertz was applied to the head for 30 minutes. The EEG was evaluated by looking at its power spectra in terms of percent of alpha, beta, delta and theta activity. No alternations were found in either the patients or the control subjects. In agreement with Bartko, Takashima et al.[53] using rabbits found no acute exposure effects. However, they did find EEG changes in rabbits exposed for 2 hours daily for six weeks. The exposures were at 1-10 megahertz (15 hertz modulation) and 60 to 500 volts/meter. At the level of 90-150 volts/meter in the chronic exposed group, there was stimulation of the EEG high frequency components (10-15 hertz) while irradiation at 500 volts/meter enhanced the low frequency waves (4-5 hertz). They reported that the rabbits' EEGs returned to normal within several days following the discontinuation of the irradiation. Goldstein and Sisko[30] used male New Zealand rabbits sedated with a low dose of pentobarbital. They exposed the rabbits to 9.3 gigahertz, power density from 0.7 to 2.8 $mW/cm^2$. The exposure time was for 5 minutes. No changes were seen in the EEG patterns during exposure. Following a latent period post-exposure, the rabbits showed a sudden arousal which lasted about 3 minutes. This was followed by a return of the sedated state. Again after another latent period of 3 to 5 minutes, a period of arousal of 2 to 10 minutes duration occurred. Bise[12] reported enhancement of the EEG's low frequency waves and an increase in amplitude in the alpha waves in human volunteers exposed to low level (less than 10 $mW/cm^2$) in the frequency range of 0.1 to 960 megahertz CW and 8.5-9.6 gigahertz pulsed.

Each of these studies is unique from the others in its design and the electromagnetic parameters used. Because of the differences in the research protocols, it is very difficult, if not impossible, to come to any firm conclusions concerning any generalized effects of electromagnetic radiation on the EEG. There is some evidence from the EEG research data, combined with research on the neuronal activity in the brain, to indicate that ELF radiation may produce some entrainment of neuronal firing rates.[16]

B. Behavioral Effects

The electromagnetic radiation (EMR) literature is replete with behavioral studies using a variety of behavioral measures and exposure parameters. The majority of the studies are clustered in very few frequencies. As with the rest of the literature in this area of electromagnetic radiation, one can find many reported behavioral effects and some studies that fail to find any effects. There are research reports that both confirm or deny previously reported findings. This paper covers only a representative sample of this extensive behavioral literature.

Delgado[25] reported behavioral changes in monkeys exposed to magnetic coils aimed at both cerebellar hemispheres. The coils were energized with 3 msec square pulses at 50 to 100 hertz and 1.0 amperes, producing 1.1 to 2.5 gauss at the cerebellar surface. The field was applied for 9 hours. The results showed that at frequencies below 50 hertz the monkeys were more relaxed, less reactive to external stimulation and had increased periods of sleep. When the frequency was increased to 100 hertz, the monkeys were more reactive and alert, with increased restlessness, and had decreased periods of sleep.

Many investigators questioned whether subjects could learn to escape or avoid electromagnetic radiation exposure. These behavioral tests have only shown positive effects when there was adequate power to produce thermal heating of the experimental subject. More effective measures of subtle effects can be found in operant conditioning schedule-controlled studies. Gavalas-Medici[28] used time-based (interval) schedules of reinforcement together with EEG measurements in monkeys. She noted a decrease in the interresponse time in the irradiated animals.

Thomas et al.[54-56] explored the interaction of electromagnetic fields and psychoactive drugs on time-based schedules of reinforcement in rats. Their studies showed a variety of results. Using both pulsed and continuous waves (CW) systems, they reported that the drug diazepam with both pulsed and CW EMR produced no effect upon the schedule-induced behavior of the rats. In contrast, pulsed fields combined with chlordiazepoxide effected the behavior of the experimental animals while chlordiazepoxide and CW fields showed no effect. Both diazepam and chlordiazepoxide are prescribed to reduce anxiety in humans.

The same researchers in another study using dextroamphetamine and pulsed EMR reported that a time-based schedule of reinforcement was affected. In summary of these studies, Thomas and Banvard[57] reported that pulsed EMR lowered the response rates while CW fields did not.

Lovely et. al.[38] attempted to reproduce some of Thomas' work but could not obtain the same result as reported by Thomas. However, the studies were not exact replications since they were not consistent with Thomas's exact exposure parameters. As noted previously, use of different parameters can lead to different results.

Mantle and Persinger[39] studied the subjective evaluations of college students when exposed to either a 5 or 9 hertz magnetic field. The study was well controlled with double-blind conditions of exposure. Analysis of the data showed that individuals exposed to the 5 hertz magnetic field had significant decreases in sweatiness, stomach ache, dry mouth, and an increase in the relaxed state. There was no difference in any of the parameters evaluated for the 9 hertz magnetic field, placebo, or control groups.

Persinger and Nolan[46] also had university students listen to a three minute narrative with a standardized structure that had been validated as to structure of recall after a five minute delay. A 5 hertz field was applied to either the right or left of the head over the temporal lobe during one of the three minutes of the narrative. If the field was applied during the first minute, there was a significant decrease in the recall of the information. The authors conclusion was that a electromagnetic field near the temporal lobe during the labile stage of memory consolidation for complex verbal behavior may have affected the retention of material.

Amassian and his co-workers[3-6] have studied the effects of magnetic coils on both central and peripheral nerves in humans. They found that strong magnetic fields pulsed over the motor cortex stimulated peripheral muscle movement. In addition, when the coil is placed over the visual (occipital) cortex and pulsed, there is a suppression of visual perception. They have also noted that the resultant effects, using a strong magnetic field from a coil, are different from the effects seen with electric stimulation.

While there is little similarity between the low power, low frequency signals that Persinger and Nolan[46]. have used and the high power magnetic coil used by Amassian[3-6] for his studies, they both indicate that brain function can be altered by the application of external EM fields.

C. Brain Chemistry

Many chemical and ionic changes in the central nervous system during or following exposure to EMR have been reported in the literature. Several authors have reported changes in various neurotransmitters. Merritt et al.[41] reported decreases in norepinephrine, dopamine, and serotonin in specific regions of the brain following a short exposure to 80 $mW/cm^2$ EM fields. Snyder[52] also found a reduction in the serotonin levels along with a reduction in 5-HIAA after multiple exposures to 10 $mW/cm^2$. Monahan[42] also reported finding changes in the cholinergic system of the brain following electromagnetic irradiation. Monahan hypothesized that EMR increased the levels of acetylcholine.

Adey and his co-workers[2,10,11] extensively reported on the efflux of calcium ions in the brain following low level irradiation. Blackman et al.[13-15] confirmed Adey's work on calcium ions and extended the observations showing that the parameters of the electromagnetic irradiation and the local geomagnetic fields influence the outcome of the experiment.

## CLINICAL STUDIES

Prusinski et al.[47] used large coils to the entire body with a pulsed electromagnetic field (PEMF) as a therapy for headaches. The treatment was given daily, 20 minutes/day for 15 days. The patients were evaluated one month after treatment. The authors claim good to excellent results for individuals with migraine, tension, or cervical headaches. Poor or no improvement were noted in patients with cluster or post-traumatic headaches. During the treatment they noted no adverse effects of exacerbation of the headache pain. In contrast, Giczi and Guseo[29] also using PEMF therapy for 15 days on patients with migraine reported a decrease in attack frequency, but no change in the intensity of the pain when an attack did occur. They found no change in tension headaches. In cases of headaches associated with cervical spondylosis there was a marked decrease in frequency, intensity and duration of the pain.

In the early 1970's, Dr. Wen in Hong Kong found that addicts who were hospitalized and had electroacupuncture during neurosurgical procedures did not have withdrawal symptoms.[59] Since that time, there have been well over 100 journal articles published describing the use of electroacupuncture in the treatment of withdrawal symptoms of drug addiction. Dr. Patterson worked with Dr. Wen in Hong Kong during the early days of his experimenting with electroacupuncture and addiction. Upon her return to England she started to use the therapy there. She soon found that the English did not like acupuncture and she started to use only electrical stimulation with surface electrodes attached over the mastoids. She referred to this treatment as neuroelectric therapy (NET). Dr. Patterson claims that one can treat all types of addiction (hard drugs, alcohol and tobacco) in this manner.[44]

It is very difficult to obtain the exact parameters used from the published literature. Wen in one of his papers[60] states:

> "The electrical stimulator is then turned to the 'on' position. The frequency of the current is gradually increased from 0 to 125 hertz. The intensity is increased until the patient feels the flow of the current but without causing pain to his ear. Should the patient complain of pain in the ear the current is suitably reduced for comfort. The length of treatment varies in each individual and also depends on the drug or drugs that the patient has been addicted to. Some need more stimulation and some less. On the average, we stimulate the patient for half an hour. In the first few days of treatment, we give the patient two to

three stimulations per day for two to three days, followed by one stimulation per day for the next four to five days."

Dr. Patterson is even more vague in describing the electrical parameters of the device she uses for her NET therapy. She states,[44] that "....latest model of Stimulator is fully programmed and automated for different drug groups, drug combinations and the longer term abstinence symptoms such as depression and insomnia". She fails to provide any specific data concerning what electrical parameters such as frequency, type of pulses, repetition rate etc., are used.

At present therapeutic parameters are loose and vague and it is difficult to evaluate the effectiveness of the treatment. Dr. Patterson stresses that the therapy is not a cure, but that it relieves the withdrawal symptoms without substituting one addicting drug for another (eg, Methadone treatment).[44] The therapy should be used as an adjunct to psychotherapy, which attempts to treat the root causes of the drug dependence.

Patterson[44] reports that electrotherapy has been beneficial in treating addiction to the following drugs: heroin, Methadone, morphine, barbiturates and other sedatives, tranquilizers, cocaine, amphetamines, hallucinogens, alcohol, and nicotine. However, adequate control studies evaluating NET use with all of the addicting drugs is, in many cases, sadly lacking in the published literature. In spite of the lack of sound scientific studies, there is a large body of clinical evidence that indicates a therapeutic benefit from the use of electrical stimulators.

Ng and his associates[43] at the National Institute of Mental Health studied rats addicted to morphine. They found a significant attenuation of certain naloxone precipitated morphine abstinence signs following treatment with a constant current stimulator attached to the lower part of the concha, posterior to the external auditory meatus. The stimulator delivered biphasic square-wave pulses of 7 hertz and a current of 0.2-0.3 ma.

Capel et al.[17-19] studied the effects of electrostimulation and barbiturates in rats. In one study they noted that barbiturate-induced sleeping time could be altered by electrostimulation. They also noted that the frequency producing the maximum reduction in sleeping time varied with different barbiturates. In these studies, Capel used frequencies from 0 to 1000 hertz with a intensity of 1 volt/meter. Most of his studies used hexabarbital.

A study in progress by myself and several colleagues was undertaken to attempt to repeat some of Capel's research using electrostimulation with barbiturates and sleeping time. The study is a double blind design using a computer chip generated signal of 1 volt/meter. Trial runs have been conducted using 10 and 500 hertz square waves. These preliminary studies have, to date, failed to confirm Capel's findings.

There are several factors which may account for the different findings. Currently, hexabarbital is unattainable in the United States or England so Nembutal(pentobarbital) was substituted. We are sure that our electronic equipment is not identical to the system used by Capel, but we have duplicated his pulse, frequency and voltage. At the present time, a thorough evaluation is underway to check on possible problems with our system.

## MECHANISM OF ACTION

Sjolund, Clement-Jones, Salar and their co-workers[51,21,48] have all reported that there is an increase in ß-endorphins in the cerebrospinal fluid following electroacupuncture or transcutaneous electrotherapy. Capel et al. [20] reported changes in rat brain levels of serotonin, tryptophan and hydroxyindole acetic acid levels following electrostimulation at various frequencies. As noted previously, there are numerous papers[52,42,2,10,11,13] relating to changes in calcium ion levels in the brain following irradiation.

It has been hypothesized that drug addiction turns off the body's own production of ß-endorphins and upon acute withdrawal, the major symptomatology is caused by the failure of the body's endorphin and enkephalin production. Further hypothesis[44] suggest that the electrostimulation enhances the natural production and release of encephalins and ß-endorphins by the body, thus preventing the onset of withdrawal symptoms. This concept may be valid if the beneficial results were limited to heroin and its derivatives. Patterson et al.[45] also reports that electrostimulation is effective with cocaine, barbiturates, alcohol and even nicotine. While some of these observations have not been validated, there is adequate evidence that electrostimulation is effective for drugs other than heroin and its derivatives. There is little or no evidence in the literature that β-endorphins are involved or play a part in the biochemistry or metabolism of some of these drugs such as barbiturates or cocaine. There is a large body of literature concerning the metabolism and biochemistry of heroin,[49,34,35] but very limited literature concerning the metabolism of many of the other drugs that appear to be affected by electrostimulation. For example, only limited data is available on the metabolism of barbiturates.[31]

In an extensive review of the literature, this author has been able to identify only calcium as a common factor required in the metabolism of each of the drugs considered to be effected by electrostimulation. It is this author's hypothesis that it is calcium that is the critical factor and not the endorphin system per se.

## COMMENTS ON THE FUTURE

The function and metabolism of the human body is an electrochemical system. Modern medicine has studied, analyzed, and treated mainly the chemical side of the equation. For the most part, the electrical half of human systems has been completely ignored. We have only used a few of the electrical systems for diagnostic purposes; electrocardiogram, electroencephalogram, and electromyogram are examples. Even fewer uses are found from a therapeutic standpoint; electroconvulsive shock therapy (ECT), cardiac pacemakers, defibrillators, and TENs devices are examples. Each of these devices, particularly ECT, has its detractors.[36,33]

Most of the body's natural chemicals are released by an electrical signal or an electrochemical reaction. Can these same chemicals be released by an external electrical signal? Can different electromagnetic parameters stimulate different chemical systems? These are exciting possibilities. Rather than doping the entire body with a drug in order to obtain a small effect at one specific site, an electrical signal may induce the desired effect at the specific site, without adversely effecting the rest of the body.

Our current culture is drug oriented. It is no wonder that we are facing a drug problem. We, as a culture, have endorsed drugs as a way to solve all our pains. Advertising on television can serve as an example. Drug treatments are offered as immediate relief for headaches, eating too much, staying awake, going to sleep, relieving cold symptoms, etc. The medical profession is no better. Most physicians have endorsed the use of drugs as the scientific way to cure all illness and disease and have rejected, or at least, ignored other possible modalities.

Human behavior tends to resist the perception and acceptance of new ideas. This, unfortunately, can preclude the acceptance and use of new and novel ideas or techniques. Electrotherapy offers a new way to look at treatment of disease. While it may become an effective treatment for many diseases, it must not be regarded as a panacea for all diseases. Each disease process must be evaluated on its own merit following adequate scientific studies. Conversely, electrotherapy should not be rejected out of hand for any disease condition without scientific evaluation.

## References

1. Adey WR, Bawin SM. Brain interactions with weak electric and magnetic field. Neurosci Res Prog Bull 1977; 15:7-17.

2. Adey WR, Bawin SM, Lawerence AF. Effects of weak amplitude modulated microwave fields on calcium efflux from awake cat cerebral cortex. Bioelectromagnetics 1982; 3:295-307.

3. Amassian VE, Cracco RQ. Human cerebral cortical response to contralateral transcranial stimulation. Neurosurg 1987; 20:148-155.

4. Amassian VE, Anziska BJ, Cracco JB, Cracco RQ, Maccabee PT. Focal magnetic coil excitation of frontal cortex activates laryngeal muscles in man. J Physio 1988; 398:41.

5. Amassian VE, Cracco JB, Cracco, RQ, Eberle L, Maccabee PT, Rudell A. Suppression of human visual perception with the magnetic coil over occipital cortex. J Physio (Lond) 1988; 398-408.

6. Amassian VE, et al. See this volume: 171-204

7. Bartko D, Turcani P, Danisova J, Janco S, Traubner P, Miklasova, A, Lesincky O, Pancak J, Vestenicka V. The Effects of the pulsing magnetic field on the cerebral circulation, EEG power spectra and some properties of the blood. A preliminary data. J Bioelectricity 1988; 7(1):131-132.

8. Bauer, W. Neuroelectric medicine. J Bioelectricity 1983; 2(2 & 3):159-180.

9. Bawin SM, Kaczmarek LM, Adey WR. Effects of modulated VHF fields on the central nervous system. Ann NY Acad Sci 1975; 247:74-81

10. Bawin SM, Adey WR. Sensitivity of calcium binding in cerebral tissue to weak environmental electric fields oscillating at low frequency. Proc Nat Acad Sci USA 1976; 73:1999-2003.

11. Bawin SM, Adey WR, Sabbot IM. Ionic factors in release of $^{45}Ca^{2+}$ from chicken cerebral tissue by electromagnetic fields. Proc Nat Acad Sci USA 1978; 75:6314-6318.

12. Bise W. Low power radio frequency and microwave effects on human electroencephalogram and behavior. Physiol Chem Phys 1978; 10:387-398.

13. Blackman CF, Benane SG, Elder JA, House DE, Lampe JA, Faulk JM. Induction of calcium ion efflux from brain tissue by radiofrequency radiation: Effects of sample number and modulation frequency on the power-density window. Bioelectromagnetics 1980; 1:35-43.

14. Blackman CF, Benane SG, House DE, Joines WT. Effects of ELF (1-120 Hz) and modulated (50 Hz) fields on the efflux of calcium ions from brain tissue in vitro. Bioelectromagnetics 1985; 6:1-11.

15. Blackman CF, Benane SG, Joines WT, Hollis MA, House DE. Calcium-ion efflux from brain tissue: power-density versus internal field-intensity dependencies at 50 MHz RF radiation. Bioelectromagnetics 1980; 1:277-283.

16. Blackwell RP. Effects of extremely-low-frequency electric fields on neuronal activity in rat brain. Bioelectromagnetics 1986; 7:425-434.

17. Capel ID, Pinnock MH, Patterson MA. The influence of electrostimulation on hexobarbital induced loss of righting reflex in rats. Acupuncture & Electro Therapeut Res Int J 1982; 7:17-26.

18. Capel ID, Pinnock MH, Withey NJ, Williams DC, Patterson MA. The effects of electrostimulation on barbiturate-induced sleeping times in rats. Drug Development Res 1982; 2:73-79.

19. Capel ID, Pinnock MH, Williams DC. The influence of differing forms of electrostimulation on the hexobarbital-induced sleeping time of rats. ICRS Med Sci 1980; 8:545.

20. Capel ID, Goode AG, Smallwood AE, Patterson MA, Tryptophan, serotonin, and hydroxyindole acetic acid levels in rat brain following slow or fast frequency electrostimulation. ICRS Med Sci 1982; 10:427-428.

21. Clement-Jones V, Tomlin S, Rees LH, McLoughlin L, Besser GM, Wen HL. Increased ß-endorphin but not met-enkephalin levels in human cerebrospinal fluid after acupuncture for recurrent pain. Lancet 1980; ii:946-949.

22. Compositions Medicae, CLXII, Paris, 1528.

23. Davis WE. The healing power of electricity. In: The Cottage Physician. King-Richardson Co, Springfield, MA. 1897; pp. 510.

24. Delgado MR, Leal J, Monteagudo JL, Gracia, MG. Embryological changes induced by weak extremely low frequency electromagnetic fields. J Anat 1982; 134; (3):533-551.

25. Delgado JMR. Biological effects of extremely low frequency electromagnetic fields. J Bioelectricity 1985; 4(1):75-91.

26. De Loecker W, Cheng N, Delpert PH. Effects of pulsed electromagnetic fields on membrane transport. 1989; See this volume pp. 45-58.

27. Ganong WF, Shepard MD, Wall JR, Van Burnt E, Clegg MT. Penetration of light into the brains of mammals. Endocr 1963; 72:962-63.

28. Gavalas-Medici R, Day-Magaleno S. Extremely low frequency weak electric fields affect schedule controlled behavior in monkeys. Nature 1976; 261:256-258.

29. Giczi J, Guseo A. Treatment of headache by pulsating electromagnetic field: A preliminary report. J Bioelectricity 1988; 7(1):125-126.

30. Goldstein L, Sisko Z. A quantitative electroencephalographic study of the acute effects of X-Band microwaves in rabbits. In: Biologic Effects and Health Hazards of Microwave Radiation. Polish Medical Publishers, Warsaw. 1974; 128-133.

31. Goodman LS, Gilman A. The Pharmacological Basis of Therapeutics 2 ed. Macmillan Co, NY. 1955; 123-155.

32. Gordon ZV. Biological effects of microwaves in occupational hygiene. 1966 Izdalestvo Medicina Leningrad Otdilemie (TT 70-50087, NASA TT F-633, 1970).

33. Hamilton M, Electroconvulsive therapy. Indications and contraindications. Ann N Y Acad Sci 1986; 462:5-11.

34. Hartvig P, Berstrom K, Lindberg B, Lundberg PD. Kinetics of $^{11}C$-labeled opiates in the brain of rhesus monkeys. J Pharmacol Exp Ther 1984; 230 (1):250-255.

35. Lacomblez l, Puech A. Pharmacology of diacetylmorphine (heroin). Rev Prat. 1987; 37(29):1691-1694.

36. Leszek J, Kiejna A, Wasik A, Kubiszewski M. Electroconvulsive therapy; pro's and con's in the light of the review of the literature. Psychiatr Pol 1987; 21(5):421-427.

37. Kellaway P. The part played by electric fish. Bull History Med 1946; 20:112-137.

38. Lovely RH, Lundstrom DL, Phillips RD. Dosimetric and behavioral analysis of microwave drug synergistic effects on operant behavior in the rat. Contract No. N00014-79-C-0819 Doc # AD-A115115 NTIS, Springfield VA. 1981.

39. Mantle ER, Persinger MA. Alterations in subjective evaluations during acute exposures to 5-Hz but not 9-Hz magnetic field devices. J Bioelectricity 1983; 2:5-14.

40. Menaher M. Light perception by extraretinal receptors in the brain of the sparrows. Proceedings 76th Annual Conv Amer Psychol Assoc 1968; 3:299-300.

41. Merritt JH, Hartzell RH, Frazer JW. The effect of 1.6 GHz radiation on neurotransmitters in discrete areas of the rat brain. In: Biological Effects of Electromagnetic Waves. 1974; HEW(FDA)77-8010:290-298.

42. Monahan JC. The Effects of 2450 MHz radiation on the cholinergic system in the mouse. In: Biological Effects of Electropollution. Information Ventures Inc, Philadelphia, PA. 1986; 91-98.

43. Ng LKY, Douthitt TC, Thoa NB, Albert CA. Modification of morphine-withdrawal syndrome in rats following transauricular electrostimulation: An experimental paradigm for auricular electroacupuncture. Biol Psych 1975; 10(5):575-580.

44. Patterson MA, Firth J, Gardiner R. Treatment of drug, alcohol and nicotine addiction by neuroelectric therapy: Analysis of results over 7 years. J Bioelectricity 1984; 3(1&2):193-221.

45. Patterson MA. Effects of neuroelectric therapy (N.E.T.) in drug addiction: Interim report. Bull on Narcotics 1976; XXVIII (4):55-62.

46. Persinger MA, Nolan M. Partial amnesia for a narrative following application of theta frequency electromagnetic fields. J Bioelectricity 1985; 4(2):481-494.

47. Prusinski A, Wielka J, Durko A. Pulsating electromagnetic field in the therapy of headache. J Bioelectricity 1988; 7(1):127-128.

48. Salar G, Job I, Mingrino S, Bosio A, Trabucchi M. Effects of transcutaneous electrotherapy on CFS ß-endorphin content in patients without pain problems. Pain 1981; 20:169-172.

49. Sawynok J. The therapeutic use of heroin: a review of the pharmacological literature. Can J Physio Pharmacol 1986; 64 (1):1-6.

50. Schecter, DC. Origins of electrotherapy: Part I. NY State J Medicine 1971; 71:1002.

51. Sjound B, Terenius L, Eriksson M. Increased cerebrospinal fluid levels of endorphins after electroacupuncture. Acta Physiol Scand 1977; 100:382-384.

52. Synder SH. The Effect of microwave irradiation on the turnover rate of serotonin and norepinephrine and the effect of monamine metabolizing enzymes. Final Report, Contract No. DADA 17-69-C-9144 1971; U.S. Army Medical Research and Development Command, Washington, DC.

53. Takashima S, Onaral B. Schwan HP. Effects of modulated RF energy on the EEG of mammalian brains. II. Appearance of fast and slow waves after chronic irradiation. Proceedings of the 1978 Symposium on Electromagnetic Fields in Biological Systems. Ottawa, Canada, June 28-30, 1978.

54. Thomas JR, Yeandle SS, Burch LS. Modification of internal discriminative stimulus control of behavior by low levels of pulsed microwave radiation. In: Biological Effects of Electromagnetic Waves. 1974; HEW(FDA) 77-8010:201-214.

55. Thomas JR, Finch E, Fulk DW, Burch LS. Effects of low level radiation on behavioral baselines. Ann N Y Acad Sci 1975; 247: 425-432.

56. Thomas JR, Maitland G. Combined effects on behavior of low level microwave radiation and dextroamphetamine. Int Symp on Biological Effects of Electromagnetic Waves. USNC-URSI, Arlie, VA. 1977; 121.

57. Thomas JR, Banvard RA. (abstract) Changes in temporal aspects of behavior by low levels of pulsed microwaves. Bioelectromagnetics Symp. 1979; Seattle WA: 452.

58. Webb SJ. Genetic continuity and metabolic regulation as seen by the effects of various microwave and black light frequencies on these phenomena. Ann NY Acad Sci 1975; 247:327-351.

59. Wen HL, Chenf, SY. Treatment of drug addiction by acupuncture and electrical stimulation. Asian J Med 1973; 9:138-141.

60. Wen HL, Cheng SYC. Treatment of drug addiction by acupuncture and electrical stimulation. Amer J Acupuncture 1973 1:71-75.

# DEVELOPMENTS FOR STIMULATION AND ANALYSIS OF NERVE REGENERATION

Betty F. Sisken

## INTRODUCTION

Research on nerve regeneration has been devoted to developing new methods to enhance growth of axonal fibers. Both chemical (growth factors, hormones, drugs, etc) and electrical [direct current (DC) and pulsed electromagnetic fields (PEMF)] applications have proven fairly successful. In this report we describe experiments using noninvasive PEMF on a crush nerve model showing that the results obtained with this technique are equal to results obtained with any chemical method. We also discuss the latest state-of-the-art methods employed in studying basic and clinical aspects of nerve growth and pathology. Finally, we attempt to predict future technological advances to be used in this research field.

### A. Administration of Electric/Electromagnetic Fields In Vitro

Growth stimulation of developing neurons in vitro and lesioned adult nerve in vivo have been studied following application of low level electric and electromagnetic fields.[1,4,5,7,9,10,20-22,24-31] This technology is relatively simple, adapted directly from the techniques used for promotion of bone healing (see review in Black[3]). For in vitro studies, whole explants or dissociated neurons of sensory and sympathetic ganglia or slices of spinal cord from embryonic or fetal embryos are cultured in conventional culture medium. In dc (electric field) experiments, metal or agar electrodes are placed directly into culture dishes[10,25,27-30] and fields of high strength (ie, 10 V/cm), down to fields of low strength (~10 μV/cm), are applied for various time periods. The higher electric fields can be used for only short time periods while the lower fields can be applied for many days. For experiments testing electromagnetic fields the culture dishes are placed between a pair of Helmholtz coils for various time periods.[7,27,28]

Growth processes are scored by assessing numbers and branching patterns of neurites (axons and dendrites) using scoring methods taken from studies on nerve growth factor; neurite length alone may be measured directly.[7] Neurite outgrowth from sensory ganglia is illustrated in a scanning electron micrograph in Fig. 1 and presented as a function of time in culture in Table I.

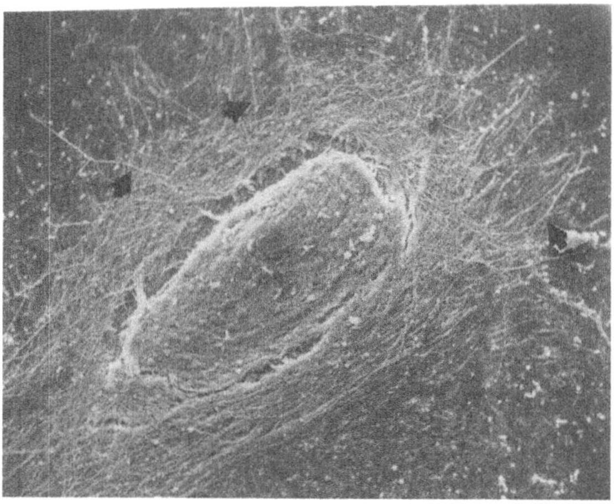

**Figure 1.** Scanning electron micrograph of dorsal root ganglia exposed to 10 nA direct current for 3 days in culture. X50. Arrows indicate neurite outgrowth emanating from the sensory neurons in the central ganglionic mass.

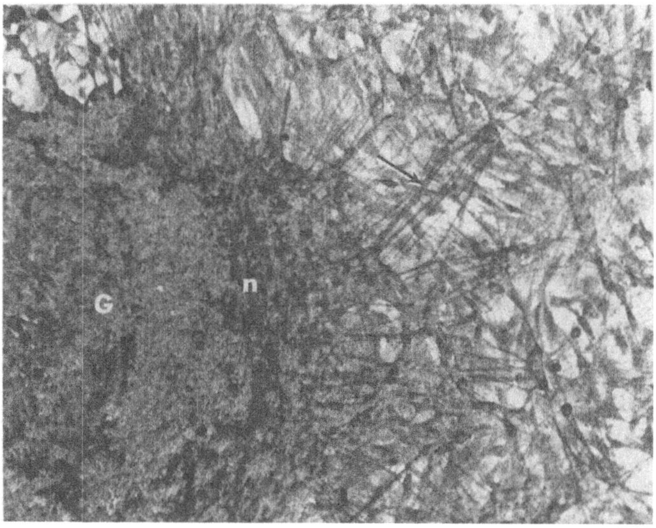


**Figure 2.** Autoradiograph of cultured dorsal root sensory ganglia exposed to PEMF, 2 Hz. The cultures were incubated for 6 days; 3H- proline was added to the medium in the last 20 hours of incubation. X150. The neurites (arrows) are extensions from neurons (n) located in the central ganglionic (G) mass.

Using liquid photographic emulsion which is sensitive to isotope radiation, incorporation of isotopically-labeled amino acid into new proteins is visualized in autoradiographs of cultured sensory neurons (Fig. 2). The dense labeling

TABLE I

**Neurite Outgrowth in Dorsal Root Ganglia after 6 Days in Culture**

| | Control, untreated | PEMF, 72 Hz |
|---|---|---|
| Number of Culture Dishes | 16 | 11 |
| Total Number of Ganglia | 73 | 67 |
| Growth Score ± S.D. | 1.28 ±0.49 | 2.16* ±0.76 |

*$p<0.05$

exhibited in neurons and neuritic processes provides qualitative and quantitative data indicating that: 1) the nerve cells are metabolically active and 2) the extent of growth stimulation can be assessed. Such information from in vitro studies has provided baseline data in planning and developing our subsequent in vivo experiments.

B. Administration of Electric/Electromagnetic Fields In Vivo

Electric and electromagnetic fields have been used to stimulate nerve regeneration in many animal preparations.[1,9,11,19-22,24,26,31] Pomeranz, et al.[21] and Kerns et al.[11] applied direct current to the lesioned nerve, assessing regeneration by counting the numbers of axons in cross-sections of the nerve or by using various functional tests. Alternatively a large number of investigators have used non-invasive electromagnetic fields to stimulate regenerative processes.[2,9,20,22, 26,27,28,31,32]

In our laboratory, and in collaboration with other laboratories, we have been exploring the in vitro and in vivo effects of a specific PEMF signal on nerve cell cultures and on two different animal nerve injury models. In the in vitro experiments cultures of dorsal root ganglia from chick embryos were exposed to a 20 millisecond magnetic pulse with a repetition rate of 2 Hz at an amplitude of 0.4 gauss for 2 hours per day for 2 days. Neuritic outgrowth scores at 3 days were consistently 30-35% higher than control cultures.[27]

TABLE II

**PEMF effects on Regeneration Rate of Rat Sciatic Nerve after a Crush Lesion**

| Groups | Number | Regeneration Rate |
|---|---|---|
| Control | 26 | 3.2 mm/day |
| PEMF, 2 Hz<br>4 hours/day | 11 | 3.9** mm/day |

** $p<0.001$

In one animal model , the sciatic nerve of adult male rats was crushed and the animals placed between two Helmholtz coils for 6 days for 1, 4, or 10 hours and in several orientations relative to the magnetic vector. The pulsed electromagnetic fields were pulsed for 20 milliseconds at an amplitude of 3.0 gauss with a repetition rate of 2 Hz. Experiments using PEMF for 4 hours/day for 3, 4 or 6 days produced an approximate 22% increase in the rate of axonal sprouting relative to controls as measured by the pinch test,[26] (Table II). Interestingly, animals pretreated with the field for four hours per day for seven days before the crush injury with no treatment subsequent to the injury also showed a response not significantly different from the post-injury treated animals (Table III). This seems to suggest that the effect of the field is not at the injury site per se, but rather on a systemic function as yet unidentified.

In the other animal model the rat sciatic nerve was transected and subjected to the same 3 gauss field. Rats whose sciatic nerves were repaired immediately after transection and treated for four hours per day for five days gave the clinical impression after one month of advanced healing and a reduced neuroma formation relative to controls. In another laboratory, gait analyses of animals with sciatic nerves that had been transected and repaired nerves were conducted at 35, 75, 105, 140 and 165 days. Animals treated for four hours/day for five days starting immediately after transection and repair did not show a long-term gain relative to controls. However, animals whose repair was delayed for five days and then treated showed a sustained improvement in gait 15-20% above controls (M.Orgel, personal communication). The results of treatment with such fields indicate that this signal has the potential to be used as adjunctive therapy for various nerve injuries.

TABLE III

**Regeneration Distance of Rat Sciatic Nerve after Pre- or Post-Exposure to 2 Hz PEMF**

| Group | Number | Regeneration Distancc |
|---|---|---|
| Control | 9 | 4.7 mm ±0.7 |
| Treat with PEMF for 3 days after crush | 9 | 6.5** mm ±0.7 |
| Treat with PEMF for 7 days before crush | 5 | 6.9* mm ±0.54 |

* p<0.05 ** p<0.01

Other techniques under exploration for treating nerve injuries are those employing hormones and growth factors to the repair site,[12,13,14,17] insertion of piezoelectric guides to provide a physical-bioelectric matrix[1,15] and treatment of nerve injuries with a He-Ne laser.[6]

## NEW TECHNOLOGY FOR ASSESSMENT

In the discussion above we emphasized studies of electric/electromagnetic techniques to enhance regeneration. However, conditioning lesions[17,18] and infusion of growth factors, hormones, ions and drugs are also being tested. In all cases, irrespective of the chemical, mechanical or electrical nature of the experimental method, these agents are assessed as a function of experimental paradigm and elapsed time. One of the most important problems confronted in this research, as well as in research in most other areas, is the limited ability to determine non-invasively the morphological and physiological changes that occur in closely-spaced time sequences (seconds, minutes, days) in the same preparation. In actuality, these determinations should be carried out in a parallel fashion in control and experimental animals; this would allow one to compare rate constants of all of the reactions being monitored.

New technologies presently available, as well as those on the horizon, indicate that this capability is, or will become, a reality using such equipment as: magnetic resonance imaging (MRI), magnetic spectroscopy, computed

**Figure 3.** Left: Image of cultured dorsal root ganglia exposed to PEMF showing extensive neuritic growth surrounding the ganglion (G). Right : Image after processing on the Kontron/Sun System.

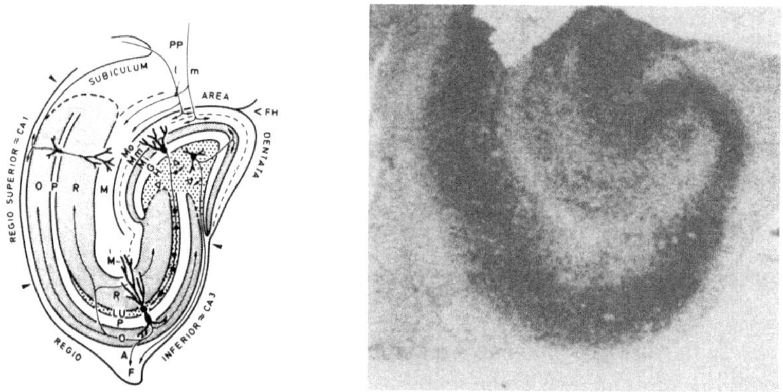


**Figure 4.** Localization of glutamate receptors in human hippocampal slices using autoradiographic and quantitative image analysis techniques.

tomography scans (CT), optical spectroscopy, positron emission tomography (PET), ultrasound sonography, and real-time video imaging with quantitative and 3-D reconstruction capability. Relative to our own interest in examining nerve growth, W. Levy[15] has developed a video-microscope system specifically adapted for continuous visualization and recording of nerve regeneration in a special in vivo chamber. At present, this system is very expensive and difficult to use on a routine basis.

In the Image Analysis Consortium at the University of Kentucky, we have the capability of using a Kontron/Sun system for quantitative analyses of images,[8] contrast enhancement, discrimination of specific objects,[16] 3-D

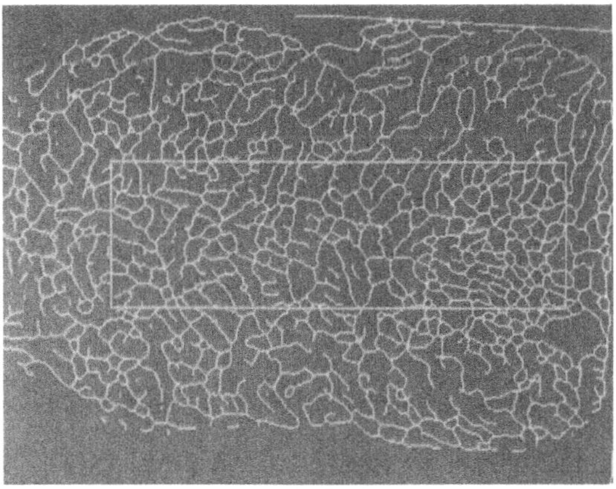

**Figure 5**. Distribution of blood vessels within a circumscribed area of a dorsal skin flap after reducing all vessels to a diameter of one pixel.

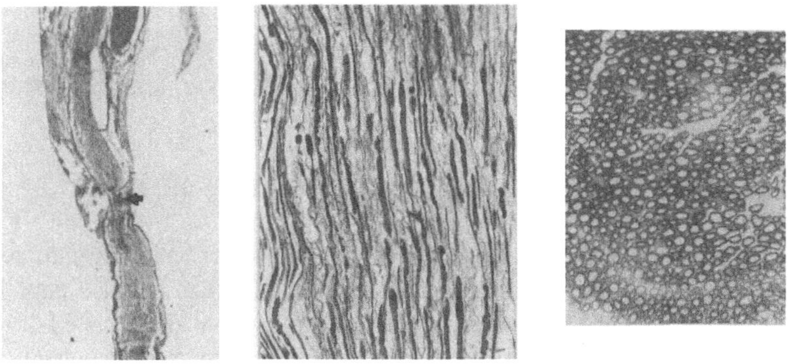

**Figure 6**. Left : Longitudinal section of rat sciatic nerve one day after a crush lesion (arrow). Center : Higher power photograph of the crushed sciatic nerve indicating the longitudinal arrangement of nerve axons stained with silver nitrate, many of which are edematous. Right : cross-section of sciatic nerve after image processing.

reconstructions, etc. In our research studies we have used this system for discriminating neurite outgrowth of dorsal root ganglia exposed to PEMF using segmentation techniques (Fig. 3), to determine the localization and concentration of glutamate receptors in human hippocampus using grey level density discrimination (Fig. 4), illustrate and quantitate blood vessel distribution in a dorsal rat skin flap[8] (Fig. 5), automatically analyze fiber size and number in longitudinal (A) and cross-sections (B) of sciatic nerves (Fig. 6) and perform 3-D

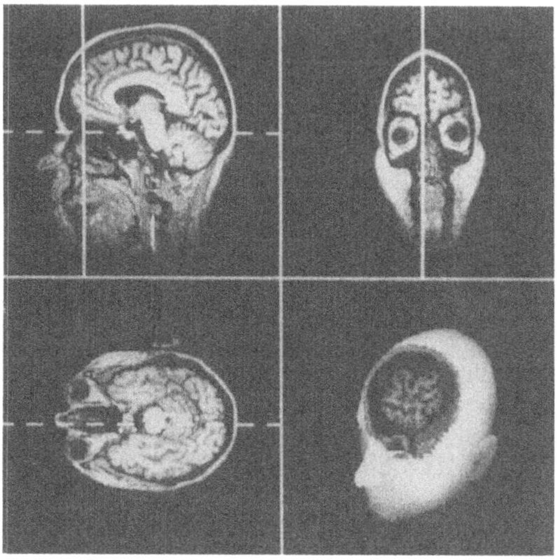

**Figure 7.** Series of images obtained from 3-dimensional reconstruction of MRI scans (compliments of the Kontron Imaging Co.).

reconstructions of the human brain/skull and head (Fig. 7) for teaching, research and clinical studies.

## PREDICTIONS FOR TECHNOLOGIES IN THE FUTURE

Sophisticated versions of techniques such as MRI or optical spectroscopy will be capable of not only obtaining 3-D images in time, but also have the ability to co-determine metabolic and chemical alterations (i.e., a "PET- four-dimensional" technology), or what Robb[23] refers to as "tomochemistry (multidimensional biochemical imaging). Eventually, this will enable us to determine non-invasively closely-timed, progressive changes in treated and non-treated experimental models. With such tools we will be able to define more accurately the sequences of biochemical, physiological and anatomical reactions that occur and also have the potential to assess mechanisms at the tissue, cell and perhaps atomic levels.

## ACKNOWLEDGMENT

Supported in part by the Office of Naval Research #N00014-86- K0221. We acknowledge support of the University of Kentucky Major Research Instrumentation Bond Program in the purchase of equipment used in this study (Bond I.D. #7E-8W34-57).

## References

1. Aebischer P, Valentini PD, Domenici C, Galletti PM. Piezoelectric guidance channels enhance regeneration in the mouse sciatic nerve after axotomy. Brain Res 1987; 436:165-168.

2. Bassett CAL, Pawluck RJ, Becker RO. Effects of electric current on bone in vivo. Nature 1964; 204: 652-655.

3. Black J. Electrical Stimulation: Its Role in Growth, Repair, and Remodeling of the Musculoskeletal System. Praeger, NY. 1987; 225 pp.

4. Borgens RB, Blight AR, Murphy DJ, Stewart L. Transected dorsal column axons within the guinea pig spinal cord regenerate in the presence of an applied electric field. J Comp Neur 1986; 250:168-180.

5. Freeman JA, Manis PB, Snipes GJ, Mayes BN, Samson PC, Wikswo JP Jr, Freeman DB. Steady growth cone currents revealed by a novel circularly vibrating probe: A possible mechanism underlying neurite growth. J Neurosci 1985; 13:257-283.

6. Friedman JE, Belkin M, Schwartz M. Low energy He-Ne laser irradiation enhances potassium-buffering capability of glia cells. Neur 1988; 14:1054.

7. Gunderson RW, Greenbaum B, Sutton, C. (abstract) Effects of pulsed magnetic fields, neurite elongation from chick dorsal root ganglia. Ann Meet Bioelectromagnetics Society 1989; 11:60.

8. Herbst E, Sisken BF, Wang H. (abstract) Assessment of vascular network in rat skin flaps subjected to sinusoidal EMFs using Image Analysis Techniques. 1988 Trans Bioelect Repair and Growth Soc 1988; 8:40.

9. Ito H, Bassett CAL. Effect of weak, pulsing electromagnetic fields on neural regeneration in the rat. Clin Orthop and Related Res 1983; 181:283-290.

10. Jaffe L, Poo MM. Neurites grow faster toward the cathode than the anode in a steady field. J Exp Zool 1979; 209:115-127.

11. Kerns JM, Pavkovic IM, Fakhouri AJ, Wickersham KL, Freeman JA. An experimental implant for applying a DC electrical fields to peripheral nerve. J Neurosci Methods 1987; 19:217-223.

12. Kanje M, Skottner A, Lundborg G. Effects of growth hormone treatment on regeneration of the rat sciatic nerve. (under editorial review).

13. Kilmer SL, Carlsen RC. Forskolin activation of adenylate cyclase in vivo stimulates nerve regeneration. Nature 1984; 307:455-457.

14. Kilmer SL, Carlsen RC. Chronic infusion of agents that increase cyclic AMP concentration enhances the regeneration of mammalian peripheral nerves in vivo. Exp Neurol 1987; 95:357-367.

15. Levy W, Humphrey MF, Dietrich D. A comparison of saline and Matrigel as substrates for peripheral nerve regeneration in an in vivo window chamber. Soc Neur 1988; 14:499.

16. Mann DM, Brandt P, Sisken BF, Vanaman TC . Identification of a synaptic membrane-localized isoform of the calcium-pumping ATPase. In: Calcium Protein Signaling 1988 (Ed. H Hidaka). Plenum Press, NY. 1989; 71-78.

17. McQuarrie IG, Grafstein B, Gersham MD. Axonal regeneration in the rat sciatic nerve: effect of a conditioning lesion and of dbcAMP. Brain Res 1977; 132:443-453.

18. McQuarrie IG. Structural protein transport in elongating motor axons after sciatic nerve crush. Effect of a conditioning lesion. Neurochem Pathol 1986; 5:153-164.

19. Nix WA, Hope HC. Electrical stimulation of regenerating nerve and its effects on motor recovery. Brain Res 1983; 272:21- 25.

20. Orgel MG, O'Brien WJ, Murray HM. Pulsing electromagnetic field therapy in nerve regeneration: An experimental study in the cat. Plast and Reconstr Surg 1984; 73:173-183.

21. Pomeranz B, Mullen M, Markus H. Effect of applied electrical fields on sprouting of intact saphenous nerve in adult rat. Brain Res 1984; 303:331-336.

22. Raji ARM, Bowden REM. Effects of high-peak pulsed electromagnetic field on the degeneration and regeneration of the common peroneal nerve in rats. J Bone and Joint Surg 1983; 65:478-492.

23. Robb RA. Four dimensional display technology reveals power of biomedical imaging. Res Resources Reporter 1989; XIII:1-5.

24. Roederer E, Goldberg NH, Cohen MJ. Modification of retrograde degeneration in transected spinal axons of the lamprey by applied DC current. J Neurosci 1983; 3:153-160.

25. Sisken BF. Effects of electromagnetic fields on nerve regeneration. In Modern Bioelectricity (Ed. A Marino). Marcel Dekker Inc, New York. 1988; 497-52.

26. Sisken BF, Kanje M, Lundborg G, Herbst E, Kurtz W. Stimulation of rat sciatic nerve regeneration with pulsed electromagnetic fields. Brain Res 1989; 485:309.

27. Sisken BF, Kanje M, Lundborg G, Herbst E, Zienowicz R, Orgel M, Kurtz W. Stimulation of nerve regeneration in vitro and in vivo using pulsed electromagnetic fields. Soc Neur Ann Meet 1988; 14:497.

28. Sisken BF, McLeod B, Pilla AA. PEMF, direct current and neuronal regeneration: Effect of field geometry and current density. J Bioelectricity 1984; 3:81-101.

29. Sisken BF, Roberts E, Goetz I. Triethanolamine, tris, hepes, and cytosine arabinoside show neuritogenic activity in cultured chick embryo ganglia. Exper Neurol 1985; 88:27-43.

30. Sisken BF, Smith S. The effects of minute direct electrical currents on cultured chick embryo trigeminal ganglia. J Embryol Exp Morphol 1975; 33:29-41.

31. Wilson DH, Jagadeesh P. Experimental regeneration in peripheral nerves and the spinal cord in laboratory animals exposed to a pulsed electromagnetic field. Paraplegia 1976; 14:12-20.

32. Young W. (abstract) Pulsed electromagnetic fields alter calcium in spinal cord injury. Central Nerv Sys Trauma 1984; 1:100.

# PERSPECTIVES IN STIMULATION OF HUMAN NERVOUS SYSTEM WITH THE MAGNETIC COIL

Vahe E. Amassian, Roger Q. Cracco and Paul J. Maccabee

## INTRODUCTION

### A. Development of the Field

Neural stimulation by powerful magnetic fields has been sought since the turn of this century, but Bickford and Fremming [26] were the first to demonstrate magnetic coil (MC) excitation of human peripheral nerve. A major advance occurred when Barker et al.,[23,24] introduced an MC stimulator powerful enough to elicit peripheral movement by transcranial stimulation of the motor cortex. Earlier, Merton and Morton[52,53] activated peripheral muscles by transcranial stimulation with individual electrical pulses. For the first time it became feasible to add the study of the central motor system in normal persons to that of patients with brains exposed during surgical operations. Despite improvements in a design of the stimulating electrodes[2,41,61] considerable discomfort with electrical stimulation remained, thus limiting its usefulness, especially in patients. However, unlike an electrical pulse, the MC pulse penetrates skin, muscle, bone and membranes covering the brain with little attenuation. Excitation of neurons by the MC pulse was presumed to result from the electric currents induced in the volume conductor.

Conventionally, the MC was applied tangentially (ie, flat) at the vertex. This orientation is the most efficient in transferring magnetic flux to the volume conductor. However, the flow of current induced in the brain is also tangentially orientated, resulting in widespread activation of motor cortex and consequent widespread activation of limb muscles. Paradoxically, despite the powerful activation of muscle, the latency of discharge was at least 2 ms longer than with electrical transcranial stimulation.[31,42,55,62] This and other findings[32] suggested that MC and electrical stimulation differed fundamentally in mode of action. Specifically the MC pulse was believed incapable of exciting directly the corticospinal system, activation of this system occurring indirectly through its synaptic inputs. Subsequently, the MC orientation was shown to be crucial in determining the type of activation of the corticospinal system.[4,8,12,18]

B. Clinical Utility of Magnetic Coil Stimulation in the Adult

The lack of discomfort with MC transcranial stimulation facilitated its use in patients. A number of careful studies from the United Kingdom established that muscle responses in patients with multiple sclerosis had an increased latency to electrical and to MC stimulation.[21,22,29,55,61,65] The increased latency was unaccompanied by an increase in peripheral motor axonal conduction time, implying an increased delay in the central motor pathways. Possibly the increased delay is due to demyelination of the fastest corticospinal fibers, although other possibilities require evaluation.

Past intraoperative monitoring of responses to transcranial stimulation have used electrical transcranial stimulation,[44] any discomfort to the patient being obviated by the general anesthesia. Any operations performed under local or very light anesthesia could utilize MC stimulation to advantage, because of the reduction in local stimulation.

In general, MC transcranial stimulation when combined with other measures including: reflex amplitude, ratio of M wave amplitude to transcranially evoked motor response, peripheral motor conduction velocity and motor unit and fiber properties, provides a powerful tool for quantifying and analyzing failure within the human motor system.

C. Utility of Magnetic Coil Stimulation in Developmental Studies

Although electrical transcranial stimulation has been used in humans to study central motor pathways in the newborn,[34] the advantage of the less uncomfortable MC in stimulating the young is obvious. Studies initiated by Eyre and Koh [35] disclosed that latency of arm muscle compound motor action potentials (CMAPs) is rapidly reduced between birth and two years of age. When these latencies are divided into height, a conduction velocity is derived that slowly increases after two years until adult values are achieved at ten to fourteen years. Because peripheral conduction velocity reaches adult values by two to five years,[68] the maturation in conduction velocity after three years is attributed to maturation of central pathways. Later observations[37] used MC stimulation to distinguish between children two to twelve years of age, with spastic hemiparesis but normal intelligence, from those eight to fifteen years of age with spastic quadriparesis, accompanied by diffuse cortical damage. The two groups of children showed remarkable differences; those with spastic hemiparesis on the affected side had either delayed or no motor evoked responses, but the quadriplegics had normal responses. Such findings suggest a possible use of transcranial MC stimulation in distinguishing between lesions of the corticospinal system and inputs thereof, related to the apraxias.

## MATCHING PROPERTIES OF MC STIMULATION TO THE NERVOUS SYSTEM: GENERAL PERSPECTIVES

The contemporary view of functional interconnections between cortical areas and either other cortical areas or subcortical nuclei is of small subsets of neurons connected specifically with other subsets. The classical concepts of topographical mapping of sensory fields on cortical receiving areas and of motor cortical areas on muscles are now generalized to connectivities within higher order systems. Therefore, any technique of transcranial stimulation should have the capability of focally exciting functional subsets of neurons, or at least small groupings thereof, to elicit specific responses. Focal stimulation can be achieved either by appropriate orientation of the standard, round MC [4,8,12,18] or by specially designed MCs.[10,11,12,18]

In contrast with the need for focal stimulation is the need in two contexts for maximum stimulation of a neural set or system. In the first, the maximum output of the system is sought (eg, the maximum amplitude of the CMAP, so important to the electromyographer, or maximum amplitude or the shortest latency of the CMAP obtainable with maximum stimulation of the motor cortex). Secondly, maximum stimulation may be required to engage a neural system whose threshold is elevated under pathological conditions (eg, following section of the medullary pyramids in the monkey the motor cortex can still be mapped, but requires stronger electrical stimulation).[45] Patients with multiple sclerosis may demonstrate greatly increased central motor delays with transcranial stimulation. Such increased delays could reflect: 1) slowed conduction in demyelinated corticospinal axons; 2) delays in intact, but slower conducting members of the corticospinal population; 3) the need for increased temporal summation of corticospinal volleys, the increased delay then being attributable to the need for summation of later, indirect corticospinal volleys; 4) a more complex, non-corticospinal pathway subserved the delayed motor responses. While contemporary transcranial stimulation techniques cannot distinguish between each of these possibilities, determining the minimal central motor delay,[4,7,8,12,18] by powerfully exciting directly with the MC corticospinal neurons whose membrane potentials are close to firing is a first step.

Related to the need for maximum stimulation of a neural system is the requirement for stimulation over a large area, where the representation is an unfolded, or a folded two dimensional sheet, (eg, visual cortex) or because several areas process information in parallel and must therefore be simultaneously affected. The cortical areas related to processing an array of visually-presented, language symbols exemplify the need for MC effects over a wide area of cerebral cortex if perception of the entire array is to be suppressed.[6] The need for MC action over a wide area is easily met by a tangential orientation such that the coil windings overlie the relevant areas. Unfortunately, the MC is likely to overlie other areas also; thus, the development of a multiple, focal MC technology should be fostered.

Repetitive stimulation of human brain has long been known to interfere with complex behaviors, such as speech.[56,59] Interference with tactile placing has been described with single electrical pulses to feline thalamic nucleus ventralis lateralis and anterior [1] and in human reaction time following a transcranial stimulus to motor cortex.[33] Nevertheless, the weakness of the effects of individual MC pulses on higher cognitive functions in the human underlines the need for MC stimulators that can deliver short trains of MC pulses. In principle, the electrical engineering requirement is for reproducing the current circuitry subserving single pulses, but adding sequential switching of separate, charged capacitors into the coil. The increased heat generation may require additional cooling, but in all probability far less current/pulse would be required in an effective train of pulses.

Probably the most difficult future requirement to meet is the capability of stimulating focally at depth, but not close to the MC. Such capability could bring the striatum within the orbit of the MC stimulator. The depth of penetration is a function of the diameter of the MC, but the larger the diameter, the larger the area excited under the windings. Possibly, the development of multiple, wide diameter MCs, separately directed at the deep nuclei may provide a useful approach.

## HOW DOES THE MC EXCITE NERVE FIBERS?

It is generally held that the MC pulse excites nerve fibers by inducing a flow of current in the volume conductor.[22] A simple test of this assertion is provided by comparing the MC stimulus intensity required for just suprathreshold excitation of human ulnar motor axons with the voltage gradients developed in a restricted, 3 dimensional volume conductor. In one of us, the threshold for a motor response of the hypothenar muscles was 22-25% of maximum output of the Cadwell stimulator; the standard round MC was held orthogonal (ie, on edge) to the ulnar nerve just above the medial epicondyle and with the plane of the MC parallel to the extended arm (Fig. 1). The volume conductor model was a trough, 22.5 cm long by 7 cm wide and filled to a depth of 4 cm with isotonic saline. The arm above and below the elbow joint has a larger volume, but is partly occupied by bone, a poor conductor. At 25% output, a similarly orientated MC induces a voltage difference of 7 V in

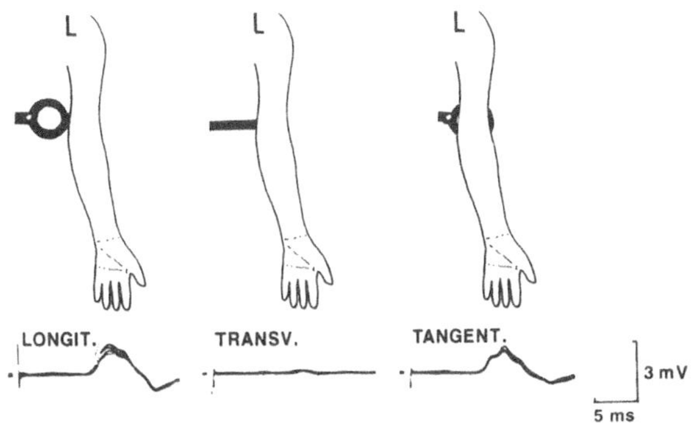


**Figure 1**. Effect of MC orientation on ulnar nerve excitation. Top, 3 orientations of MC energized at 60% maximum output of Cadwell MES 10 stimulator. Bottom, superimposed CMAPs recorded with bipolar disc electrodes on skin overlying hypothenar muscles. The ulnar nerve was stimulated submaximally to reveal any differing efficacies of stimulation, the largest CMAPs being 1/4 maximum amplitude with electrical stimulation.

**Figure 2**. The polarity of the electrical field induced in the arm by the MC modeled in an isotropic volume conductor. MC energized at 25% maximum stimulator output throughout. Top, positions 1-4 show differing relationships between the MC and the volume conductor model, with polarity of the induced electric field indicated by (+) and (-) signs. Bottom, left, the potential differences recorded between bipolar electrodes, 2 cm apart, at the corresponding MC positions (1-4). Reversal of polarity between positions 1 and 2 indicates the direction of current flow in the volume conductor is anticlockwise (ie, opposite to that in the MC). Position 4 corresponds to an orthogonal-longitudinal MC orientation over the right ulnar nerve (bottom right).

longitudinally orientated bipolar recording electrodes, that were separated by 2 cm and located near one edge of the volume conductor (Fig. 2). The recorded voltage gradient of 35 mV/mm is then compared with the theoretical threshold value for excitation.

Tasaki[67] measured values of 15-25 MΩ/mm length of axis cylinder and 41 MΩ for nodal resistance of frog myelinated axons. From Kirchoff's Law, approximately 40 % of an imposed external voltage gradient would be dropped across the stimulated nodal membrane, hyperpolarization occurring at another node.[17] If a 15 mV depolarization at a node is required for excitation, then the external, longitudinal voltage gradient required is estimated at 37.5 mV/mm. A 15 μm motor fiber has an internodal distance of approximately 15 x 100 m = 1.5 mm,[66] requiring an external gradient of 37.5 x 1.5 = 56 mV/1.5 mm. But at 25% stimulator output, the MC induces a 35 x 1.5 = 52.5 mV/1.5 mm voltage gradient in the volume conductor. Given the anisotropy of the arm volume conductor and differences between human and frog axons, the agreement between observed and predicted values for voltage gradient in the volume conductor and those predicted for excitation is surprisingly close.

Thus, the hypothesis is supported that the MC pulse excites nerve fibers through the electric field induced in the volume conductor, rather than by some alternative action (eg, directly on axonal membrane).

## EFFECTS IN ANIMAL MODELS OF A TRANSIENT ELECTRICAL STIMULUS ON A COMPLEX, INTERCONNECTED NEURAL SYSTEM (THE CEREBRAL CORTEX): EFFECTS OF REPETITIVE STIMULATION

The conclusion that the MC pulse excites neural elements through the electric field it induces in the volume conductor permits us to interpret effects of the MC on complex neural systems such as cerebral cortex by drawing on previous analyses performed with electrical stimuli. The effect of a brief electrical stimulus applied to the pial surface of cerebral cortex was first studied in monkeys and cats by analyzing the evoked responses of a particular set of output neurons from cortex, those projecting into the pyramidal or corticospinal tracts.[57,58] The electrical stimulus excited pyramidal tract (PT) or corticospinal tract (CT) neurons directly and, or indirectly through transynaptic action of excited inputs to these neurons. Intracellular recording close to the site of electrical stimulation similarly disclosed both direct and indirect excitation of PT neurons and other, unidentified motor cortical neurons.[60] Indirect excitation usually occurred through a monosynaptically delayed excitatory postsynaptic potential (EPSP). A later study disclosed corticocortical afferent inputs and the deep tangential plexus as important sources of the monosynaptic excitation of corticospinal neurons, with excitation of specific thalamocortical afferents from thalamic nucleus ventralis lateralis and anterior playing a lower, dispensable role. A vertically orientated chain of interneurons with excitation spreading from the surface towards lamina V accounted for the later indirect CT discharges.[20]

The EPSP was followed by a prolonged postsynaptic inhibitory potential (IPSP), the increased latency implying that the IPSP was mediated by small inhibitory interneurons (ie, by a polysynaptic circuit).[60] The inhibitory neurotransmitter is most likely GABA.[43]

The effects of repetitive electrical stimulation are more complex. Both direct[57] and monosynaptic indirect[20] discharge can follow each stimulus in repetitive trains well above 60-100 Hz, common frequencies used when exposed human cortex is stimulated.[56,59] During high frequency trains, facilitation of synaptic activation of corticospinal neurons occurs in the motor cortex.[20] Thus, repetitive stimulation of a cortical area is likely to engage distant areas through corticocortical connection more effectively than would individual stimuli. A monotonic relationship between stimulus frequency and neuronal output would not be expected if reverberation through interneuronal chains contributed to the facilitation. Optimal responses at particular frequencies (ie, 'tuning') could occur if the network has stable reexcitatory delays.

## FOCAL MAGNETIC COIL EXCITATION OF NERVOUS SYSTEM

### A. Peripheral Nerve

The simplest system in which to study the influence of MC orientation on nerve excitation is a set of approximately longitudinally orientated, peripheral axons, there being no complications by synaptic interactions. Either the ulnar nerve just above the elbow of an extended arm[17] or the median nerve[48] afford appropriate analytical systems. The best signal-to-noise ratio is given by recording the CMAP from muscles innervated by these nerves. Figure 1 illustrates the effect of symmetrical tangential, orthogonal-longitudinal and transverse orientations of the MC on ulnar nerve excitation. The symmetrical tangential orientation couples the maximum fraction of the magnetic flux to the arm volume conductor, the induced current flowing in annular rings deep to the MC windings. At the left of the arm, a portion of the induced electric field approximately parallels the ulnar nerve and therefore is capable of exciting ulnar motor axons. Unfortunately, nerve or possibly muscle membranes elsewhere in the arm with a similar parallel relationship to the other portions of the electric field, will also be excited. Thus, excitation of ulnar motor axons, as evidenced by the hypothenar CMAP, is accompanied by excitation of many other muscles, with a resulting unpleasant sensation. By contrast, the electric field set up by

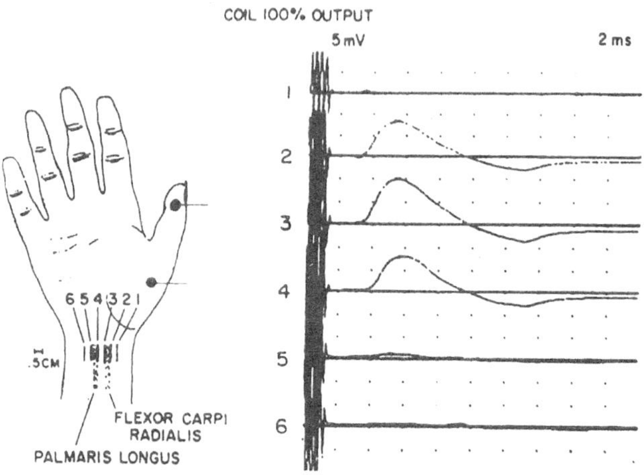


**Figure 3**. The effect on median nerve excitation of stimulating with an orthogonal-longitudinally orientated MC at different positions in the radial-ulnar axis. Thenar CMAPs (1-6) recorded with MC at indicated wrist positions differing by 0.5 cm intervals. (Figure modified from Maccabee et al,[48] with permission.)

the orthogonal-longitudinal orientation has its major component parallel to the ulnar nerve, resulting in its focal excitation.

Further evidence of focal excitation with the orthogonal-longitudinal orientation is provided by stimulating over the median nerve and, at 0.5 cm intervals, on either side. Although the thickness of the MC used was 12 mm, the thenar CMAP (mainly abductor pollicis brevis) was markedly reduced 10-15 mm lateral to the optimum location (Fig. 3). This finding implies that with the orthogonal-longitudinal orientation, the more distant, ulnar nerve was not excited; otherwise, the ulnar-innervated adductor pollicis would have contributed to the thenar CMAP (eg, at positions 5 and 6).

With MC stimulation, in addition to the failure in some individuals to excite all motor axons in a nerve, the site of axonal excitation is not as precisely defined as with electrical stimulation. For example, by comparing thenar CMAP latencies with cathodal and MC stimulation of the median nerve, the distally propagating motor impulses were estimated to arise 2 to 15 mm distal to the midpoint of the contacting edge of the MC.[48] Such imprecision in the site of excitation would be a disadvantage in clinical electromyography.

B. Cranial Nerve

Stimulation by the MC has a major advantage over electrical stimulation in the ease with which the facial nerve can be transcranially excited.[49,63] It was

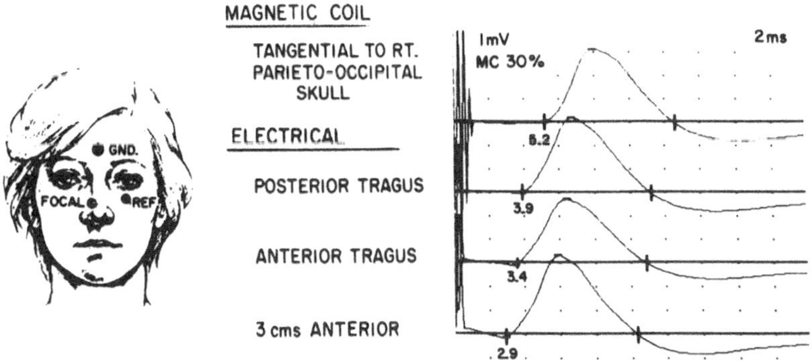


**Figure 4.** Comparison of CMAPs with right-sided MC versus electrical stimulation of facial nerve. Near-maximal, nasalis CMAPs recorded with bipolar discs at locations shown on left. MC and cathodal stimulus locations indicated for each trace; in bottom trace, cathode was 3 cm anterior to anterior tragus. (Figure modified from Maccabee et al,[48] with permission.)

noted that when the MC was placed over the occiput to elicit visual suppression,[5] a twitch of the facial muscles occurred, with a latency significantly greater than that elicited by cathodal stimulation of the facial nerve after it emerges from the stylomastoid foramen (Fig. 4). After measuring the peripheral conduction velocity (V) of facial motor axons in the same subject,the extra CMAP delay (t) with MC stimulation can be converted (Vxt) into an extra conduction distance (d). Interpolating this extra conduction distance proximally from the region of the stylomastoid foramen yielded a site of excitation of the facial nerve close to its emergence from the brainstem.[49] This result is not only of clinical interest in the differential diagnosis of facial palsy, but also poses a question as to why an MC placed flat on the occiput, or even more anteriorly, can excite the distant facial nerve with a small fraction (<50%) of the maximum output of the stimulator. By contrast, when eliciting visual suppression, close to 100% output delivered into the same MC is required to block the much closer, striate cortex. Evidently, the tangential orientation of the MC over occipito-parietal scalp is appropriate for a selective effect on cranial nerves such as the facial. Presumably, the lateral limb of an induced current loop parallels the course of the facial nerve between brainstem and internal auditory meatus. A large potential gradient is secured by the high resistance barrier of the tentorium just superior to the facial nerve. Possibly, the transition from relatively high resistance brainstem to low resistance CSF surrounding the facial nerve as it emerges produces a large change in voltage gradient that may promote excitation at this site. Summarizing, the intracranial excitation of facial motor axons by a tangentially applied MC is a rare example of focal activation at depth.

C. Motor Cortex: Localized Peripheral Movement

The importance of inducing an electric field parallel to motor cortical neurons whose focal excitation is required has led to two developments: 1) the appropriate orientation of the standard round MC; 2) the design (eg, by Cadwell Laboratories) of special MCs, such as the 'double square', the smaller, figure of '8' and the small pointed MC (Fig. 5). Presumably, focal activation of corticospinal neurons is promoted if the MC pulse activates a small clump of such neurons that are related preferentially to a muscle. If the muscle is relaxed, direct corticospinal activation will usually not suffice to bring the motorneurons to firing level; therefore, some excitation of inputs to corticospinal neurons is necessary, eliciting multiple, indirect corticospinal discharges and subsequent motorneuron discharge.[18,62] Theoretically, with the standard, round MC in the lateral-sagittal orientation, the induced electric field is appropriately orientated to activate preferentially, radially orientated corticospinal neurons and their projections (Fig. 6, left, position2). In addition, the terminal portions of corticocortical trajectories where they turn superficially into area 4 would also be stimulated. Both direct and indirect corticospinal activation would be expected over a restricted volume. By contrast, tangential stimulation at the vertex would be expected to excite corticocortical inputs to corticospinal neurons over a wide area, thereby eliciting contraction of many muscles (Fig. 6, left, position 1).

The expected differences between lateral-sagittal and vertex-tangential MC orientations have been confirmed in several ways including: 1) recording the magnetic flux intensities in the 3 axes; 2) recording the voltage gradient modeled

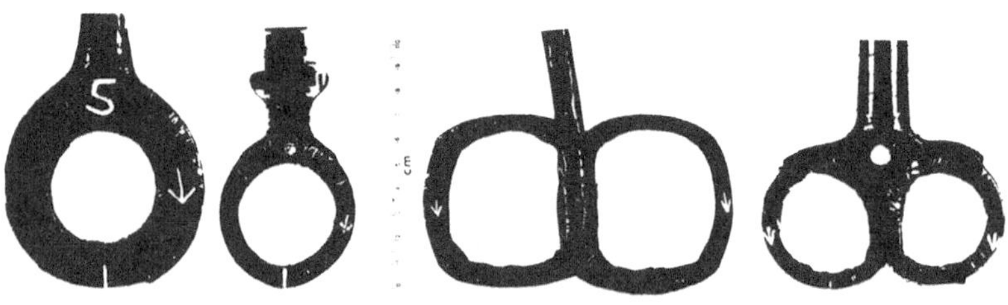

**Figure 5**. Four Cadwell designs of MCs used in our studies. From left to right, standard round, small pointed, double square and figure of '8' MCs. The cm scale applies throughout. All MC's shown are wrapped in plastic tape. The standard round MC generally supplied by the manufacturer is epoxy-covered, which slightly increases the distance from the windings, with a consequent minor reduction in MC output.

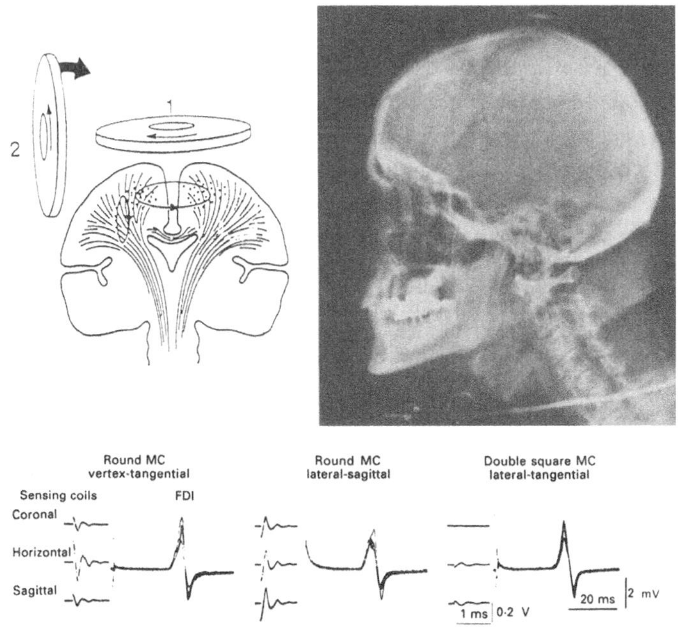


**Figure 6.** Diagram of differing directions of current flow in brain with vertex-tangential versus tilted, lateral-sagittal MC orientations; effect of MC orientation and design on magnetic fields recorded at a distance below the brain. Top left, diagrammatic coronal section of cerebrum; radially directed fibers represent corticofugal projections to brainstem and spinal cord. Stippled regions represents corticocortical fibers coursing in antero-posterior axis. Major directions of current flow induced by vertex-tangential (position 1) and lateral-sagittal (position 2) orientations indicated by arrows and are opposite to the directions in the MC. Top right, Xray of skull of one of us showing position of sensing coil assembly below the chin and anterior to the larynx. Bottom, voltages induced in each sensing coil in the indicated 3 planes with the round MC orientated vertex-tangentially (left) and slightly tilted, lateral-sagitally (middle). Right, the double square MC is located tangentially and laterally. Stimulator outputs were 55% (tangential), 100% (sagittal) and 60% (double square MC). Five superimposed CMAPs from the first dorsal interosseous (FDI) of the hand are shown corresponding to each type of MC stimulation (Figure modified from Amassian et al,[15,18] with permission.)

in an isotropic volume conductor as a function of MC orientation and shape of MC; 3) recording the corticospinal responses in the monkey related to the different MC orientations; and 4) recording the movements and, or the CMAPs related to the different MC orientations.

In reference to 1) above, the magnetic fields were detected by 3 'sensing' coils of wire, each orthogonal to one another, 2 cm in diameter and containing 18 turns.[15] The sensing coils were orientated in sagittal, coronal and horizontal planes,the assembly being placed below the chin and against the larynx (Fig. 6, right). The sensing assembly was not placed in the mouth closer to the brain, because it could not be placed sufficiently posteriorly to lie under the MC, without causing gagging. As a measure of the intensity of motor cortical activation by the different MC orientations, the CMAPs of the relaxed first dorsal interosseous (FDI) muscle of the hand were recorded. The intensities of MC stimulation at the various orientations were adjusted to elicit comparable, submaximal FDI responses, the underlying assumption being that the responses by corticospinal neurons projecting to FDI motorneurons were equivalent. As expected, a vertex-tangential orientation yielded the largest magnetic flux component in the horizontal plane, sensing coil. Shifting the MC to the slightly tilted, lateral-sagittal position significantly increased the sagittal component relative to the horizontal component; the horizontal component may show an absolute increase because of the MC is now much closer to the sensing assembly. However, there was no reduction in CMAP latency of the relaxed FDI, a background of voluntary activity being essential for demonstrating minimal motor latencies. More remarkably, equivalent sized FDI responses were elicited by the double square MC despite a many-fold reduction in magnetic flux recorded by the sensing assembly. Evidently, equivalent FDI responses are elicited by the double square MC with much less total magnetic flux traversing the brain. Put another way, a higher fraction of the magnetic flux injected into the brain is usefully employed in eliciting responses by the desired muscle, rather than eliciting responses in many other muscles and elsewhere in the brain.

In reference to 2) above, an inference drawn from 1) that the distribution of induced currents in the brain is a function of orientation of the standard round MC and of the design of special MCs such as the double square, or figure of '8', has been tested by measuring the voltage gradients in an isotonic saline, volume conductor. As expected, a round MC tangentially and symmetrically applied to the volume conductor induces a much broader profile of potential gradient than that induced by the figure of '8' MC (Fig. 7 left and right); that is, with the round MC the voltage gradient falls off more slowly with distance. However, this is not true when the round MC is orientated orthogonally to the volume conductor (Fig. 7, middle). Although, both round and figure of '8' MCs were equally energized at 20% of maximum stimulator output, the peak voltage gradient with the round MC was greatly reduced, evidencing the inefficiency of the orthogonal orientation in coupling magnetic flux to the volume conductor.

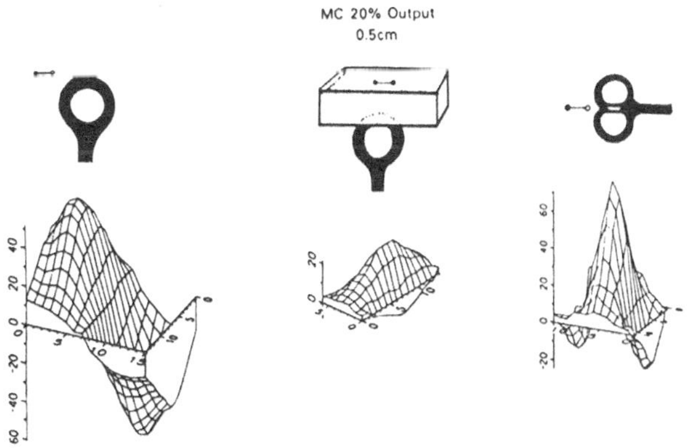


**Figure 7.** Voltage gradients induced in an isotropic volume conductor by different orientations of a round MC and by a figure of 8 MC. Bipolar recording electrodes, 1 cm apart, were moved in the XY plane 0.5 cm above the MCs. Vertical axis indicates mV/mm recorded at 20% of maximum stimulator output. Left and middle, round MC tangentially and orthogonally orientated, respectively. Right, figure of '8' MC tangentially orientated (Figure modified from Maccabee et al,[50] with permission. © 1989 IEEE)

In reference to 3) above, and by analogy with the effects of a single electrical stimulus, an MC pulse would be expected to induce an electrical field, which would excite appropriately orientated neurons or their projections. The effects would include; 1) direct excitation of some neurons, including those projecting from the cortical area; 2) indirect (transynaptic) excitation of neurons, including those projecting from the cortical area; and 3) a prolonged inhibition of cortical neurons, including those projecting from the cortical area.[60] Unfortunately, the large amplitude of the stimulus artefact has precluded directly testing these inferences by intracellular recording near the MC. Nevertheless one sample of motor cortical output, the corticospinal tract, can readily by recorded in the monkey. Such recordings have disclosed both direct and indirect discharge (ie, excitation of corticospinal neurons when the MC is orientated laterally and sagittally over arm area of motor cortex).[19] Significantly, in the anesthetized monkey at least, with this MC orientation, direct discharge had a lower threshold than indirect. This finding is consistent with the similarity in trajectory between one component of the electric field induced by the lateral-sagittal MC and the radial orientation of the bulk of the corticospinal neurons and their projections (Fig. 6, left, position 2). By contrast, a vertex-tangential MC orientation in the monkey could elicit indirect corticospinal discharge at lower threshold than direct discharge, which required a larger MC pulse (Fig. 6, left, position 1). Again, such findings are interpretable in terms of the orientation of the electric field induced by the tangential site of the MC at the vertex. This orientation induces a series of annular rings of current, opposite in direction to that in the MC and with diminishing intensity at increasing distance from the MC. Such field orientation would largely be transverse and therefore be relatively ineffective in directly exciting corticospinal neurons. However, the lateral components of the current loop would approximately parallel the trajectory of corticocortical inputs to area 4, leading to preferential, indirect excitation of corticospinal neurons. Corticocortical inputs capable of indirectly discharging corticospinal neurons arise both from premotor and from postcentral gyri.[57,58] Similarly, the deep tangential plexus in grey matter would be optimally excited by a tangentially orientated MC.[20]

In reference to 4) above, a lateral-sagittally orientated round MC, appropriately positioned and powerfully energized to compensate for the reduced coupling of magnetic flux to the volume conductor, should be able to activate focally motor cortex and, or its projections. It must be emphasized that focality of cortical excitation by the MC can only be inferred when a small subset of possible motor outcomes results from the MC stimulus. Thus, it is essential to demonstrate either differential movements (eg, of digit) or CMAPs by recording from multiple motor sources that lie close to one another. Theoretically, although contraction of flexors and extensors with equal force could result in no net movement, in practise the timing of agonist and antagonist excitation are usually clearly separated.[12] Therefore, recording digit movement rather than CMAPs from separate muscles yielding at least 4 movements (flexion, extension, abduction, adduction) appears a justifiable economy, especially when 4 or 5 digits are tracked. However, even at 100 % Cadwell stimulator output, a lateral-sagittal orientation of the round MC will not usually elicit movement in the relaxed hand. Some tilting towards the midline is required to couple sufficient magnetic flux to the volume conductor to elicit any movement. Given such tilt, movement predominantly of a single digit is readily elicited.[14,18] In Fig. 8, the movements were tracked by sticking fluorescent markers onto the tips of the digits, illuminating them with UV light and recording every 1/60 sec the X,Y coordinates of each marker by a TV camera-computer system. The stored positions of each digit tip can then be displayed as X and Y components of the movement vector as a function of time. The example shown illustrates the ability to elicit predominant movement in digit V, a difficult digit for which to obtain such predominant movement.

An alternative to focally stimulating with the appropriately orientated standard round MC is to use one of the 3 special MC designs (Fig. 5, right 3 MCs). These MC's have important advantages in that part, if not the whole coil, can easily be made to conform to the skull. Also their asymmetry facilitates stimulation at the same angular rotation about the axis of the coil.

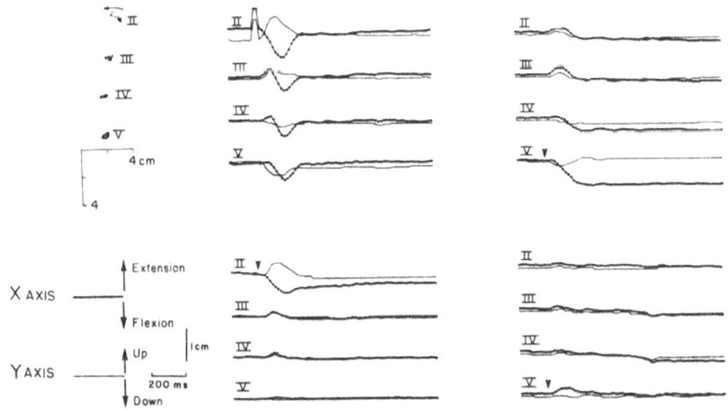


**Figure 8.** Contralateral digit movements elicited by stimulation of motor cortex with a laterally positioned, round MC. Fluorescent spots glued to tips of digits II-V, illuminated with UV light and tracked at 1/60 sec intervals with a TV camera-computer system. Top left inset shows camera view of left hand during the response shown at its immediate right. Responses to 4 stimuli shown; in each, the thick line shows the X axis and the thin line the Y axis components respectively, of the movement vector as a function of time. Left bottom key, downward movement of thick line indicates digit flexion. Artefact and arrows indicate incidence of MC stimuli. (Figure from Amassian et al,[12] with permission.)

Thus, Fig.9 illustrates a predominant digit II movement with stimulation at just above threshold (52% of maximum output) with the small pointed MC.

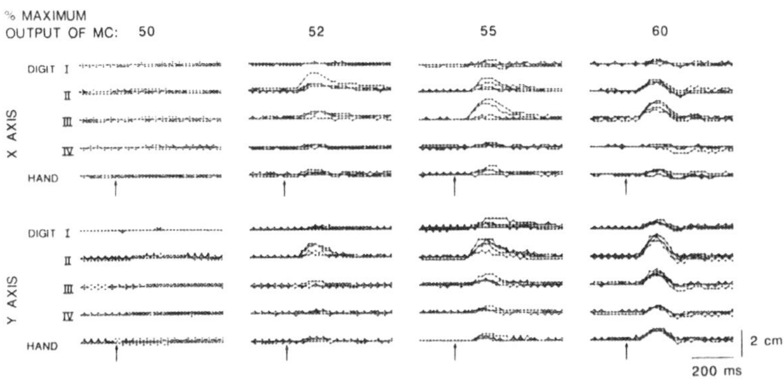


**Figure 9.** Contralateral digit movements elicited by stimulation of motor cortex with a laterally positioned, focal double square MC. X and Y axis components separately displayed. Responses to 4 stimuli superimposed in each record. Downward movement in X axis records of digits II-V and in Y axis of digit I, indicates flexion. (Figure from Amassian et al,[11] with permission.)

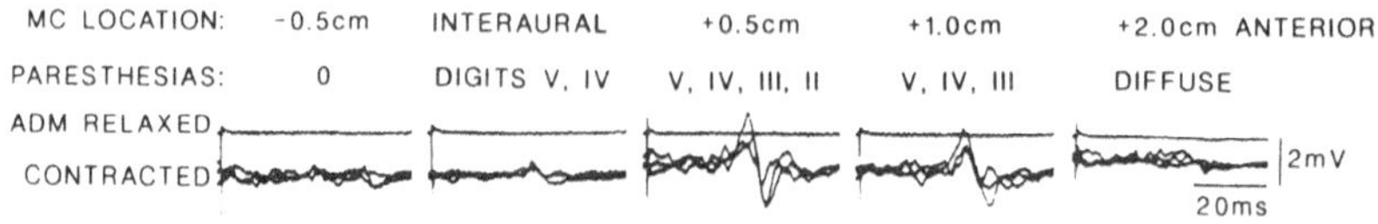


**Figure 10**. Contralateral paraesthesias and CMAPs of abductor digiti minimi elicited by focal stimulation of motor cortex with a figure of '8' MC. Top, location in antero-posterior axis of the MC energized at 70% of maximum stimulator output; the junction region lay horizontally over frontal cortex with the handle lying anteriorly and the point of divergence of the two coils lying posteriorly at the indicated location. Therefore, optimal stimulation probably occurred approximately 2-3 cm anterior to the site, as indicated by Figure 7, right. Middle, location of paraesthesias reported by subject; "diffuse" included all digits and portion of hand. Bottom pair, CMAPs with muscle relaxed and voluntarily contracted, respectively, prior to the MC stimulus. (Figure from Amassian et al,[7] with permission.)

The spatial gradient in the antero-posterior axis is illustrated in Fig. 10; using the most focal MC available (the small figure of '8'), moving the MC as little as 0.5 cm markedly attenuated the response by the contracted abductor digit minimi muscle.[16]

D. Motor Versus Somatosensory Cortex: Source of Projected Somatosensory Sensations

Despite repeated attempts, we never elicited projected sensations by individual pulses when the MC clearly lay over the parietal cortex. Such failure is entirely consistent with the careful parametric study of Libet et al.,[47] who found the adequate stimulus for projected paraesthesias was a repetitive train. Such findings resemble our failure to elicit bright phosphenes by single pulses of the Cadwell round MC located over visual cortex;[14] these were described with the Sheffield MC,[23] which is much wider (14 cm versus 9.2 cm) and therefore is effective at greater depths.

By analogy with our finding of visual suppression by individual MC pulses delivered to visual cortex, somatosensory suppression was sought by delivering single MC pulses to the parietal lobe. An electrical stimulus was delivered through ring electrodes to a digit; the MC pulse was delivered over contralateral parietal lobe at various times (eg, 20, 50, 90 ms) after the stimulus to the digit. No depression was observed unless the MC pulse also elicited a peripheral movement, probably from spread to motor cortex. Any somatosensory depression observed with movement or motor cortical stimulation could be secondary to inhibition from afferent input, or inhibition by the corollary discharge elicited from motor cortex.[11] Possibly, the failure to elicit depression with somatosensory cortical stimulation reflects a multiple afferent representation, only part of which falls within the orbit of the MC.

By contrast, definite sensory responses were elicited by single MC pulses delivered to motor cortex. The commonest response was a sense of movement, projected even to a single contralateral finger. However, such sensation could result from afferent feedback from concomitant movement. Distal movement can reversibly be abolished by ischemia of the arm for 25-35 mins, which is easily produced by a cuff inflated above arterial pressure on the upper arm.[45] The experiment then requires that the motor cortical stimulation elicit movement only in distal limb, which subsequently can be paralyzed. Furthermore, the MC must be reliably replaceable at the appropriate scalp site for the duration of the experiment (approx. 1 hour). These requirements could only be met after the focal MCs become available.

The experiments were done in two ways:[10,11] 1) A scalp site was selected that elicited a finger movement (eg, extension of the index finger). When the distal arm was ischemically paralyzed, focal MC stimulation at the same site elicited an unambiguous sense of movement. The intensity of the sensation and its projected location were both altered by changing the intensity and site of MC stimulation, respectively. Position sense of the digit was lost during the ischemia, rendering it unlikely that any relevant afferent input had escaped block. 2) The distal limb was first ischemically paralyzed; the motor cortex was then mapped for sensation of movement (Fig. 11). After restoring circulation and function in the arm, the motor cortex was remapped for movement and sensations thereof. The conclusion from these studies is that motor cortex is a source of a central sense of movement, supporting the concept of corollary discharge.[39,51] Presumably, the central sense of movement, reflecting that intended, is centrally compared with afferent feedback resulting from the motor performance. Whatever the role of the central sense in regulating on-going motor activity, its refinement and correction by repeated comparison with afferent feedback would seem crucial when skilled movements unfold too rapidly to be regulated by afferent feedback.

Less frequent somatosensory effects of MC stimulation of motor cortex include paraesthesias, projected usually to the contralateral hand. In an initial study,[16] paraesthesias were encountered in only one out of eight physicians tested. The subject, a 56 year old neurologist carefully distinguished the 'tingling' paraesthesias from a sense of movement. The figure of '8' MC proved particularly useful in dissociating pre- from post-central gyral sources of the

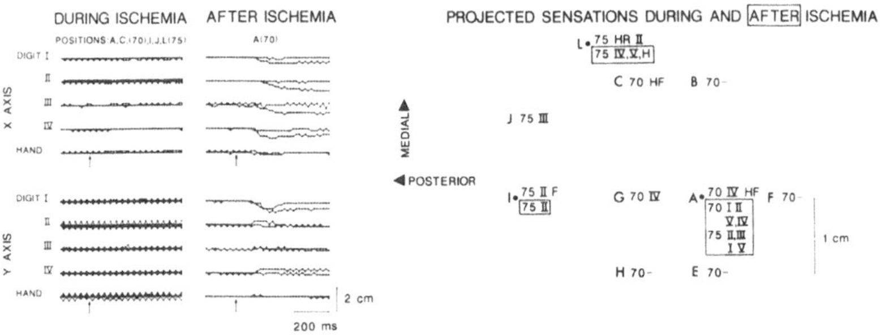


**Figure 11**. The sense of distal movement elicited during and after arm ischemia by focal stimulation of motor cortex with the small pointed MC. Left, superimposed traces show absence of movement elicited by MC stimulation at various sites during ischemia and return of movement after blood readmitted. Stimulus intensity at each site (A-L) indicated as % (eg, 70, 75) of maximum stimulator output. Sensations of digit or hand movement reported in order of diminishing intensity when separated by commas. (-) indicates no sensation; HR and HF indicate sense of radial deviation and flexion, respectively of hand; H, undefined sense of movement of hand. Data enclosed in rectangles obtained after blood readmitted. (Figure from Amassian et al,[11] with permssion.)

paraesthesias (Fig. 10). Scalp sites from which the paraesthesias were elicited in the relaxed hand yielded no movement at intensities of MC stimulation adequate for the projected sensation. Voluntary contraction, known to enhance the effects of transcranial stimulation,[42,53,55,62] was used to detect whether the MC lay over motor rather than sensory cortex. This procedure revealed that paraesthesias were elicited by MC stimulation clearly over motor cortex, although additional participation by somatosensory cortex is not completely excluded.

As with the centrally elicited sense of movement, it is unclear whether the paraesthesias reflect the effect of orthodromic projections from motor cortex, or antidromic stimulation of projections to motor cortex, with distribution of effects through axonal collaterals located elsewhere. An additional problem, the rarity of individuals exhibiting paraesthesias implies either an abnormal response by motor cortex, such as a long train of impulses following a single MC stimulus, or an abnormal response by the 'perceptual' system to the volley from motor cortex. We favor the latter explanation because the muscle responses of the subject to MC stimulation were normal in duration (Fig. 10). The finding of such paraesthesias, although rare, raises the possibility in patients that hallucinatory disturbances, which are not necessarily confined to the somatosensory system, might be precipitated and be analyzed by using focal transcranial MC stimulation.

E. Frontal Lobe Representation of Speech Motor Functions

Focal MC stimulation proved useful in distinguishing between separate areas of frontal cortex related to speech motor activity.[3,8,18] Our studies were largely completed before the special MCs became available; therefore, the standard round MC was orientated so as to 'beam' the magnetic flux away from temporalis muscle inferiorly and from portions of frontal lobe, such as area 4, that are hierarchically lower in the system related to speech. The compromise between these requirements was achieved by an approximately 45$^{o}$ orientation of the MC to the skull (ie, tilted antero-medially). The standard 10-20 electrode positions used in electroencephalography did not define the MC position sufficiently precisely; therefore, an XY coordinate system with 1 cm subdivisions was used with 0, 0 at the interaural line and the outer canthus of the eye at Y = 2.5-3.0 cm. Surface CMAPs were recorded with bipolar discs from muscles acting on the larynx, mainly sternohyoid, sternothyroid and probably cricothyroid. Surprisingly, such surface recordings often reveal one or more distinguishable motor units, which responded specifically to articulating some alphabetical letters, but not others. The laryngeal muscle response was greatly enhanced when the MC stimulus was triggered by the start of the articulation.[18] Articulating 'W' was particularly useful in demonstrating differential effects from focal frontal lobe stimulation.

The findings may be classified as 'simple', motor cortically elicited responses and 'higher' level motor responses. Motor cortically elicited responses had CMAP latencies of 6-8 ms and were obtained in close proximity to sites yielding arm responses, thus confirming their attribution to classical area 4 (Fig. 12, top). Typically, MC stimulation of motor cortex did not prolong the utterance 'W'. By contrast, stimulation at three other sites yielded CMAPs with longer latencies of 13-20 ms (Fig 12, bottom). Furthermore, the da-b-l-u sound of 'W' was converted into daaaaa-b-l-u. This implies that the motor program generating the sequence of phonemes constituting 'W' had undergone a temporary interference. The first, antero-medial area, presumably the supplementary motor area, was easily differentiated from the second, more antero-lateral area, Broca's area. The third area lay at the lateral extremity of the precentral gyrus, which is cytoarchitecturally area 6 and not area 4.[28] Although, we believe that the effects from the lateral extremity of precentral gyrus and Broca's area are separate from one another the areas are so close to one another that the question can only be settled definitely by use of still more focal stimulation.

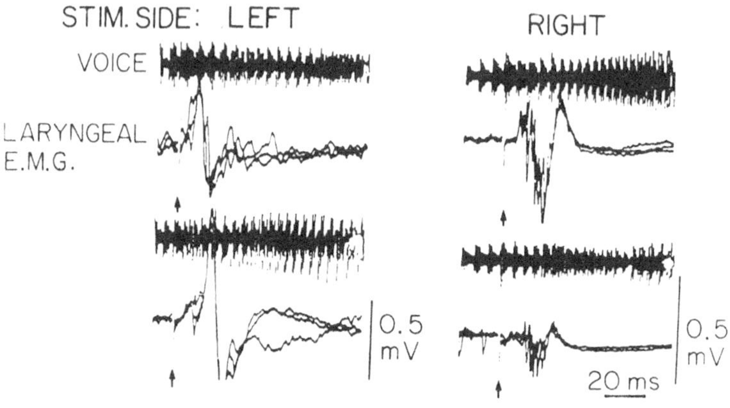


**Figure 12.** Patterns of contralateral laryngeal muscle activation by focal MC stimulation of frontal cortical areas. Paired superimposed traces: top, microphone recording of utterance "Aaaaa-ah"; bottom, surface recording by two discs from paramedian muscles below thyroid cartilage (mainly sternohyoid and sternothyroid). MC stimulus triggered soon after start of utterance. Top pairs elicited by MC stimulation of classical motor cortex; bottom pairs after MC moved to sites of 'higher representation` 2.5 cm anteriorly and 4.5 cm inferiorly, respectively on left and right sides. (Figure from Amassian et al,[3] with permission.)

F. Suppression of Visual Perception

Initially, we thought suppression of visual perception was an example of a 'maximal' effect because almost 100% output of the stimulator was required with the round MC placed tangentially over occipital cortex. However, the topographical effect of changing the MC position and the ability to elicit the suppression with an appropriately orientated focal MC, the double square, imply that this suppression is better classified among focal MC effects. To demonstrate suppression of visual perception, three random letters are briefly flashed on a screen with the MC pulse delivered to visual cortex at various times later.[54,18] In Fig. 13, virtually none of the letters were correctly reported by the three subjects when the MC pulse was delivered 80-100 ms later. The subjects reported either that they "saw nothing" or a "blur" at such test intervals. However, when the MC pulse was delivered 0-40 ms after the visual presentation, no clear effect resulted, nor was a pulse delivered at 140 ms or later effective.

There are a number of possible mechanisms that might account for suppression of visual perception, but the possibilities were narrowed when it was shown that moving the MC to the right spared the right most letter in a

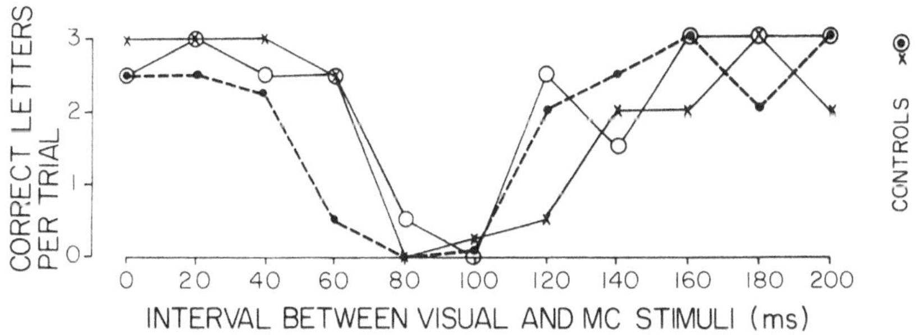


**Figure 13**. Perceptual suppression in three subjects of random letters by MC stimulation of visual cortex. Three 'dark' letters (against light background) presented at visual angle of 0.4-0.8°; at varying times later, MC pulse delivered with lower edge of coil approximately 2 cms above the inion and symmetrically tangentially applied to the occiput. For each subject, MC trials: N= 54(•) 26(0); 26(x). Controls N = 9(•); 13(0); 11(x). (Figure from Amassian et al,[5] with permission.)

horizontal array and vice versa; furthermore, moving the MC anteriorly spared the top letter in a vertical array. Thus, the suppression is related to the topography of the geniculo-striate system. Our more recent findings with the double square MC disclosed a powerful suppression when the junction region lay symmetrically and horizontally over occipital cortex; however, at best, the middle letter of a horizontal array was suppressed when the junction region was orientated in the midline sagittal position.

By analogy with one of the effects of a local electrical pulse on motor cortex,[60] we attribute the suppression to an IPSP in occipital cortex elicited by the MC induced electric field. Regardless of mechanism, our findings imply that the information subserving letter recognition has been transmitted to higher processing centers by 140 ms after the visual stimulus.

G. Tracing CorticoCortical Connectivities by Evoked Potential Recording

We previously recorded transcallosal responses in homologous cortex to focal anodic stimulation of frontal cortex.[2] The cortical responses were initially positive waves with latencies of 9-14 ms, but were almost invariably preceded by a large CMAP from directly activated muscle. The contamination by muscle response was a consequence of the large electrical stimuli that had to be applied to the scalp to excite cortex underlying high resistance bone, etc. The important advantage of MC stimulation is that transcallosal responses can be elicited with minimal activation of muscle.[9,30] The responses are quite focal, movement of the sagittally orientated MC by as little as 1 cm markedly attenuating the response (Fig. 14). Provided the stimulus artefact can be controlled, perhaps by

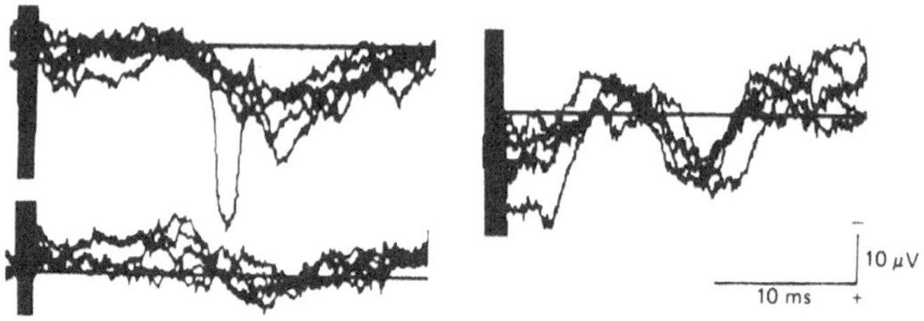


**Figure 14**. Transcallosal responses elicited in two subjects by focal MC stimulation of frontal cortex. Responses recorded by focal disc on scalp between C3 and F4 over contralateral homologous cortex versus temporal reference disc. Left, in this subject, location of lateral edge of MC in top records was 1 cm posterior to site in bottom records. Increased proximity to central sulcus probably accounts for the superimposed, brief diphasic corticofacial EMG deflection. (Figure from Amassian et al,[9] with permission.)

the development of 'switched off' amplifiers during the stimulus duration, there appears to be no technical barrier to using the MC to trace other important cortical connectivities in humans, such as the arcuate fasciculus, occipitofrontal projections and corticopontine projections.

## MAXIMAL STIMULATION WITH THE MAGNETIC COIL

### A. Peripheral Nerve

The need to elicit a maximum CMAP is imperative in clinical electromyography, but conflicts with the need for focality and predictability of the site of stimulation. Thus, an orthogonal-longitudinal orientation of the MC focally stimulates peripheral nerve, but with this orientation much less magnetic flux is coupled to the volume conductor. Thus, in some subjects, excitation of all ulnar motor axons may not be possible without tilting the MC.[17] Such compromise, although permitting excitation of all the ulnar motor axons, then increases the uncertainty as to the point of excitation and would tend to spread excitation to other nerves.

### B. Motor Cortex

Analysis of CMAP latencies following MC stimulation of motor cortex requires assurance that corticospinal neurons are powerfully activated directly. By utilizing a tilted, lateral-sagittal MC orientation and near maximum output of the stimulator, latencies of CMAPs in voluntarily contracted forearm and hand muscles can be made as brief as with focal anodic stimulation, implying that the central excitation pathways are identical. In Fig. 15, the CMAP latency of digit

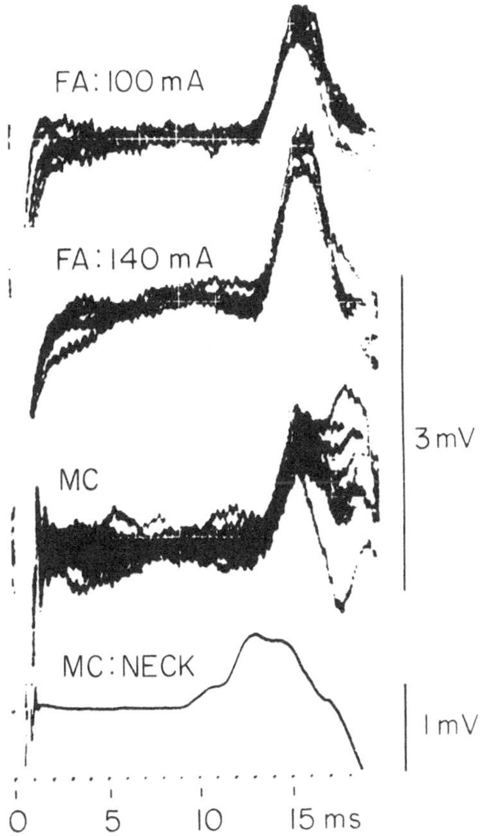


**Figure 15.** Latencies of contralateral digit flexor CMAPs matched by focal anodic and MC stimulation of motor cortex; peripheral conduction time determined by orthogonal (edge) MC stimulation at neck. The electrical pulse lasted 200 μs and was delivered between the focal anode (5 $cm^2$) and the semicircular cathode (26 $cm^2$), separated by several cm beyond the 3-4 cm used to elicit the transcallosal response.[2] To stimulate the peripheral motor axons as centrally as possible, the MC was applied orthogonally to the lower neck, slanted rostro-caudally and was moved in one cm increments from contralateral to ipsilateral positions. The CMAP appeared abruptly at 1 cm ipsilaterally and showed (bottom record) a further marked increase at 2 cm. (Figure from Amassian et al,[9] with permission.)

flexors with focal anodic stimulation had already reached a minimal value (13.3 ms) at 100 mA, an increase in stimulus intensity to 140 mA producing only a modest increase in amplitude. Appropriate MC stimulation yielded a more complex CMAP, but with a similar latency. An MC orthogonally orientated on the lower neck was used to elicit CMAPs (latency 9.0 ms) in the same muscle. After adding a delay (0.5 ms) for conduction between the point of peripheral stimulation and the ventral horn, the total peripheral delay is estimated at 9.5 ms. Subtracting this delay from the CMAP latency with motor cortical stimulation (13.3 ms) yielded a minimal central delay of 3.8 ms. Subtracting a further 0.7 ms for monosynaptic delay for discharge of an active motoneuron close to firing level, yields a conduction time of 3.1 ms in the corticospinal tract. A conduction delay of 3.1 ms over a conduction distance of 210 mm from motor cortex to C6-C7 yields a maximum conduction velocity of 68 m/s, a value consistent with other measurements.[62,65]

C. Higher Cortical Processing Areas for Visual Perception

The finding of visual suppression by an MC pulse delivered to visual cortex suggested the possibility of tracking the flow of visual information through higher cortical processing areas by appropriately placed and timed MC pulses. However, even with near maximum output of the stimulator and the MC laid tangentially over presumed Wernicke's area, the supramarginal and angular gyri, correct reporting of a word flashed before the subject could not be prevented. Such failure may be contrasted with the blocking effects on various aspects of language reported by Ojeman and Mateer,[56] who used repetitive stimulation lasting many sec of much smaller cortical areas, as did Penfield and Roberts.[59] Possible explanations of the discrepancy are that repetitive stimulation: 1) by temporal facilitation, accesses other cortical, or subcortical components of the overall circuit (eg, thalamic association nuclei) that are not adequately influenced by the corticofugal response to a single MC pulse; 2) outlasts a temporary storage modality (eg, chemical) which is inaccessible to the brief changes in membrane potential that are induced by a single MC pulse; and 3) outlasts 'refreshing' by visual cortex of the input to the higher processing centers.

Our first successful attempts to influence higher level processing were designed to hinder possible refreshing by visual cortex.[6] Two random words were presented at intervals of 50-83 ms, with the MC pulse given 90-150 ms after the first word. Statistically significant reductions (proportionality test; $P<0.05$) in word retrieval or correct sequencing were demonstrated by MC stimulation of the areas listed above, but the explanation for these findings need not be related to the previous possibility (3) above. Significant suppression was also obtained by MC stimulation of Broca's area, supporting a language receptive function of this area additional to the conventional motor expressive function. The suppression from all areas tested was clearly weaker than that obtained with the MC over visual cortex. A further difficulty was that MC effects were manifested by the failure of the subject to make a correct report, without the concommittant topographical relationships seen with visual cortical stimulation, which were important in eliminating nonspecific effects (eg, on attention). To improve the paradigm by requiring a positive response by the subject, two random words or three letter arrays were presented; the second stimulus (S2) was approximately 4 X brighter than S1 and was delivered 100 ms later. Under these conditions, powerful backward masking occurred (Fig. 16) (ie, the first word was not retrieved).[54] The representation of S2 does not arrive in

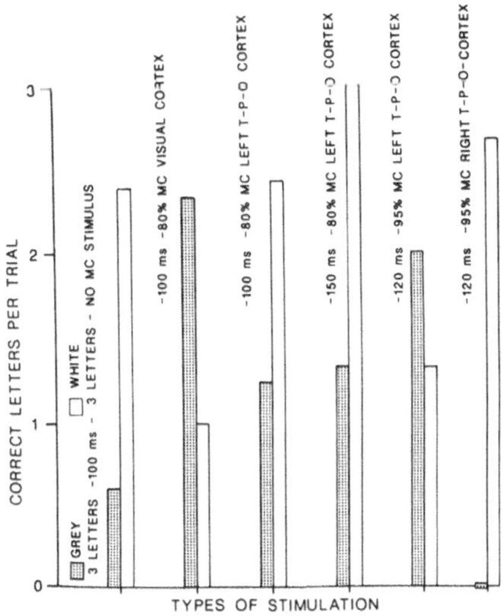


**Figure 16.** Unmasking of visual perception as a function of site, intensity and timing of the MC pulse. The visual stimuli were 3 grey letters (stippled rectangles) followed after a 100 ms interval by 3 white letters (open rectangles). 80% MC, VC, 100 ms means that the MC was energized by 80% of maximum stimulator output applied to visual cortex 100 ms after the white letters. T-P-O temporo-parieto-occipital cortex was stimulated on left (dominant) side except at extreme right. (Figure from Amassian et al,[6] with permission.)

visual cortex until 100 ms plus at least 60 ms transmission time, the earliest latency of the visual cortical evoked potential with the type of stimuli we used. However by 160 ms, the representation of S1 has already been transmitted to the higher processing centers, where at an S1-S2 interval of 100 ms, the masking presumably occurs. If an MC pulse is delivered to visual cortex 100 ms after S2, the response to S1 is readily retrieved (unmasked), most likely because the response to S2 is not transmitted beyond visual cortex. In Fig. 16 (second from left rectangles) the increased retrieval of the first set of letters is approximately matched by the reduced retrieval of the second set. In two of us who were subjects, the unmasked response to S1 had the brightness attribute of S2, as though the MC had suppressed letter recognition, but not the associated brightness, which may have been represented in cortex outside the orbit of the MC. Stimulation with the MC approximately covering Wernicke's area, supramarginal, angular gyri and inferior portions of the occipital lobe (ie, on the dominant side) also resulted in statistically significant ($P<0.05$), but less robust unmasking (Fig. 16). The unmasking was partially manifested by incorporation of S1 letters into the responsc to S2. No unmasking could be demonstrated in this subject on the non-dominant side.

It seems clear that despite the use of double visual stimulation, near maximum stimulator output and the MC overlying a large area of 'relevant higher level' processing cortex, the effects are still weak. Major advances might acrue from the use of a short train 3-4 of MC pulses with a period of 50 ms (ie, each period approximating the duration of visual cortical suppression and the presumed duration of the cortical IPSP). The use of multiple focal MC's would also be an advance over an MC covering a wide area. Given these technical improvements, further attempts to track the flow of information from visual cortex would be aided. A design problem that would arise with repetitive stimulation would be overheating of the MC. The possibility of kindling might also be increased.[38] Our initial attempt to use two MCs for stimulating two cortical areas resulted in failure of one stimulator, probably from a high, reversed voltage induced in the MC by the other MC. Protection from interaction between multiple MC stimulators would appear to be necessary.

## SAFETY OF MAGNETIC COIL STIMULATION

Virtually from its inception, the question of safety of the MC stimulator has been of concern to investigators. There are three major safety concerns: 1) Is an MC pulse of the intensity used safe in the short and in the long term? 2) Is the total number of MC pulses used per session (lasting about 1-2 hours),or over an extended period of many sessions, a factor in tissue damage? 3) Is a particular stimulator capable of delivering MC pulses safely (eg, without either exploding from mechanical or thermal stress, or the insulation breaking down, with the danger of an electric shock to the subject).

Attempts at answering such questions related to the safety of MC pulses are derived from several sources, including: 1) Calculations of the amount of energy transferred by the MC to the brain. Barker et al.[22] estimate the amount of heating is 1/300 of the international standard for continuous thermal energy. Similarly, they calculate the maximum charge transferred by the MC as 50 μC/pulse, which is 0.05-0.005% of that used in electroconvulsive therapy. Cadwell[27] shows very little mechanical energy is imparted by the MC to objects provided they are small (eg, a scalpel blade). 2) Personal experiences with the MC of several groups in Great Britain (from 1985) and in North America (from 1987), although admittedly anecdotal, have not evidenced that the MC can cause brain damage. The recent extension of such studies to the newborn,[35] increases confidence that MC stimulation does not affect development, a confidence requiring direct testing on animals. A number of investigators, including members of our group, have had thousands of such stimuli to the head with no clear short term or long term effect, other than a local, cranial muscle ache lasting a few hours. The absence of reports of a throbbing headache is particularly significant because the MC pulse might directly excite nerve fibers supplying cerebral arterioles causing either constriction, with subsequent reactive dilatation or, less likely, initial dilatation. An MC effect directly on smooth muscle membrane is still less likely. Increased regional cerebral blood flow is known to accompany physiological activities that are expected to increase neuronal activity. Not surprisingly, Iofetamine HCl 123 (Spectamine) brain scans reveal increased activity at the site of focal motor cortical stimulation by the MC. The increased blood flow could be either secondary to the increased neuronal activity or a direct consequence of the MC pulse on cortical blood vessels. However, contralateral, homologous cortex and cerebellum also show increased activity;[64] the MC pulse is unlikely to act directly on blood vessels at such distance. Because the distant regions receive direct or indirect projections from the stimulated frontal lobe, the increased blood flow in these distant regions most likely reflects increased neuronal activity. Curiously, a brief report of the effect of MC pulses in the cat failed to disclose any changes in cerebral blood flow, although the stimuli elicited responses in the ulnar nerve.[36] Until more details are available (eg, of the effects of the general anesthesia on metabolic responses) further discussion of these data is unwarranted.

As yet, there is no clear evidence that seizures can be precipitated in otherwise normal individuals by an MC pulse. Obviously, it is wise routinely to exclude from conventional MC stimulation subjects with a personal or family history of epilepsy. It should be emphasized that conventional EEG recordings, while potentially valuable in demonstrating the possible development of focal spiking, can not record events within many ms of the MC pulse because of the large stimulus artefact. By contrast, a transient electrical stimulus to motor cortex of a seizure prone cat elicits a marked prolongation of indirect corticospinal discharge.[25] Therefore, an early warning sign of pathologically increased excitability of motor cortex may be prolongation of the CMAP following single MC pulses.

Other possible tests for 'damage' by MC stimulation include hormonal levels of prolactin, cortisol and noradrenalin. Unfortunately, the anxiety of an untrained subject requires that elaborate controls be included in the protocol to avoid spuriously attributing changes in hormonal level to the direct effect of the MC pulse on the brain. Thus, a minimal control would be to stimulate orthogonally with the MC over an interposed spacer, several cm wide, shaped like the MC and pressing on the skull. Preferably some local contraction of muscle would be elicited to mimic further the effect of the MC. The anxiety manifested even by some neurologist volunteers undergoing MC stimulation underlines the importance of appropriate controls in any study based on hormonal levels.

We suspect that the most sensitive indices of possible neural damage will be manifest by changes in cognition and skilled motor performance. The study of suppression of visual perception implies that the effect of a series of MC pulses to visual cortex lasts for a brief period of approximately 50 ms after each pulse and is not cumulative over many trials. The two word or three letter array presentations require the closest attention by the subject. No systematic trend in control performance occurred after many intense MC stimuli were delivered to temporo-parieto-occipital cortex, nor was a deterioration in performance manifest during experimental sessions over a period of months.

It must be emphasized that if repetitive MC stimulators are introduced, concern about a possible kindling effect[38] would have to be addressed directly. The potential dangers of repetitive stimulation are greatest among self-experimenting investigators, who generally receive many more stimuli than any patient would receive. The risks would be ameliorated if, as expected, the use of repetitive MC stimulation permits a substantial reduction in intensity per pulse and the number of pulses in the train is kept, small (eg, 4). However, the risk will have to be weighed by choice of an appropriate animal model. The problems with such models include: 1) devising an appropriately scaled down version of the MC to deliver pulses, equivalent to those used in the human to smaller brains; 2) animal brains have different shapes and nearby resistance barriers, such as a bony tentorium; and 3) susceptibility to seizures differs among animal models as compared with the human and in a given species (eg, the humans in infancy compared with adulthood).

## SUMMARY

Percutaneous magnetic coil (MC) stimulation of human nervous system is a relatively painless method of exciting peripheral nerves, cranial nerves (especially VII and V), cerebral cortex and, or its projections. Although MC stimulation has little advantage over conventional electrical stimulation in peripheral nerve studies, the importance of orientation of the MC was best demonstrated on this relatively simple, neural system. Thus, a longitudinal-orthogonal orientation of the MC relative to the ulnar or median nerves just above the elbow and wrist, respectively, yielded the most focal excitation of motor axons, presumably because the induced current flows most intensely parallel and close to the axons. Comparison of the MC threshold for exciting human ulnar motor axons and the longitudinal voltage gradients developed in a simple trough model of the elbow (containing isotonic saline) implied that axonal excitation by the MC was accounted for by the electric field induced in the volume conductor. Given this premise, a past intracellular analysis of electrical excitation of cat motor cortex was used to predict that the effects of an MC pulse would be: 1) direct excitation of motor cortical, including corticospinal neurons; 2) indirect excitation of 1) through transynaptic action; and 3) a delayed inhibitory postsynaptic potential through polysynaptic action. Recordings from monkey corticospinal tract confirmed that the MC elicited both direct and indirect corticospinal excitation, direct preferentially by lateral-sagittal orientations and indirect preferentially by midline-tangential orientations.

Our main thesis is that the MC stimulator has multiple uses, some of which imply conflicting aims in designing coils and stimulators. Thus, focal activation (eg, of cerebral cortex) is an important physiological objective, which has been achieved either by special orientation of the standard round MC or by specially designed MCs which can be conveniently laid tangentially over the local area to be stimulated. Focal stimulation of the motor cortex has been used: to elicit individual digit movements; identify and map the central sense of movement and projected paraesthesias (present in rare individuals); to delineate four frontal areas related to speech, including Broca's area, supplementary motor area and two precentral areas; to demonstrate the topographical nature of suppression of visual perception by MC stimulation of visual cortex; and to trace interhemispheric connectivities by recording the evoked cortical potential inhomologous cortex to MC stimulation of the frontal lobe.

By contrast, 'maximal' MC stimulation was needed for some purposes with a trade off against focality: to excite all motor axons in a peripheral nerve; to elicit a powerful direct discharge by corticospinal neurons, which, against a background of voluntary contraction, yielded an estimate of minimal central motor delay; and to stimulate powerfully several cortical areas, including Wernicke's area with supramarginal and angular gyri and Broca's area, leading to perceptual suppression of letters or words.

The safety of neural stimulation by individual MC pulses is discussed, with the conclusion that, given the exclusion of inappropriate subjects, no clear danger so far has been identified. However, the need for repetitive stimulation in exploring higher cognitive functions makes it important to evaluate the safety of such stimulation, especially if repeated in the experimental laboratory.

## ACKNOWLEDGMENTS

Our thanks are due to Dr. John Cadwell for supplying us with a variety of special designs of magnetic coils and to our collaborators whose names appear in the reference section.

### References

1. Amassian VE. The use of contact placing in analytical and synthetic studies of the higher sensorimotor control system. In: Integration in the Nervous System. A Symposium in Honor of David PC Lloyd and Rafael Lorente de No (Eds. H Asanuma, VJ Wilson). Igaku-Shoin Press, Tokyo. 1979; 279-304.

2. Amassian VE, Cracco RQ. Human cerebral cortical responses to contralateral transcranial stimulation. Neurosurg 1987; 20:148-155.

3. Amassian VE, Anziska BJ, Cracco JB, Cracco RQ, Maccabee PJ. Focal magnetic coil excitation of frontal cortex activates laryngeal muscles in man. J Physiol (Lond) 1988; 398:41P.

4. Amassian VE, Cadwell J, Cracco RQ, Maccabee PJ. Focal cerebral and peripheral nerve stimulation in man with the magnetic coil. J Physiol (Lond) 1987; 390:24P.

5. Amassian VE, Cracco JB, Cracco RQ, Eberle L, Maccabee PJ. Suppression of human visual perception with the magnetic coil over occipital cortex. J Physiol (Lond) 1988; 398:40P.

6. Amassian VE, Cracco RQ, Eberle L, Maccabee PJ, Rudell A. Unmasking visual perception by magnetic coil stimulation of human cerebral cortex. J Physiol (Lond) (in press).

7. Amassian VE, Cracco RQ, Maccabee PJ. Focal magnetic coil activation of human motor cortex elicits a sense of movement in ischaemically paralyzed, distal arm. J Physiol 1988; 403:75P.

8. Amassian VE, Cracco RQ, Maccabee PJ. Basic mechanisms of magnetic coil excitation of nervous system in humans and monkeys and their applications In: IEEE/EMBS 10th Annual Conference, Maturing Technologies and Emerging Horizons in Biomedical Engineering. McGregor & Werner Inc, Washington, DC. 1988; 10-17.

9. Amassian VE, Cracco RQ, Maccabee PJ. Transcallosal responses elicited by focal magnetic coil stimulation of human frontal cortex. J Physiol 1988; 407:69P.

10. Amassian VE, Cracco RQ, Maccabee PJ. Focal magnetic coil activation of human motor cortex elicits a sense of movement in ischemically paralyzed, distal arm. J Physiol (Lond) 1988; 403:75P.

11. Amassian VE, Cracco RQ, Maccabee PJ. A sense of movement elicited in paralyzed distal arm by focal magnetic coil stimulation of human motor cortex. Brain Research 1989; 479:355-360.

12. Amassian VE, Cracco RQ, Maccabee PJ. Focal stimulation of human cerebral cortex with the magnetic coil: A comparison with electrical stimulation. Electroenceph Clin Neurophysiol (in press).

14. Amassian VE, Cracco RQ, Maccabee PJ, Cracco JB, Rudell A, Eberle L. Suppression of visual perception by magnetic coil stimulation of human occipital cortex. Electroenceph Clin Neurophysiol (in press).

15. Amassian VE, Cracco RQ, Maccabee PJ, Marsh R. Comparing magnetic fields when stimulating human brain with different coils. J Physiol (Lond) 1989; 412:7P.

16. Amassian VE, Cracco RQ, Maccabee PJ, Somasundaram M. Focal magnetic coil stimulation near human central sulcus exceptionally elicits localized contralateral paraesthesias. J Physiol (Lond) (in press).

17. Amassian VE, Maccabee PJ, Cracco RQ. Focal stimulation of human peripheral nerve with the magnetic coil: A comparison with electrical stimulation. Exp Neurol 1989; 103:282-289.

18. Amassian VE, Maccabee PJ, Cracco RQ, Cracco JB. Basic mechanisms of magnetic coil excitation of nervous system in humans and monkeys: Applications in focal stimulation of different cortical areas in humans. In: Magnetic Stimulation in Clinical Neurophysiology (Ed. S Chokroverty). Butterworth, Stoneham. (in press).

19. Amassian VE, Quirk GJ, Stewart M. Magnetic coil versus electrical stimulation of monkey motor cortex. J Physiol (Lond) 1987; 394:119P.

20. Amassian VE, Stewart M, Quirk GJ, Rosenthal JL. Physiological basis of motor effects of a transient stimulus to cerebral cortex. Neurosurg 1987; 20:74-93.

21. Barker AT, Freeston IL, Jalinous R, Jarratt JA. Clinical evaluation of conduction time measurements in central motor pathways using magnetic stimulation of the human brain. Lancet 1986; i:1325-1326.

22. Barker AT, Freeston IL, Jalinous R, Jarratt JA. Magnetic stimulation of human brain peripheral nervous system: an introduction and the results of an initial clinical evaluation. Neurosurg 1987; 20:100-109.

23. Barker AT, Freeston IL, Jalinous R, Merton PA, Morton HB. Magnetic stimulation of the human brain. J Physiol (Lond) 1985; 369:3P.

24. Barker AT, Jalinous R, Freeston IL. Non-invasive magnetic stimulation of the human motor cortex. Lancet 1985; i:1106-1107.

25. Berlin L, Amassian VE. Pyramidal tract responses during seizures. Electroenceph Clin Neurophysiol 1965; 19:587-597.

26. Bickford RG, Fremming BD. Neuronal stimulation by pulsed magnetic fields in animals and man. Digest 6th Int Conf Med Electronics Biol Eng 1965; 112.

27. Cadwell MES-10 Operator's Manual, 1-16, appendix 12.

28. Campbell AW. Histological Studies on the Localization of Cerebral Function. Cambridge University Press, Cambridge, England. 1905.

29. Cowan JMA, Rothwell JC, Dick JPR, Thompson PD, Day BL, Marsden CD. Abnormalities in central motor pathway conduction in multiple sclerosis. Lancet 1984; ii:304-307.

30. Cracco RQ, Amassian VE, Maccabee PJ, Cracco JB. Comparison of human transcallosal responses evoked by magnetic coil and electrical stimulation. Electroenceph Clin Neurophysiol (in press).

31. Day BL, Dick JPR, Marsden CD, Thompson PD. Differences between electrical and magnetic stimulation of the human brain. J Physiol (Lond) 1986; 378:36P.

32. Day BL, Dressler D, Maertens de Noordhout A, Marsden CD, Nakashima K, Rothwell JC, Thompson PD. Differential effect of cutaneous stimuli on responses to electrical or magnetic stimulation of the human brain. J Physiol (Lond) 1988; 399:68P.

33. Day BL, Maertens de Noordhout A, Marsden CD, Nakashima K, Rothwell JC, Thompson PD. Temporary interruption of brain processing by an electrical or magnetic cortical shock in man. J Physiol (Lond) 1987; 390:197P.

34. Duron B, Khater-Boidin J. Percutaneous stimulation of spinal cord and brain: central motor pathways, conduction velocities in human newborn. J Physiol (Lond) 1988; 406:35P.

35. Eyre JA, Koh THHG. Maturation of descending motor pathways in man from birth to adulthood. J Physiol (Lond) 1988; 396:58P.

36. Eyre JA, Flecknell PA, Kenyon BR, Koh THGG, Miller S. Effects of electromagnetic stimulation of the brain on cortical activity, cortical blood flow, blood pressure and heart rate in the cat. J Physiol (Lond) 1988; 396:154P.

37. Eyre JA, Gibson M, Koh THHG, Miller S, O'Sullivan MC, Ramesh V. Corticospinal transmission excited by electromagnetic stimulation of the brain is impaired in children with spastic hemiparesis but normal in those with quadriparesis. J Physiol (Lond) 1989; 414:9P.

38. Goddard GV, McIntyre DC, Leech CK. A permanent change in brain function resulting from daily brain stimulation. Exp Neurol 1969; 25:295-330.

39. Goodwin GM, McCloskey DJ, Matthews PBC. The contribution of muscle afferents to kinaesthesia shown by vibration induced illusions of movement and by the effects of paralyzing joint afferents. Brain 1972; 95:705-748.

40. Harrington DO. The Visual Fields: A Textbook and Atlas of Clinical Perimetry. CV Mosby, St Louis, MO, 1976.

41. Hassan NF, Rossini, Cracco RQ, Cracco JB. Unexposed motor cortex excitation by low voltage stimuli. In: Evoked Potentials: Neurophysiological and Clinical Aspects (Eds. C Morocutti, PA Rizzo). Elsevier, Amsterdam. 1985; 107-113.

42. Hess CW, Mills KR, Murray NMF. Responses in small hand muscles from magnetic stimulation of the human brain. J Physiol (Lond) 1987; 388:397-419.

43. Krynjevic K, Phillis JW. Iontophoretic studies of neurones in the mammalian cerebral cortex. J Physiol (London) 1963; 165:274-304.

44. Levy WJ, Jr. Clinical experience with motor and cerebellar evoked potential monitoring. Neurosurg 1987; 20:169-182.

45. Lewis R, Brindley G. The extrapyramidal motor map. Brain 1965; 88:397-406.

46. Lewis T, Pickering GW, Rothschild . Centripetal paralysis arising out of arrested blood flow to the limbs, including notes on a form of tingling. Heart 1931; 16:1-32.

47. Libet B, Alberts WW, Wright EW Jr, Delatrre LD, Levin G, Feinstein B. Production of threshold levels of conscious sensation by electrical stimulation of human somatosensory cortex. J Neurophysiol 1964; 27:564-578.

48. Maccabee PJ, Amassian VE, Cracco RQ, Cadwell JA. An analysis of peripheral motor nerve stimulation in humans using the magnetic coil. Electroenceph Clin Neurophysiol 1988; 70:524-533.

49. Maccabee PM, Amassian VE, Cracco RQ, Cracco JB, Anziska BJ. Intracranial stimulation of facial nerve in humans with the magnetic coil: Electroencephal Clin Neurophysiol 1988; 70:350- 354.

50. Maccabee PJ, Eberle L, Amassian VE, Cracco RQ, Rudell A. Studies of 3-dimensional voltage distributions induced by homogeneous media volume conductors by round and butterfly magnetic coils. In Proc 11th Ann Int Conf IEEEE/EMB. IEEE, Piscataway, NJ. 1989: 1259-1260.

51. McCloskey DI, Colebatch JG, Potter EK, Burke D. Judgements about onset of rapid voluntary movements in man. J Neurophysiol 1983; 49:851-863.

52. Merton PA, Morton HB. Stimulation of the cerebral cortex in the intact human subject. Nature 1980; 285:227.

53. Merton PA, Morton HB. Electrical stimulation of human motor and visual cortex through the scalp. J Physiol (Lond) 1980; 305:9-10P.

54. Michaels CF, Turvey MT. Central sources of visual masking: indexing structures supporting seeing at a single brief glance. Psychol Res 1979; 41:1-61.

55. Mills KR, Murray NMF, Hess FMH. Magnetic and electrical transcranial brain stimulation: Physiological mechanisms and clinical applications. Neurosurg 1987; 20:164-168.

56. Ojemann G, Mateer C. Human language cortex: Localization of memory, syntax and sequential motor-phoneme identification systems. Science 1979; 205:1401-1403.

57. Patton HD, Amassian VE. Single- and multiple-unit analysis of cortical stage of pyramidal tract activation. J Neurophysiol 1954; 17:345-363.

58. Patton HD, Amassian VE. The pyramidal tract: Its excitation and functions. In: Handbook of Physiology. Neurophysiology. American Physiology Society, Washington, DC. 1960; II(1):837-861.

59. Penfield W, Roberts L. Speech and brain mechanisms. Princeton University Press. Princeton, NJ, 1959.

60. Rosenthal J, Waller HJ, Amassian VE. An analysis of the activation of motor cortical neurons by surface stimulation. J Neurophysiol 1967; 30:84458.

61. Rossini PM, Caramia MD, Zaroli F. Mechanisms of nervous propagation along central motor pathways: non-invasive evaluation in healthy subjects and in patients with neurological disease. Neurosurg 1987; 20:183-191.

62. Rothwell JC, Day BL, Thompson PD, Dick JPR, Marsden CD. Some experiences of techniques for stimulation of the human cerebral motor cortex through the scalp. Neurosurg 1987; 20:156-163.

63. Schriefer TN, Mills KR, Murray NMF, Hess CW. Evaluation of proximal facial nerve conduction by transcranial magnetic stimulation. J Neurol Neurosurg Psychiatry 1988; 51:60-66.

64. Shafran B, Maccabee PJ, Amassian VE, Cracco RQ, Strashun A, Vaquer R, Singman R. Spectamine imaging of cerebral blood flow during voluntary movement and focal stimulation of motor cortex with magnetic coil. Neurosc Abstr 1989; 15:789.

65. Snooks SJ, Swash M. Motor conduction velocity in the human spinal cord: Slowed conduction in multiple sclerosis and radiation myelopathy. J Neurol Neurosurg Psychiatry 1985; 48:1135-39.

66. Tasaki I. Nervous Transmission. Thomas, Springfield, IL, 1953.

67. Tasaki I. Conduction of the nerve impulse. In: Handbook of Physiology. Amer Physiol Soc, Washington, DC. 1959; I(1):75-121.

68. Thomas JE, Lambert EH. Ulnar nerve conduction velocity and H-reflex in infants and children. J Appl Physiol 1960; 15:1-9.

# SECTION III

# CLINICAL APPLICATIONS

# RECENT TECHNICAL DEVELOPMENTS IN CANCER HYPERTHERMIA

Chung-Kwang Chou, and Kenneth H. Luk

## INTRODUCTION

Historically, treatment of cancer with hyperthermia can be traced back to 3,000 B.C. when smoldering sticks of wood were inserted in tumors.[42] More recently in 1866 Busch described the disappearance of a large sarcoma after high fever. Since then, sporadic reports of using heat to treat cancers have appeared in Western journals. During the last 15 years, technological advances have revived interest in using hyperthermia in combination with other forms of therapy.[1,7,18,23,25,29,33,43,59] Renewed clinical interest in hyperthermia was due in part to a plateau in treatment outcomes with both radiation therapy and chemotherapy that led scientists and physicians to look for other methods to combat cancer. Although, there were earlier reports published of success in treating animal and human tumors by heat alone, there are many papers emphasizing the synergistic effects of heat and radiotherapy or heat and chemotherapy.[35] The effective temperature range of hyperthermia is small (41-45°C). At lower temperatures, the effect is quite minimal. For typical treatment times, at temperatures higher than 45°C, unacceptable normal tissue damage occurs. Due to the difference in blood flow in tumor tissue, as compared to normal tissues, tumor temperatures are usually higher than those in the surrounding normal tissue during hyperthermia treatment.[57] In addition, it is generally believed that tumors are more sensitive to heat. This is thought to be due to the hypoxic (lack of oxygen), acidic, and poor nutritional state of tumor cells. The synergism of radiation and hyperthermia is accomplished by the thermal killing of hypoxic and S phase (DNA synthesis period) cells which are resistant to radiation. Hyperthermia has been used in combination with chemotherapy because heating increases the membrane permeability and the potency of some drugs.[28] Although there are many encouraging clinical results, in the United States hyperthermia remains an experimental treatment, usually applied to late stage patients. This is due largely to the difficulties clinicians encounter in raising and keeping the tumor temperature at therapeutic level. Heating methods include whole body heating using hot wax, hot air, hot water suits, or infrared irradiation; partial body heating utilizes radio frequency (RF) current or field, microwave (MW), ultrasound (US), hot blood, or fluid perfusion.[15,22,25]

The temperature increase in tumors and tissues is determined by the energy deposited and the physiological responses of the patient. When electromagnetic (EM) methods are used, the energy deposition is a complex function of frequency, intensity, polarization of the applied fields, and geometry and size of applicator as well as the dielectric property, geometry, size and depth of the tumor.[8] The final temperature depends not only on the energy deposition but also on blood flow and thermal conduction in tissues. In present hyperthermia research, thermal dosimetry and treatment planning are far from adequate. Further clinical applications cannot be designed without better knowledge of how heat is to be delivered in various clinical situations. The development of advanced hyperthermia equipment and techniques will allow successful treatment of cancers that are resistant to other therapies.

During the last 15 years, there has been significant progress in the methods of heat delivery, temperature monitoring, and thermal dosimetry. In this paper, we summarize technical advances in hyperthermia and then discuss possible future developments. The importance of interdisciplinary team work is emphasized.

## HEAT DELIVERY

### A. Basic Methods

The superficial cooling mechanism of tissue makes deep heating by the conductive method difficult. Penetrating methods such as MW, RF and US are necessary to bring EM or acoustic energy to tissue beneath the subcutaneous fat layers.[36] The following heating methods, used for diathermy in Rehabilitation Medicine for four decades, have been modified and refined for heating tumors.

#### 1. Resistive Heating

Tissues can be heated by RF currents through the use of needle or plate electrodes. The operating frequency should be higher than 100 kHz to prevent nerve excitation.

#### 2. Capacitative Heating

Tissues can be placed between two capacitor plates and heated by displacement currents. This method is simple, but overheating the fat remains a major problem.[22] In planar tissue models, the rate of temperature rise is about 17 times greater in fat than in muscle due to the large differences in dielectric property and specific heat. In addition, blood flow is significantly less in fat, so the final temperature is much higher than that in muscle. In Japan, the 8 MHz Thermotron system uses a water cooled bolus to minimize heating fat tissues.[32] The sizes of the two electrodes are adjustable for controlling the heating patterns during treatments.

3. Inductive Heating

Magnetic fields generated by solenoidal loops or pancake magnetic coils induce eddy currents in tissue. Since the induced electric fields are parallel to the tissue interface, heating is maximized in muscle rather than in fat. However, the heating pattern is generally toroidal with a null at the center of the coil.

4. Radiative Heating

The above three heating methods use frequencies in the RF band where a quasi-static condition applies. In the MW frequency range, energy is coupled into tissues through waveguides, dipoles, microstrips, or other radiating devices.[25] The shorter wavelengths of MWs make it possible to direct and focus the energy into tissues by direct radiation from a small applicator. Loading an applicator with dielectric material can reduce its aperture size and provide more flexibility in controlling the amount of energy deposited at tumor sites.

5. Ultrasound Heating

Compared to EM waves, US can penetrate more deeply. Because its wavelength is shorter than RF and MW, focusing is easier. Frequencies in the range of 0.5 to 3.0 MHz are generally used for US hyperthermia. Since fat and muscle have similar acoustic impedances, the reflections between soft tissue structures are minimum in contrast to those that occur in EM heating. However, due to the large impedance mismatches at tissue/air or tissue/bone interfaces, clinical applications of US at various sites are limited.

B. Clinical Applications

1. Interstitial Hyperthermia

Interstitial techniques for radiation implants as primary or boost treatments have been practiced successfully by radiation oncologists for many years. Once hyperthermia was known to be cytotoxic and synergistic with radiation, it was natural to consider the combination with conventional interstitial radioactive implantation.[11,67] Other advantages of this technique include better control of heat distribution within the tumor as compared with external hyperthermia, and ability to spare overlying normal tissue, particularly the skin.

a. Local current fields (LCF) techniques:

At the City of Hope National Medical Center interstitial hyperthermia equipment utilizing 0.5 MHz RF current was used for Phase I and II clinical trials. Most of the patients treated had advanced primary breast and cervical cancers which were considered uncontrollable by conventional methods.[67] A number of technical improvements have occurred in the last few years in the development of the LCF unit. Multiple thermocouples can be built into a sensor probe so that up to five measurement points can be made. The aggregate temperature information gives an excellent representation of the temperature distribution across a plane or volume of tissue. Our present unit (Oncotherm LCF 2032) can record up to 32 temperature readings, has 20 electrode pairs and is able to display simultaneously these temperatures on a video terminal. Feedback control of RF power and current dwell time (period when current flows) across the selected needle pairs can be adjusted by computer, based on the temperature monitored by a reference sensor. Current research involves the refinement and clinical testing of a multi-point feedback (MPF) system. The MPF system provides adjustments of individual dwell time across each needle pair, and theoretically should provide more homogeneous temperature distribution than single point control can achieve.

b. Microwave technique

Small MW antennae inserted into hollow tubing can produce satisfactory heating patterns between 300 MHz and 2450 MHz. A common frequency used in the United States is 915 MHz. A small coaxial antenna can irradiate a volume of approximate 60 cc. With a multi-node coaxial antenna, the length of the heating pattern can be extended to about 10 cm in a 3 node antenna. Strohbehn has calculated the isotherms for an array of antennas, taking into account the absorbed power in tissues and relating that to the bio-heat equation.[61] Assumptions with or without blood flow were included in the theoretical modeling. As in LCF hyperthermia, the control of MW power radiating from these antennas is important in order to achieve homogeneous heating. Multiple point feedback control would be important as well. It was found that the heating pattern of multiple dipole antenna array is dependent on insertion depth.[6] A fixed hot spot relative to the surface of the body was observed. This characteristic limits its use for treatment. Various other forms of antenna design have been developed to minimize this problem.[31,45,54]

c. Ferromagnetic seed implants

Burton et al.[4] used thermally self-regulating implants for the production of brain lesions. This technique is applicable for delivering thermal energy to deep-seated tumors. When exposed to magnetic fields, the implants absorb power and become heated, but when a critical temperature (Curie point) is reached, the implants become non-ferromagnetic and no longer produce heat.[58] The surrounding tissues are heated by thermal conduction. One can compensate for the influence of blood flow and tissue inhomogeneities of the tumor, which may affect the temperature distribution, by the self-regulation of the implants, thus maintaining a temperature close to the Curie point. The rate of energy absorption by the implant is strongly dependent on the permeability of the material, the frequency of the magnetic field, the implant diameter, and the orientation of the implants with respect to the magnetic fields. Since the temperature of the tissues between the implants is dependent on thermal diffusion, the spacing between the implants must be small enough to achieve therapeutic temperature. In animal studies the spacing was 1 cm or less. In areas with high blood flow, the spacing may need to be reduced. This technique is being clinically tested at the University of Arizona, University of California San Francisco, and City of Hope National Medical Center.

d. Intracavitary/intraluminal hyperthermia

Certain tumor sites at hollow viscera or cavities may be treated with these techniques: 1) gastrointestinal (esophagus, rectum); 2) gynecological (vaginal, cervix, uterus); 3) genitourinary (prostate, bladder); 4) pulmonary (trachea, bronchus); and 5) miscellaneous sites where it is technically feasible to insert such applicators.[69] MW and RF applicators are most commonly used for this application.[38,40] Recently US intracavitary applicators also have been developed.[14]

2. Local External Hyperthermia

Heating of small volume tumors, usually up to 50 $cm^2$ in area and up to 4 cm deep, located near the surface of the body can be quite easily achieved. The majority of studies involve the use of MW, usually at 915 MHz. The others involve US (see below). If there is no evidence of superficial tumor, skin cooling is employed. Engineering development has been concentrated on the design of new MW applicators.[25] Most of them are dielectrically-loaded, with water bolus for surface cooling. Low profile, lightweight and easy to use microstrip applicators have been developed. To reduce the applicator size and weight, high permittivity dielectric material, electric wall boundary, and magnetic material have been used. For large treatment areas such as recurrent breast cancer, a MW blanket developed by the Stanford group and the current sheet applicator described by Hand are being tested (personal communication).

3. Regional Hyperthermia

To heat deep-seated tumors noninvasively is difficult. RF energy can be deposited into the center of the body but a large region is affected. Differential increases of blood flow in the normal and tumor tissues may result in higher temperature in tumor than normal organs. However, this temperature differential cannot be assured. The annular phased array systems (APAS) made by BSD Corporation (Salt Lake City, Utah) and the Magnetrode made by Henry Medical Electronics, Inc. (Los Angeles, California) have been evaluated by several institutions.[51,53,55]

The APAS consists of four sets of dual radiating apertures which operate in transverse electromagnetic (TEM) mode over the frequency range of 50 to 110 MHz.[63] The maximum power is 2 kW. The patient is placed inside the octagonal aperture. Distilled water bolus bags fill the air space within the aperture to improve energy coupling, reduce stray radiation and provide surface cooling. Due to Federal Communication Commission regulations, the system has to be operated inside a shielded room. In clinical evaluations for achieving therapeutic temperatures in the abdomen and pelvis, the safety, feasibility, and limitations of the APAS have been explored. Treatment of the chest can cause excessive systemic heating and also unusual hot spots in the head and neck due to spurious fields. In addition, the invasive thermometry currently employed can puncture lungs or major vessels. More pilot studies are needed before this modality can be evaluated by randomized clinical trials. A new version of the APAS which is more compact and has amplitude and phase control is being clinically evaluated.

Several experimental designs for deep heating have been described. The basic principle is to generate an electrical field parallel to the body axis. Helical coils have been shown to produce transversely uniform heating in cylindrical phantoms.[46] A toroidal transformer can provide uniform heating in a cylindrical structure inside the toroidal ring.[62] Ring capacitors have shown good heating capabilities in small cylindrical phantoms such as arms and lower legs.[66] However, there are hot spots on the body trunk surface near the applicator. The coaxial TEM system produces a circumferential electrical field around the body.[12] The patient is immersed in water inside a hollow inner conductor. Phantom (simulated tissue) tests have shown good center heating. Harrison and Storm[27] modified the TEM design into a loosely coupled applicator. In this design, the body is at a distance from the electrode plates (about 10 cm), the primary electrical field in the body is parallel to the body surface. While good deep heating was measured, no excessive fat heating was seen in either pigs or human patients.

4. Phased-Array Hyperthermia

An array of applicators with variations in phase, frequency amplitude, and orientation of the applied fields can add more dimensions to controlling the heating patterns during the treatment. The APAS radiates 16 RF fields in phase toward the patient. When the electrical field in the tissue is increased by a factor of N fold, the specific absorption rate (SAR), and therefore the temperature rise, is $N^2$ times higher. By changing the phase and amplitude of the applied fields incident from different directions, the SAR pattern can be controlled. Theoretically it is possible to achieve the temperature elevation at the tumor only. A few researchers have shown theoretically and/or experimentally that the focusing effect can be achieved in models.[19,39,41] A number of papers on this subject were published in a special issue of the IEEE Transaction on Microwave Theory and Techniques (May 1986).

5. Ultrasound Hyperthermia

For clinical applications, various US transducers have been designed. They can be classified as single vs. multiple, planar vs. focused, and stationary vs scanned transducers. Corry et al.[10] have reviewed the clinical applications of single, planar, and stationary transducers on superficial tumors and Fessenden et al.[17] described their experience with multi-transducer system for deep heating. The variables for treatment are the size, frequency, orientation of the transducer, applied power, and the temperature of the cooling bolus. Because of the complicated near field, there are localized hot spots near the surface. Bone heating often occurred when using these transducers. Mechanically scanning the transducers over the skin is also used to distribute the power over a large area and smooth out the peak heating. Alternatively, multiple transducers with individual power control can produce desired power deposition patterns.[64] A commercial system based on the multi-transducer array is marketed by the Labthermics Technologies, Inc.

To avoid heating superficial tissues, multiple planar transducers can be mounted on a spherical shell to produce focused heating for deep-seated tumors. Lele[37] used multiple transducers but at high frequencies. A diagnostic transducer mounted at the center of an array of six therapeutic transducers was used to provide depth information so that focused heating could be achieved by controlling the tilt of these transducers. The focusing mode concentrates the US energy into a small focal region, typically between 0.5 to 5 mm in diameter and

2.5 to 25 mm in length. A diagnostic transducer was used to constantly monitor the location of the focal point relative to blood vessels, bone, and air pockets. The focused beam was also computer controlled to scan the tumor in a specified trajectory. Therefore, focusing, scanning, optimization of frequency and aperture size as well as spatial and temporal intensity modulation are essential for uniform or controlled distribution of hyperthermia without overheating the surrounding normal tissues. Hynynen et al.[30] have been testing a similar system on animals. Clinical evaluations are in progress by the same group. Roemer[48] cautioned that while scanned focused US has great potential and is being implemented by several institutions, much work remained to be done by mechanical engineers to optimize the engineering designs of clinical systems for actual patient treatments.

## TEMPERATURE MONITORING

The therapeutic temperature range for cancer treatment is very narrow, with 42.5°C as the minimum temperature[13] while 44°C and above is damaging for normal cells. It is very important to monitor the tissue temperature accurately. It would be ideal if three dimensional temperature information could be obtained noninvasively. However, at the present time, only invasive techniques with limited measurement points are practical.

### A. Invasive Thermometry

Cetas[5] gave excellent reviews on thermometry. Sensors such as thermocouples and thermistors are good for general temperature measurements because of their sensitivity, accuracy, stability, response time, and size. However, they must be used with care for accurate temperature monitoring during hyperthermia treatments. This is due to interactions of the sensor probes with the heating modality. In EM fields, metallic probes can cause three problems: perturbation of the SAR pattern, intensification of field at the tip of the metallic probe, and interference of electronic readout. Proper shielding and filtering can minimize interference. Orienting the probe perpendicular to the applied field can also reduce the perturbation and intensification. However, it is very difficult to keep the probe perpendicular to the electric field during clinical treatment.

Several nonperturbing temperature probes have been developed during the last 15 years. The probes have utilized liquid crystal,[50] small thermistors connected with high resistance leads,[2] gallium arsenide chips at the end of optical fibers,[9] and fluorescent decay time of a rare earth phosphor.[68] There are reports and presentations of other types of probes, (eg, birefringence of crystal, MW integrated circuit, photo luminescence, viscosity of fluid, and non- metallic thermocouple). However, they are not commercially available or have not gained wide acceptance in this country.

Since the heating may vary considerably over a three-dimensional volume, a single point measurement with the above described probes will not define the temperature pattern. Multiple probes can provide more information about the heating pattern.[65] An alternative technique is to move the temperature probe with a stepping motor.[16,20]

For US hyperthermia, small size needle thermocouples are used to minimize perturbation of the field. However, the heating of the probe sheath can cause considerable measurement errors. Kuhn and Christensen[34] found that polyethylene catheter tubing results in much smaller artifacts than Teflon in planar fields. In focused heating, shearing forces can cause large artifacts due to viscosity heating which is associated with the relative motion between fluid or tissue and the probe. A practical method to read the correct temperature is to turn the power off momentarily and extrapolate the slow decaying temperature back to the onset of the US power.

B. Non-invasive Thermometry[24]

Thermographic techniques have been used to sense the infrared portion of black body radiation. Since the penetration depth of this frequency band is very small, it can only be used for surface temperature measurements. Its application in clinical hyperthermia is limited to skin temperature measurements. However, this technique has been widely used in the laboratory for energy deposition studies in phantoms.[21]

Microwave radiometry is based on the MW energy radiated by the warm body. For higher temperatures under the skin at depths of a few centimeters, MW radiation in the range of 0.5 to 2 GHz can be detected. Since tumors are usually about 1°C warmer than surrounding normal tissue, this technique has been used for cancer detection. The feasibility of using MW radiometry for temperature control during hyperthermia has been studied and found to be possible based on a one-dimensional simple model.[44] However, it will need more research before it is clinically applicable.

Ultrasonic radiometry has much better depth penetration than MW radiometry due to the shorter wavelengths in tissue. Since the velocity of US propagation in tissue is proportional to temperature, time of flight (US propagation time) tomography has been explored. Although many tissues show similar slopes with increasing sound velocity at higher temperatures, fat shows negative slopes in certain temperature ranges. Due to the complex composition of tissues, the negative slopes make the interpretation of the data difficult.

It is also possible to use the changes in CT scans and electrical conductivity at different temperatures as noninvasive methods for temperature monitoring. However, none of these are close to being useful in the clinic. The method of using nuclear magnetic resonance (NMR) for temperature measurement has become a popular topic recently, especially in Japan. Since the decay of relaxation time of proton spin orientation with respect to spin lattices (T1) is a function of temperature, the localized temperature can be determined by measuring the tissue relaxation in an NMR system. However, there are two problems. The first is the heating of tissues from the strong magnetic fields, because the exposure time in the NMR system will be in the order of several minutes to obtain the necessary resolution. The second is that the presence of the large metallic hyperthermia system is not compatible with the metal sensitive NMR system. Apparently, the use of noninvasive methods for clinical application will still take years of research.

## THERMAL DOSIMETRY

### A. Four Areas of Thermal Dosimetry

Roemer and Cetas[49] subdivided the thermal dosimetry into four areas: 1) comparative (equipment evaluation); 2) prospective (treatment planning); 3) concurrent (treatment control); and 4) retrospective (treatment evaluation). Bowman[3] has also reviewed this subject.

The purpose of comparative thermal dosimetry is to compare the ability of different heating modalities or different configurations for treating a particular tumor. This is an equipment evaluation and selection process.

The prospective dosimetry is about individual treatment planning. Detailed power distribution, tumor size, location, blood flow, etc., are needed to determine a complete temperature distribution. The goal of simulation is to optimize the proposed thermal treatment by determining the power deposition parameters which maximize the therapeutic effects in tumors and minimize normal tissue damage and patient discomfort.

Concurrent thermal dosimetry consists of monitoring of power and temperatures during treatment and, through computer feedback control, optimizing the treatment. At the present time, control based on discrete point temperature measurements are used. It will be the goal of concurrent thermal dosimetry to control the complete temperature fields in an optimal manner.

Mathematical models of treatment process and advanced control algorithms will be needed. The retrospective thermal dosimetry involves post-treatment evaluation of a completed therapy. Temperature distribution data and thermal dose calculation can be obtained. These data are needed for clinical evaluation of the efficacy of the treatment modality or protocol. To date, most of the hyperthermia work has been on comparative thermal dosimetry. A few techniques on prospective thermal dosimetry have appeared in the literature. Concurrent and retrospective thermal dosimetry are just emerging. The major problem is the lack of blood flow data for the prospective thermal dosimetry. The other problem is the computer time required for calculations. With the rapid progress in computer sciences, that may not be a problem in the future.

### B. Heating Patterns

For treatment planning, the heating patterns of a particular modality are needed before the treatment. The heating patterns can be predicted by numerical modelling. After the pattern is found, the next step is to use the heat transfer equation to calculate the temperature distribution. Numerous papers on mathematical modeling have been published.[15,60] Numerical techniques, such as finite differences, finite element, moment, and finite-difference time domain methods, have been used for numerical modelling.

For heating pattern measurements in phantoms heated by EM energy, measurements by temperature sensors and thermographic technique have been described in the literature.[8] The absorption of EM energy in tissue is determined by many factors. Among them are amplitude, frequency, duration and polarization of applied fields; dielectric properties, size, geometry, and depth of tissues; size and shape of applicator; as well as spacing or coupling between tissue and applicator. Therefore, for effective treatment in hyperthermia, in order to apply penetrating MW or RF energy for tissue heating, measurements of energy absorption in tissues are required. Phantoms with dielectric properties and geometry similar to those of real tissues have been used for EM heating studies.

## DISCUSSIONS AND FUTURE DEVELOPMENTS

In reviewing 15 years of efforts in clinical hyperthermia it is clear that much progress has been made. Improvements of treatment equipment have been made based on medical demands, and the technological progress, in turn, fulfilled many of the clinical requirements. It is unlikely that a single piece of equipment will satisfy all the clinical requirements for patient treatments. There are advantages in having different types of equipment to treat different tumors. Depending on the location and vascularity of the tumor and adjacent tissues, and the general physical condition of the patient, a physician may wish to have the option of choosing the most appropriate equipment. Hence, a Comprehensive Hyperthermia Clinic, where a patient can be evaluated and treated with the optimal techniques available, is necessary.

In the authors' opinions, the equipment in a Comprehensive Hyperthermia Clinic would include the following:

1. Local external MW operating at frequencies of 2450 MHz, 915 MHz and perhaps 433 MHz with various sizes of applicators

2. MW interstitial antennae suitable for a large volume implant which can deliver heat in an array configuration

3. RF local current field interstitial implant capability with multi-point feedback control of power based on temperature measurements

4. External MW and/or RF annular phased array for treatment of deeper tumors

5. US, single or multiple transducers with scan and focusing capabilities for treatment of superficial or deep tumors in suitable locations.

6. MW and US applicators for intracavitary or intraluminal hyperthermia at anatomical sites such as the vagina, rectum, and esophagus; and

7. Ferromagnetic seed implant hyperthermia which allows for temperature control and multiple heating possibilities.

Naturally there will be other equipment suitable for treating certain types of tumors. They are either currently under evaluation, or will be developed in response to clinical impetus. For example, the whole body hyperthermia technique of a combination of convective heating and control of body heat loss appears to be promising.[47] After further testing of its toxicity, and proving the efficacy of combining hyperthermia and chemotherapy, one might wish to include such a piece of equipment in a comprehensive hyperthermia clinic, for those patients with metastases.

In a Comprehensive Hyperthermia Clinic, a team approach will be needed to care for patients. The Hyperthermia Treatment Team should consist of one or more physicians who have special training and a primary interest in this field, and who offer consultations, prescribe, and supervise hyperthermia treatments of various kinds. There should be at least one oncology nurse to assist the physicians in the care of these patients before, during and after hyperthermia, and to provide supportive care during follow-up examinations. Technologists who operate the equipment require special didactic and practical training, and can be recruited from similar professional backgrounds. Medical physicists also play a key role in treatment planning and delivery, maintenance of the equipment, and in ongoing quality assurance programs. Biomedical engineers are an indispensable and integral part of the treatment team as well, ensuring a smooth and trouble-free operation. Their contributions have been well highlighted in the foregoing discussions. This interdisciplinary team is vital for the success of a hyperthermia program.

Future developments in the field of hyperthermia cannot be fully anticipated as the foundation of this field is still being laid by present researchers. These developments will respond to the demands of the medical environment, whether they will be clinical trial results, discoveries in other branches of medicine, or socio-economical trends. If one were to predict the accomplishments in hyperthermia in the foreseeable future, one might see much more automation and robotics, more rapid and fuller computer controls, noninvasive thermometry, dynamic treatment planning with immediate feedback, and a comprehensive selection of applicators and accessories of advanced designs to suit specific situations and clinical presentations.

Even though the ultimate role of hyperthermia in cancer therapy cannot be ascertained at this time, biological studies point to expectations of successful clinical trials. Technological shortcomings of today are being overcome rapidly. The entry of commercial interests in the form of venture capitalists and entrepreneurs certainly gives great impetus to the continuing progressions described above. The past 15 years has seen the revival of hyperthermia based on biology, and the move on to clinical trials. In the next decade developments are likely to include:

1. Better tools and detailed dosimetry for site specific treatment

2. Better integrated systems

3. Multi-parameter control with faster computer

4. Better temperature monitoring and, it is hoped, noninvasive

5. Practical three dimensional treatment planning

6. Better treatment scheduling

7. Improved thermal dose[52] which is adopted in clinics

8. Standardized data format for documentation and evaluation

9. Advances in quality assurance[26,56]

10. More intracavitary treatment

11. More interstitial heating

12. Whole body hyperthermia in combination with local hyperthermia to improve temperature uniformity

13. More local and whole body hyperthermia in combination with chemotherapy

14. Other medical applications such as a combination with tumor necrosis factor and interferon, to reduce multi-drug resistance, and to treat benign conditions such as hemorrhoids, sinusitis and prostatic hypertrophy; and

15. Japanese contributions in this field will continue to increase.

It is hoped that hyperthermia will be recognized as a legitimate division of medical science, and be permitted to mature as an important modality in cancer therapy.

## ACKNOWLEDGMENT

This work was supported in part by NCI under Grant CA33572.

## References

1. Hyperthermia in Cancer Treatment. (Eds. LJ Anghileri, J Robert), CRC Press, Boca Raton, FL, 1986.

2. Bowman RR. A probe for measuring temperature in heated material. IEEE Trans Microwave Theory Tech 1976; 24:43-45.

3. Bowman HF. Thermal dosimetry. In: Hyperthermia in Cancer Treatment (Eds. LJ Anghileri, J Robert), CRC Press, Boca Raton, FL. 1986; 2:155-178.

4. Burton CV, Hill M, Walker AE. The RF thermal seed - A thermally self-regulating implant for the production of brain lesions. IEEE Trans Biomed Eng 1971; 18:104-109.

5. Cetas TC. Thermometry. In: Therapeutic Heat and Cold, 3rd edition (Ed. JF Lehmann), William and Wilkins, Baltimore, MD. 1982;35-69.

6. Chan KW, Chou CK, McDougall JA, Luk KH, Vora NL, Forell BW. Changes in heating patterns of interstitial microwave antenna arrays at different insertion depths. Int J Hyper 1989; 5(4):499-507.

7. Chou CK. Application of electromagnetic energy in cancer treatment. IEEE Trans Instru Meas 1988; 37(4):547-551.

8. Chou CK. Electromagnetic energy deposition patterns in phantoms. In: Biological, Physical, and Clinical Aspects of Hyperthermia (Eds. B Paliwal, FW Hetzel, MW Dewhirst), American Institute of Physics, AAPM Monograph 16, New York. 1988;132-151.

9. Christensen DA. A new nonperturbing temperature probe using semiconductor band edge shift. J Bioeng 1977; 1:541-555.

10. Corry Pm, Spanos WJ, Tilchen EJ, Barlogie B, Barke HT,Armour EO. Combined ultrasound and radiation therapy treatment of human superficial tumors. Radiology 1982; 145:165-169.

11. Cosset JM, Dutriex J, Haie C, Gerbaulet A, Janoray P, Dewar JA. Interstitial thermoradiotherapy: A technical and clinical study of 29 implantations performed at the Institute Gustave-Roussy. Int J Hyper 1985; 1:3-13.

12. De Leeuw AA, Lagendijk JJ. Design of a clinical deep-body hyperthermia system based on the 'coaxial TEM' applicator. Int J Hyper 1987; 3(5):413-421.

13. Dewhirst MW, Sim DA, Sapareto SA, Connor WG. Importance of minimum temperature in determining early and long-term response of spontaneous canine and feline tumors to heat and radiation. Cancer Res 1984; 44:43-50.

14. Diederich CJ, Hynynen K. Induction of hyperthermia using an intracavitary multielement ultrasonic applicator. IEEE Trans Biomed Engr 1989; 36(4):432-438.

15. Durney CH. Calculation of electromagnetic power deposition. In: Physics and Technology of Hyperthermia. (Eds. SB Field, C Franconi), Martinus-Nijhoff Dordricht Publisher, Amsterdam, Netherlands. 1987;152-158.

16. Engler MJ, Dewhirst MW, Winget JM, Oleson JR. Automated temperature scanning for hyperthermia treatment monitoring. Int J Radiat Oncol Biol Phys 1987; 13:1377-1382.

17. Fessenden P, Lee ER, Anderson TL, Strohbehn JW, Meyer JD, Samulski TV, Marmor TB. Experience with multi-transducer ultrasound system for localized hyperthermia of deep tissues. IEEE Trans Biomed Engr 1984; 31:126-135.

18. Physics and Technology of Hyperthermia. (Eds. SB Field, C Franconi), Martinus-Nijhoff Dordricht Publisher, Amsterdam Netherlands, 1987.

19. Gee W, Lee SW, Bong NK, Cain CA, Mittra R, Magin RL. Focused array hyperthermia applicator: theory and experiment. IEEE Trans Biomed Engr 1984; 31:38-46.

20. Gibbs FA. Thermal mapping in experimental cancer treatment with hyperthermia: description and use of a semi-automatic system. Int J Rad Oncol Phys 1983; 9:1057-1063.

21. Guy AW. Analysis of electromagnetic fields induced in biological tissue by thermographic studies in equivalent phantom models. IEEE Trans Microwave Theory Tech 1971; 19:205-215.

22. Guy AW, Chou CK. Physical aspects of localized heating by radiowaves and microwaves. In: Hyperthermia in Cancer Therapy (Ed. FK Storm), GK Hall Medical Publishers, Boston, MA. 1983;279-304.

23. Hahn GM. Hyperthermia and Cancer. Plenum Press, New York, 1982.

24. Hahn GM. Non-invasive methods of temperature measurement. In: Physics and Technology of Hyperthermia. (Eds. SB Field, C Franconi), Martinus-Nijhoff Dordricht Publisher, Amsterdam, Netherlands. 1987;509-516.

25. Physical Techniques in Clinical Hyperthermia. (Eds. JW Hand, RJ James), Research Study Press, Latchworth, Hartfordshire, England, 1986.

26. Hand JW, Lagendijk JJW, Anderson JB, Bolomey JC. Quality assurance guidelines for ESHO protocols. Int J Hyper 1989; 5(4):421-428.

27. Harrison WH, Storm FK. Loosely coupled coaxial TEM applicators for deep heating. Int J Hyper 1989; 5(3):297-306.

28. Hetzel FW, Mattiello J. Interactions of hyperthermia with other modalities. In: Biological, Physical and Clinical Aspects of Hyperthermia, (Eds. BR Paliwal, FW Hetzel, MW Dewhirst), American Institute of Physics, AAPM Monograph 16, New York. 1988;30-56.

29. Hyperthermia and Cancer: Human Clinical Trial Experience. (Ed. NB Hornback) CRC Press, Boca Raton, FL, 1984.

30. Hynynen K, Roemer R, Anhalt D, Johnson C, Xu ZX, Swindell W, Cetas T. A scanned, focused, multiple transducer ultrasonic system for localized hyperthermia treatments. Int J Hyper 1987; 3(1): 21-35.

31. Iskander MF, Tumeh AM. Design optimization of interstitial antennas. IEEE Trans Biomed Engr 1989; 36(2):238-246.

32. Kato H, Hiraoka M, Nakajima T, Ishida T. Deep-heating characteristics of an RF capacitive heating device. Int J Hyper 1985; 1(1):15-28.

33. Koga S (Ed.). Hyperthermia Oncology in Japan '87. Imai Printing and Publishing, Yonago, Japan, 1988.

34. Kuhn PK, Christensen DA. Influence of temperature prove sheathing material during ultrasound heating. IEEE Trans Biomed Engr 1986; 3:536-538.

35. Leeper D. Molecular and cellular mechanisms of hyperthermia alone or combined with other modalities. In: Hyperthermia Oncology 1984 (Ed. J Overgaard), Taylor & Francis, London. 1985;9-40.

36. Lehmann JF (Ed), Therapeutic Heat and Cold, 3rd edition. Williams and Wilkins, Baltimore, MD, 1982.

37. Lele PP. Scanned, focused ultrasound for controlled localize heating of deep tumors. In: Hyperthermia in Cancer Treatment (Eds. LJ Anghileri, J Robert), CRC Press, Boca Raton, FL. 1986;75-98.

38. Li DJ, Luk KH, Jiang HB, Chou CK, Hwang GZ. Design and thermometry of an intracavitary microwave applicator suitable for treatment of vaginal and rectal cancers. Int J Rad Oncol Biol Phys 1984; 10:2155-2162.

39. Loane J, Ling H, Wang BF, Lee SW. Experimental investigation of a retro-focusing microwave hyperthermia applicator: conjugate-field matching scheme. IEEE Trans Microwave Theory Tech 1986; 34:490-494.

40. Luk KH, Jiang HB, Chou CK. SAR patterns of a helical microwave intracavitary applicator. In: Hyperthermia Oncology 1984 (Ed. J Overggard), Taylor and Francis, London. 1984; 1:591-594.

41. Magin RL, Peterson AF. Noninvasive microwave phased arrays for local hyperthermia: A review. Int J Hyper 1989; 5(4):429-450.

42. Overggard J (Ed.). Hyperthermic Oncology 1984; Volume 2. Review Lectures, Symposium Summaries and Workshop Summaries. Taylor & Francis, London, 1985.

43. Paliwal BR, Hetzel FW, Dewhirst MW (Eds.). Biological, Physical and Clinical Aspects of Hyperthermia. American Institute of Physics, AAPM Monograph 16. New York, 1988.

44. Plancot M, Prevost B, Chive M, Fabre JJ, Ledee R, Giaux G. A new method for thermal dosimetry in microwave hyperthermia using microwave radiometry for temperature control. Int J Hyper 1987; 3(1):9-19.

45. Prevost B., Roos D, Hugander A. Design and test of microwave interstitial applicators with improved longitudinal heating pattern. Int J Hyper 1988; 4(6):609-615.

46. Ruggera PS, Kantor G. Development of a family of RF helical coil applicators which produce transversely uniform axially distributed heating in cylindrical fat-muscle phantoms. IEEE Trans Biomed Engr 1984; 31:98-106.

47. Robins HI, Neville AJ. Biology and methodology of whole body hyperthermia. In: Hyperthermia in Cancer Treatment (Eds. LJ Anghileri, J Robert), CRC Press, Boca Raton, FL. 1986; 3:183-206.

48. Roemer RB. Ultrasonic heating techniques. In: Physics and Technology of Hyperthermia (Eds. C Franconi, S Field) Martinus Nijhoff Dordricht Publisher, Amsterdam, Netherlands: 1987;390-402.

49. Roemer RB, Cetas TC. Applications of bioheat transfer simulations in hyperthermia. Cancer Res (Supply) 1984; 44:4788S-4798S.

50. Rozzell TC, Johnson CC, Durney DH, Lords JL, Olsen RG. A non-perturbing temperature sensor for measurements in electromagnetic fields. J Microwave Power 1974; 9:241-248.

51. Samulski TV, Kapp DS, Fessenden P, Lohrbach A. Heating deep seated eccentrically located tumors with an annular phased array system: a comparative clinical study using two annular array operating configurations. Int J Radiat Oncol Biol Phys. 1987; 13(1): 83-94.

52. Sapareto S. Thermal Dose. In: Biological, Physical and Clinical Aspects of Hyperthermia (Eds. BR Paliwal, FW Hetzel, MW Dewhirst), American Institute of Physics, 1988, AAPM Monograph, New York: 16;57-70.

53. Sapozink MD, Gibbs FA Jr, Thompson JW, Eltringham JR, Stewart JR. A comparison of deep regional hyperthermia from an annular array and concentric coil in the same patients. Int J Rad Oncol Phys 1985; 11:179-190.

54. Satoh T, Stauffer PR. Implantable helical coil microwave antenna for interstitial hyperthermia. Int J Hyper 1988; 4(5):497-512.

55. Shimm DS, Cetas TC, Oleson JR, Gross ER, Buechler DN, Fletcher AM, Dean SE. Regional hyperthermia for deep-seated malignancies using the BSD annular array. Int J Hyper 1988; 4(2):159-170.

56. Shrivastava P, Luk K, Oleson J, Dewhirst M, Pajak T, Paliwal B, Perez C, Sapareto S, Saylor T, Steeves R. Hyperthermia quality assurance guidelines. Int J Radiat Oncol Biol Phys 1989; 16:571-587.

57. Song CW. Blood flow in tumors and normal tissues in hyperthermia. In: Hyperthermia in Cancer Therapy (Ed. FK Storm), GK Hall Medical Publisher, Boston: 1983;187-206.

58. Stauffer PR, Cetas TC, Fletcher AM, DeYoung DW, Dewhirst MW., Oleson JR, Roemer RB. Observations on the use of ferromagnetic implants for inducing hyperthermia. IEEE Trans Biomed Engr 1984; 31:76-90.

59. Storm FK (Ed), Hyperthermia in Cancer Therapy. GK Hall Medical Publisher, Boston: 1983.

60. Strohbehn JW, Roemer RB. A survey of computer simulations of hyperthermia treatments. IEEE Trans Biomed Eng 1984; 31:136-149.

61. Strohbehn JW. Interstitial techniques for hyperthermia. In: Physics and Technology of Hyperthermia (Eds. SB Field, C Franconi), Martinus-Nijhoff Dordricht Publisher, Amsterdam, Netherlands: 1987;211-39.

62. Tiberio CA, Raganella L, Banci G, Franconi C. The RF toroidal transformer as a heat delivery system for regional and focused hyperthermia. IEEE Trans Biomed Eng 1988; 35(12):1077-1085.

63. Turner PF. Regional hyperthermia with an annular phased array. IEEE Trans Biomed Engr 1984; 31:106-114.

64. Underwood HR, Burdette EC, Ocheltree KB, Magin RL. A multi-element ultrasonic hyperthermia applicator with independent element control. Int J Hyper 1987; 3(3): 257-267.

65. Vanguine VA, Christensen DA, Lindley JH, Walston TE. Multiple sensor optical thermometry system for application in clinical hyperthermia. IEEE Trans Biomed Engr 1984; 31:168-172.

66. Van Rhoon GC, Visser AG, Van Den Berg PM, Reinhold HS. Evaluation of ring capacitor plates for regional deep heating. Int J Hyper 1988;4(2):133-1 42.

67. Vora N, Shaw S, Forell B, Desai K, Archambeau J, Pezner RD, Lipsett JA, Covell J. Primary radiation combined with hyperthermia for advanced (Stage III-IV) and inflammatory carcinoma of breast. Endo Hypertherm Oncol 1986; 2:101-106.

68. Wickersheim KA, Sun MH. Fiberoptic thermometry and its applications. J. Microwave Power and EM Energy 1987; 22(2):85-94.

69. Wong JYC, Vora NL, Chou CK, McDougall JA, Chan KW, Findley DO, Forell BW, Luk KH, Philbcn VJ, Beatty JD. Intracatheter hyperthermia and Iridium-192 radiotherapy in the treatment of bile duct carcinoma. Int J Radiat Oncol Biol Phys 1988; 14(2): 353-359.

# CURRENT AND EMERGING MEDICAL APPLICATIONS OF MICROWAVE AND RADIO FREQUENCY ENERGY IN THE TREATMENT OF CANCER

Carl H. Sutton, M.D.

## INTRODUCTION

The use of hyperthermia for the treatment of cancer by physicians in the modern era began with the observation by Busch[5] in 1866 that a facial sarcoma in one of his patients regressed and resolved completely following two sustained febrile episodes of erysipelas, a streptococcal infection. Two years after the high fevers, the patient remained tumor-free. In 1893, Coley[8] induced systemic hyperthermia in cancer patients, using injections of erysipelas initially and a mixed bacterial toxin later.[9] These experiments followed his observation of the complete disappearance of a recurrent sarcoma in a patient who also developed an intercurrent streptococcal infection with prolonged high fever and was found to be completely tumor-free seven years later. His first ten cancer patients were treated with toxin-induced hyperthermia that ranged from 38 to 40.2°C.[8] With Coley's studies, hyperthermia first became a part of the oncologist's armamentarium for the treatment of cancer.[10]

In 1898, Westermark conducted hot water through metallic tubes placed in proximity to the uterus in patients with endometrial carcinoma. He was able to produce selective necrosis in the tumor without damage to normal uterine tissue.[46] In 1935, Warren devised a heating cabinet with which he produced whole body hyperthermia in patients with disseminated cancer.[45] His studies followed observations of sarcoma resorption following systemic hyperthermia in a patient with both systemic and cutaneous metastases. Patients in his series were treated with systemic hyperthermia at 41.5°C for five hours with encouraging results.

Shortwave and microwave diathermy have been used for more than forty years in departments of physical medicine and rehabilitation to produce deep heating of muscle masses for the relief of spasm, for the reduction of tissue inflammation and to increase circulation locally in various areas of the body.[24] Schliephake[32] in Germany described the use of shortwave therapy in physical medicine in 1938. Soon after the introduction of radar during World War II, Krusen[20] utilized microwave diathermy for deep heating of localized areas. Since then, microwave heating has remained a standard technique in the field of physical medicine. During the last fifteen years, microwave energy has become the most widely utilized modality for producing localized adjuvant hyperthermia in the treatment of cancer.

## RATIONALE

Biochemical and clinical studies by Cavaliere and his associates[6,7] beginning in 1967 demonstrated convincingly that there is a selective vulnerability of cancer cells to hyperthermia, as compared with their tissues of origin. In vitro and in vivo studies have also confirmed the differential thermal sensitivity between cancer cells and their tissues of origin.[15,18] This differential sensitivity can be utilized therapeutically in cancer patients by heating the neoplastic tissue to a tumoricidal level that can be tolerated by the host tissue in which it is growing.[18] The difference of thermal sensitivity between normal and neoplastic cells varies with individual histological types,[18,26] and determines the "therapeutic range" for treatment with hyperthermia in a specific tissue or organ.

It has also been established that hyperthermia produces sensitization of mammalian cells to ionizing radiation.[18] Hypoxic tumor cells are relatively radioresistant, but are sensitive to hyperthermia. The rationale for the increasing clinical use of hyperthermia in combination with ionizing radiation during the past fifteen years has been based in large part upon its radiosensitizing capability.[26] The increase in the thermal enhancement ratio (TER) is defined as the ratio of the effect of the combined therapeutic regimen compared with the effect of the radiotherapy alone. It has been demonstrated that the TER obtained when hyperthermia is added to ionizing radiation treatment is approximately 1.5 for a number of tumors.[18,26]

Another factor that has strong influence on cell survival is its position in the cell cycle at the time of exposure to radiotherapy. Cancer cells are significantly more radioresistant in the late S phase.[18] The influence of cell cycle position on the hyperthermic response indicates that there is a peak in hyperthermia resistance from late $G_1$ phase to early S phase, whereas cells in late S phase are extremely sensitive to hyperthermia. It is clear that the combination of hyperthermia with radiation therapy is likely to be more effective for killing cancer cells than is either agent alone.

## BIOLOGICAL EFFECTS

In addition to its antineoplastic effects, hyperthermia has a marked effect on local blood flow in the thermal field. It induces vasodilatation in tissues with moderate levels of temperature elevation (40-43°C).[34] As long as the blood remains at 37°C, this vasodilatation has a cooling effect on tissues, requiring a greater input of microwave or RF energy to maintain a constant level of hyperthermia during the entire treatment period. The capacity for vasodilatation is substantially greater in normal tissues than in most neoplasms, in which the circulation is predominantly composed of capillary vessels. Thus, reactive vasodilatation serves to protect normal tissues selectively. Above 45°C, however, blood vessels in some normal tissues, as well as in malignant tumors, can be damaged by prolonged heating and lead to edema, hemorrhage or infarction of the tissue supplied by those vessels.[34,40]

Unfortunately for the hyperthermic oncologist, a resistance to hyperthermia can develop in cancer cells after an initial treatment. This resistance has been termed 'thermal tolerance,'[17] and is accompanied by, or perhaps is secondary to, the development of 'heat-shock proteins'[23] within treated cells. Thermal tolerance has significance in determining the optimal timing for individual hyperthermia treatments. Since thermal tolerance in cancer cells subsides after a period of approximately 72 hours, it has become customary to treat patients with ionizing radiation and adjuvant hyperthermia twice a week or at intervals of no less than 72 hours.

## CLINICAL USE OF HYPERTHERMIA

Hyperthermia has been employed clinically as an adjuvant to ionizing radiation therapy, to chemotherapy or as a primary therapy for relatively heat-sensitive neoplasms. There are three important clinical uses of hyperthermia in combination with ionizing radiation: 1) to palliate recurrences, especially with limited doses of ionizing radiation after local treatment failures; 2) to reduce the incidence of therapeutic complications without compromising the overall cure rate by adding it to a less than curative dose of ionizing radiation; and 3) to increase the survival rate beyond levels achievable with 'curative' doses of radiation alone.

## DOSAGE DETERMINATION

The temperature achieved in a tumor and in the surrounding normal tissue is determined by several factors. First, heat absorption is greater if the tumor has a high water content. Second, heating is more easily achieved if the vascularity of the area is low, so that the blood flow exerts little cooling action. In this regard, it is recognized that most tumors have lower blood flow rates than their tissues of origin.[34] Third, tissues at the site of previous surgery or ionizing radiation therapy may be relatively hypovascular, with only capillary blood flow remaining in the residual fibrous scar tissue. The capacity for vasodilatation is very limited in capillary vessels, and necrosis of scarred normal tissue can be produced with levels of hyperthermia that are required to kill tumor cells contained within it.

It has been demonstrated that excessive heat (above 45°C) is cytotoxic to most mammalian cells.[18] In general, cancer cells are more sensitive to moderate levels of hyperthermia (41-44°C) than their tissues of origin. This difference in thermal sensitivity has been exploited by elevating the temperatures in neoplastic tissues to levels above the minimum effective dose (MED) that produces a tumoricidal effect, while remaining at a temperature level lower than the maximum dose tolerated (MTD) by the tissue which is host to the cancer cells. The difference between the MED and the MTD has been termed the "therapeutic range" for safe and effective medical applications of hyperthermic therapy. Similar concepts have been used traditionally in the field of cancer chemotherapy.

The therapeutic range for hyperthermia is frequently relatively narrow. Lele has summarized observations from the literature with respect to reported observations of the MTD, as assessed histologically, for several normal human and animal tissues, including central nervous system, liver, kidney, skeletal and cardiac muscle, and skin.[22] Damage in normal host tissues surrounding the cancer can result from excessive elevation of temperature levels. Brain and spinal cord tissue, for example, can be damaged permanently by temperature levels above 42.5°C for more than 10 min.[22] In order to prevent thermal injury, periods of prolonged heating at 42°C for 40 hours[38,39] or 72 hours[25] have been used for the treatment of intracranial malignant gliomas with safety for the patients, as well as with antineoplastic efficacy.

Because of the considerations outlined above, the standard prescription for a dose of hyperthermia as an adjuvant to ionizing radiation is 43°C for 1 hour.[35] A number of centers have adopted the concept of converting thermal exposure into equivalent minutes at 43°C, in order to calculate a thermal isoeffect dose.[35] It appears that the ideal therapeutic goal for the use of hyperthermia with ionizing radiation is to administer hyperthermia at a dose of 43°C for one hour twice each week for a minimum of four sessions, and, in some cases, as many as ten.

With respect to the timing of hyperthermia treatments, it is clear that hyperthermia given either before or after irradiation inhibits the repair of potentially lethal damage by cancer cells,[26] while hyperthermia following radiation is more effective than that given prior to treatment.[18] Dewhirst has demonstrated that, for temperatures above 43°C, the rate of cell kill increases by a factor of 2 for every 1°C rise in temperature.[14] In contrast, below 43°C, the rate of cell kill drops by a factor of 4 to 6 for every 1°C drop in temperature. He has suggested that the concept of an equivalent number of minutes at a standard temperature such as 43°C might constitute a unit which, if made standard, might permit comparison of thermal dose used in one patient with that in another, and at one treatment center with that at another.[14]

## CURRENT STATUS AND CLINICAL RESULTS

### A. Local Externally Applied Hyperthermia

Externally applied hyperthermia alone or as an adjuvant to radiation therapy or chemotherapy has been employed over the past decade. The most common frequencies for the delivery of hyperthermia with external electromagnetic (EM) fields are 2450, 915, and 433 MHz. The principal advantage of external fields is their non-invasive nature. Their principal disadvantages are a limited depth of penetration (3-4 cm) and heating patterns that are frequently non-uniform.[28] It appears that externally applied hyperthermia can facilitate the control of malignancies located in a superficial position within approximately 3 cm beneath the skin surface. It can provide palliation and, in some instances, prolongation of survival.[28] The use of hyperthermia for these lesions is relatively easy, as is thermometry during treatment.

Several sites and tumor types have been delineated in which significant clinical improvement has been achieved with adjuvant microwave hyperthermia. The first of these is breast cancer which has recurred in the chest wall or in regional lymph nodes.[28,29,41] When lesions of this type were treated with hyperthermia accompanied by low dose radiation therapy (less than 30 Gy), the complete response (CR) rate was nearly doubled, as compared with radiation therapy alone.[29] When full dose radiation therapy was accompanied by hyperthermia, higher local control rates up to 86 to 94% were obtained with the combined treatment, as compared with radiation therapy alone. This was also approximately double the local control rates seen with radiation therapy alone.[29,33] Second, locally advanced extension or invasion by breast cancer in the primary site has shown improvement in two-year recurrence-free survival rates.[19] Third, when cervical lymph node and subcutaneous metastases from head and neck cancer are treated with microwave hyperthermia and low doses of radiation therapy (20 to 40 Gy), an approximate doubling of CR rates has been reported.[29] Reduced CR rates were noted for larger tumors, however.[41] When patients were treated with full dose radiation therapy plus hyperthermia, higher local control rates were obtained.[2,29] In a randomized trial, full dose radiation therapy (60 to 70 Gy) in combination with hyperthermia was compared to radiation therapy alone. Complete response rates of 82% with hyperthermia and 37% without were noted at three months.[41] Fourth, in superficial malignant melanoma, numerous clinical trials have demonstrated improved CR rates for cutaneous, subcutaneous and peripheral lymph node metastases when the combination of radiation therapy and hyperthermia was compared with radiation therapy alone.[2,16,29] Isodose thermal enhancement ratios of 1.4 to 4.0 have been reported for melanoma.[29]

A combined protocol with a number of institutions collaborating through the Radiation Therapy Oncology Group (RTOG) was designed to evaluate acute or late toxicity of microwave hyperthermia (43°C for 45-60 min) for normal tissues.[33] In addition, the efficacy of definitive doses of radiation therapy at a tumor dose of 60 to 65 Gy was studied in 41 patients with advanced malignant lesions of the breast, head and neck and soft tissues. Necrosis of skin or subcutaneous tissue occurred within six months in 2% of the patients, while 12% experienced thermal blisters. When 14 patients who received the full course of therapy were followed up after six months, complete responses of tumor were observed in 51% of the patients.[33]

The use of externally applied radio frequency energy has been explored in efforts to heat large or deep-seated tumors with either capacitive or inductive techniques at frequencies of 6.0 to 27.12 MHz. Capacitive deep heating for the treatment of cancer patients has been studied in this country[21] and in Japan[1] using a Japanese clinical hyperthermia unit with an 8 MHz RF generator. The advantage of this approach is its relatively great depth of penetration for internal heating. Its primary limitation is the excessive heating of subcutaneous fat layers. Another concern is the need for invasive thermometry in deep-seated tumors under conditions in which thermocouples may deform the electrical field. The thermal profile of the entire tumor cannot be measured accurately with currently available methodology.[21] The heat distribution within large, deeply placed tumors is not uniform, so that the interpretation of temperature data is difficult, and dose-response information is limited. Histological studies of a number of treated tumors have demonstrated massive necrosis.

Radio frequency energy at 13.56 MHz has also been employed to produce localized heating by magnetic-loop induction.[36,37] The patient or an extremity that is the site of cancer is placed in the EM field produced by the device, and heat is produced within the tissues by magnetic induction. Adjuvant hyperthermia can be produced in this manner for use with either radiotherapy or chemotherapy. Preliminary trials have confirmed the relative safety of this approach.[36]

B. Local Interstitial Hyperthermia

There are three modes of interstitial hyperthermia delivery, including (1) microwave (MW), (2) radio frequency (RF) and (3) ferromagnetic heating. In the case of MW energy, the frequencies in general clinical use are 2450, 915 and 433 MHz, although most of the recent human studies employ antennae operating at 915 MHz.[12]

Microwave energy is absorbed as heat within tissues adjacent to each of several microwave antennae, which are usually placed from 0.5 to 2.0 cm apart when used in conjunction with interstitial radiation sources. The antennas are placed into plastic or nylon catheters that have been inserted into the tumor previously. In most cases, the catheters are after-loaded with the antennas and then with radioactive seeds in sequence. Thermometry readings are desirable at as many points as feasible.[12] The principal disadvantage of implanted interstitial hyperthermia devices is the possibility of an increased incidence of infection. A major advantage of the interstitial approach is the ability to produce hyperthermia within the tumor without the need to heat intervening normal tissues as well. If the interstitial antennae are inserted with the guidance of an MRI or CT scanner, greater precision and uniformity of placement is possible. Placement of antennae into tumors in most areas of the body, including the brain,[31] is feasible.

RF energy at frequencies between 500 kHz and 13 MHz can be used to produce localized current fields (LCF), in which the tissue between pairs of steel electrodes is heated to desired temperature levels.[12] The pattern of power deposition is steerable through computer control of individual electrode pairs. Therefore, a number of thermometry probes are required to control the uniformity of heating during treatment. For use in LCF fields parallel electrodes are required for a uniform electrical field. Equipment for RF heating is relatively inexpensive and easy to use. It is ideal for use in treating large pelvic tumors, for example, with many electrodes in a template, as well as tumors of the breast and cervix.[11,42,43]

In local interstitial hyperthermia using ferromagnetic seeds, induced electromagnetic energy is converted to heat in the implants which have been placed in tumor tissue. An externally applied magnetic field is used to deposit energy into each of the seeds, which require spacing of 0.8 to 1.0 cm between the implants.[4] The heat generated then passes by thermal conduction from each point source into adjacent tumor tissues. The heating is self-adjusting, since the seeds are made from materials that lose their ferromagnetic properties at a predetermined temperature which has been termed the Curie point.[4] This property provides automatic temperature control in all implanted sites, and eliminates excessive heating and the need for wire connections. The use of ferromagnetic seeds has been relatively free from complications.

Advantages of interstitial hyperthermia produced with this technique are that (1) it is not compromised by previous therapy or by tumor mass, (2) it can provide a safe, effective adjuvant to radiation or drug therapy, and (3) it does not produce myelosuppression or mutagenesis. There is no alteration in heating patterns with the concurrent use of brachytherapy with $^{125}$Iodine. Therefore, permanent implants of this isotope can be employed in combination with interstitial ferromagnetic heating, allowing simultaneous local radiation therapy and fractionated weekly hyperthermia treatments. Pre-treatment planning is critical to position the seeds accurately. At present, relatively few clinical trials have been completed.[3] Currently, developmental and preclinical studies are underway at several universities and medical centers.

C. Whole-Body Hyperthermia

In cancer which has metastasized beyond its site of origin and involves distant organs or tissues, any use of hyperthermia must involve heating the entire body. Techniques for producing whole body hyperthermia have undergone an evolution from the heating cabinet of Warren[45] to the present state-of-the-art radiant heat chamber used at the University of Wisconsin.[30] With this radiant heat system and accurate thermometry, the investigators maintain a constant level of 41.8°C, which produces relatively few side-effects. Patients do not require general anesthesia for this therapeutic approach, which can be combined with chemotherapy, immunotherapy or total body irradiation.[30] Heating periods can last for several hours, but the preferred duration is approximately 2.5 hours. One precaution required for whole body hyperthermia is the exclusion of small subclinical brain metastases that might produce cerebral edema in the hyperthermic host.

D. Hyperthermia and Chemotherapy

When local hyperthermia is used as an adjuvant to chemotherapy, it may be able to overcome drug resistance in previously treated patients[44] and can potentiate the action of a number of cytotoxic agents, including Cisplatin, bleomycin, methotrexate and the nitrosoureas. Local hyperthermia has very little toxicity when added to standard doses of chemotherapeutic agents.[13] Results employing Cisplatin, bleomycin and methotrexate in conjunction with local hyperthermia have revealed high response rates for many tumor types, including a number that are located in previously irradiated sites[13]

The timing of adjuvant hyperthermia treatments, as well as the total thermal dose of each one, is likely to be substantially different when hyperthermia is used in combination with chemotherapy or immunotherapy, in comparison with its adjuvant role with ionizing radiation therapy. In the case of chemotherapy, the rationale is to produce vasodilatation in order to convey a greater percent of the injected bolus into the heated tumor. To attain that goal, local hyperthermia would be induced immediately prior to each parenteral dose of the chemotherapeutic agent(s), according to the time schedule for administration determined to be optimal by the medical oncologist.

For dissemated tumors, it is necessary that therapy be systemic, so that whole body hyperthermia will be the standard modality. Thermal gradients might also be expected to be effective in directing chemical or immunologic agents to tumors and, thus, to obtain the highest possible concentration of these substances in the neoplastic tissue.

## FUTURE TRENDS AND NEEDS

At the present time, most of the existing equipment for the delivery of localized hyperthermia by microwave or radio frequency techniques is located in departments of radiation oncology and is employed in regimens combining non-ionizing and ionizing radiation for cancer therapy. In departments of radiation therapy, decisions regarding adjuvant hyperthermia are usually made by the radiation oncologist, the health physicist and a technologist or nurse who administers the treatment. For regimens combining chemotherapy with adjuvant hyperthermia, either with or without radiotherapy, an appropriate treatment team would also include a medical oncologist and a surgical oncologist, if the regimen is to be combined with an operative procedure. In addition to a health physicist who has been trained in the biological effects of non-ionizing radiation, other members of the team might include an electrical engineer, a hyperthermia physiologist, and a nurse trained in monitoring cancer patients, who are often debilitated, for possible side-effects of the combined regimens during treatment. A hyperthermia team of this composition would be prepared to make recommendations with respect to all invasive techniques, dosage regimens, and timing of treatments, to assess the therapeutic results, and to cope with any medical or surgical problems that might occur.

A multidisciplinary team of this type is likely to improve the efficacy of many hyperthermic regimens. Improved results are most likely if the treatment team uses MRI or other scanning techniques to localize the tumor tissue precisely, and computer simulation to predict the thermal profiles (isotherms) desired in all parts of the target area. Simulations of this type will permit the hyperthermia treatment team to calculate and assign power levels for individual antennae, applicators or electrodes within arrays.

In addition, the need to individualize the thermal dose, the multiple factors upon which the uniformity of heating depends, and the longer time required for hyperthermia administration renders the use of this adjuvant modality more labor intensive than ionizing radiation alone for the personnel involved. Since the majority of microwave and radio frequency hyperthermia equipment now in use is under control of radiation oncologists, it can also be anticipated that increased usage of this type of equipment by medical oncologists will lead to an increased need for this type of equipment, unless interdisciplinary regimens can be developed.

The specialized techniques of interstitial hyperthermia are likely to be utilized more extensively as an adjuvant to radiation therapy during the coming decade. These include ferromagnetic seeds, intracavitary hyperthermia, intraoperative hyperthermia, and LCF fields for the treatment of tumors beneath the body wall. Individualized hyperthermia regimens are likely to prove effective for tumors located in specific sites, such as the brain, in which neoplasms are confined to a local area with no threat of metastasis, as well as in those that are relatively resistant to radiation therapy and other forms of treatment. These resistant tumors include cancer of the pancreas, melanoma, and hypernephroma.

The use of adjuvant hyperthermia in combination with radiotherapy has been explored at a number of centers, but cooperative protocols have often proved inconclusive. At present, relatively few patients with cancer are candidates for local or regional hyperthermia using current techniques. New approaches may prove effective by using adjuvant thermal gradients to improve the localization of chemotherapeutic and immunotherapeutic agents selectively in neoplastic tissue.

It is likely that within the next ten years, therapy with thermal gradients will have settled into its rightful niche in the armamentarium of the hyperthermic oncologist, whether radiological, medical, or surgical. For those tumor types and locations which are most favorable for the effective utilization of local hyperthermia, it is likely that specific regimens will have been formulated and adopted as standard among oncologists.

Funding for future research in the area of cancer hyperthermia is increasingly uncertain. At present, the federal government appears to have little interest in funding new projects in this field. Therefore, additional progress is likely to be slow as a result of limited research funding.

In summary, clinical hyperthermia has yet to reach its maximum potential, but has already become a reliable and effective adjuvant treatment modality for cancer in combination with ionizing radiation therapy, with chemotherapy or with both approaches. As treatment teams gain experience with evolving techniques for achieving uniform thermal profiles in localized tumors, more effective results are likely to be obtained, and a greater percent of patients with cancer will be able to receive benefit from therapeutic hyperthermia.

## References

1. Abe M, Hiraoka M, Takahashi M, et al. Multi-institutional studies on hyperthermia using an 8-MHz radiofrequency capacitive heating device (Thermotron RF-8) in combination with radiation for cancer therapy. Cancer 1986; 58:1589-1595.

2. Arcangeli G, Guerra A, Lovisolo GA, Cividalli A, Marino G, Mauro C. Tumor response to heat and radiation: prognostic variables in the treatment of neck node metastases from head and neck cancer. Int J Hypertherm 1985; 3:207-217.

3. Au KS, Cetas TC, Shimm DS, Sinno JS, Haider SA, Buechler DN, Lutz WR, Cassady JR. Interstitial ferromagnetic hyperthermia and brachytherapy: Preliminary report of a phase I clinical trial. Endocuriether/Hypertherm Oncol 1989; 5:127-136.

4. Brezovich, IA, and Meredith, RF. Practical aspects of ferromagnetic thermoseed hyperthermia. Radiolog Clin N Amer 1989; 27:589-602.

5. Busch W. Uber den Einfluss welche heftigere Erysipeln zuweilig auf organisierte Neubildungen ausuben. Vrh Naturhist Preuss RheinWestphal 1866; 23:28-30.

6. Cavaliere R, Ciocatto EC, Giovanella BC, Heidelberger C, Johnson RO, Margottini M, Mondove B, Moricca G, Rossi-Fanelli A. Selective heat sensitivity of cancer cells (biochemical and clinical studies). Cancer 1967; 20:1351-1381.

7. Cavaliere R, Mondovi B, Moricca G, Monticelli G, Natali PG, Santori FS, Di Filippo F, Varanese A, Aloe L, Rossi-Fanelli A. Regional perfusion hyperthermia. In: Hyperthermia in Cancer Therapy (Ed. FK Storm). GK Hall Med Publ, Boston, MA. 1983; 369-399.

8. Coley WB. The treatment of malignant tumors by repeated inoculations of erysipelas, with a report of ten original cases. Am J Med Sci 1893; 105:488-511.

9. Coley WB. A report of recent cases of inoperable sarcoma successfully treated with mixed toxins of erysipelas and Bacillus prodigiosus. Surg Gynecol Obstet 1911; 13:174-190.

10. Coley-Nauts H, Swift W, Coley B. The treatment of malignant tumors by bacterial toxins as developed by the late William B. Coley, MD: Reviewed in the light of modern research. Cancer Res 1946; 6:205-216.

11. Cosset JM, Dutreix J, Haie C, Gerbaulet A, Janoray P, Dewar JA. Interstitial thermal radiotherapy: Technical and clinical study of 29 implantations performed at the Institut Gustave-Roussy. Internat J Hypertherm 1985; 1:3-13.

12. Coughlin CT, Strohbehn JW. Interstitial thermoradiotherapy. Radiolog Clin N Amer 1989; 27:577-588.

13. Cruciani G, Salfano V, Crippa E, Morelli M, Gnani C, Marangolo M. Clinical experiences with multifrequency hyperthermia system (MHS) and chemotherapy in the treatment of superficial metastases. In: Recent Results in Cancer Research. Springer Verlag, Berlin. 1989: (in press).

14. Dewhirst MW. Animal modelling and thermal dose. Radiolog Clin N Amer 1989; 27:509-518.

15. Dickson JA, Calderwood SK. Thermosensitivity of neoplastic tissues in vivo. In: Hyperthermia in Cancer Therapy (Ed. FK Storm). GK Hall Med Publ, Boston, MA. 1983; 63-137.

16. Emami B, Perez CA, Konefal J, Pilepich MV, Leybovich L, Straube W, vonGerichten D, Hederman MA. Thermoradiotherapy of malignant melanoma. Int J Hypertherm 1989; 4:343-381.

17. Gerner EW. Thermotolerance. In: Hyperthermia in Cancer Therapy (Ed. FK Storm). GK Hall Med Publ, Boston, MA. 1983; 141-162.

18. Hetzel FW. Biologic rationale for hyperthermia. Radiolog Clin N Amer 1989; 27:499-508.

19. Hofman P, Knol, REF, Lagendijk JJW, Schippers J. Thermoradiotherapy of primary breast carcinoma. Int J Hypertherm 1989; 5:1-11.

20. Krusen FH. Medical applications of microwave diathermy. Proc Roy Soc Med 1950; 40:638.

21. Lee, CKK, Song, CW, Rhee, JG, and Levitt, SH. Clinical experience with Thermotron RF-8 capacitive heating for bulky tumors: University of Minnesota experience. Radiolog Clin N Amer 1989; 27:543-558.

22. Lele PP. Physical aspects and clinical studies with ultrasonic hyperthermia. In: Hyperthermia in Cancer Therapy (Ed. FK Storm). GK Hall Med Publ, Boston, MA. 1983; 333-367. (cf chart, p 342)

23. Li GC, Petersen NS, Mitchell HK. Induced thermal tolerance and heat shock protein synthesis in Chinese hamster ovary cells. Int J Radiat Oncol Biol Phys 1982; 8:63-67.

24. Licht S. History of therapeutic heat. In: Therapeutic Heat and Cold, 2nd ed. (Ed. S Licht). Waverly Press, Inc, Baltimore, MD. 1972; 196-231.

25. Marchosky JA, Moran C, Fearnot N. Long term interstitial hyperthermia therapy for brain tumors using implanted catheters. Proc Internat Clin Hypertherm Soc Indianapolis, IN. 1988.

26. Overgaard J. The biologic basis for clinical treatment with combined hyperthermia and radiation. In: Progress in Radio-Oncology II (Eds. KH Karcher, et al). Raven, New York, 1982.

27. Overgaard J, Overgaard M. Hyperthermia as an adjuvant to radiotherapy in the treatment of malignant melanoma. Int J Hypertherm 1987; 3:453-501.

28. Perez CA, and Emami B. Clinical trials with local (external and interstitial) irradiation and hyperthermia: Current and future perspectives. Radiolog Clin N Amer 1989; 27:525-541.

29. Perez CA, Kuske RR, Emami B, Fineberg B. Irradiation alone or combined with hyperthermia in the treatment of recurrent carcinoma of the breast in the chest wall: a nonrandomized comparison. Int J Hypertherm 1986; 2:179-187.

30. Robins, HI, Hugander, A, and Cohen, JD. Whole body hyperthermia in the treatment of neoplastic disease. Radiolog Clin N Amer 1989; 27:603-610.

31. Salcman M, Samaras GM. Interstitial microwave hyperthermia for brain tumors. J Neuro-Oncol 1983; 1:225-236.

32. Schliephake E. Les Ondes Courtes en Biologie. Paris, 1935.

33. Scott RS, Johnson RJR, Story KV, Clay L. Local hyperthermia in combination with definitive radiotherapy: Increased tumor clearance, reduced recurrence rate in extended follow up. Int J Radiat Oncol Biol Phys 1984; 10:2119-2123.

34. Song CW. Blood flow in tumors and normal tissues in hyperthermia. In: Hyperthermia in Cancer Therapy (Ed. FK Storm). GK Hall Med Publ, Boston, MA. 1983; 187-206.

35. Steeves RA. Clinical considerations in the use of external microwaves for local hyperthermia. Radiolog Clin N Amer 1989; 27:519-524.

36. Storm FK, Harrison W, Elliott BS, Kaiser LR, Silberman AW, Morton DL. Clinical radiofrequency hyperthermia by magnetic-loop induction. J Microwave Power 1981; 16:179-184.

37. Storm FK, Harrison WH, Elliott RS, Morton DL. Physical aspects of localized heating by magnetic-loop induction. In: Hyperthermia in Cancer Therapy (Ed. FK Storm). GK Hall Med Publ, Boston, MA. 1983; 305-283.

38. Sutton CH. Tumor hyperthermia in the treatment of malignant gliomas of the brain. Transact Amer Neurol Assoc 1971; 96:195-199.

39. Sutton CH. Use of local hyperthermia in brain tumor therapy. Proc Amer Assoc Cancer Res Denver, CO. 1977; 200.

43. Sutton CH, Carroll FB. Effects of microwave-induced hyperthermia on the rat blood-brain barrier. Radio Science 1979; 14(6s):299-331.

41. Valdagni R, Liu FF, Kapp DS. Important prognostic factors influencing outcome of combined radiation and hyperthermia. Int J Radiat Oncol Biol Phys 1988; 14:632-639.

42. Vora N, Forell B, Joseph C, Lipsett JA, Archambeau JD. Interstitial implant with interstitial hyperthermia. Cancer 1982; 50:2518-2423.

43. Vora N, Shaw S, Forell B, Desai K, Archambeau J, Pezner RD, Lipsett JA, Covell J. Primary radiation combined with hyperthermia for advanced (Stage III-IV) and inflammatory carcinoma of breast. Endocuriether/Hypertherm Oncol 1986; 2:101-106.

44. Wallner KE, DeGregorio MW, Li GC. Hyperthermic potentiation of cis-diaminedichloroplatinum (II) cytotoxicity in Chinese hamster ovary cells resistant to the drug. Cancer Res 1986; 43:6242-6235.

45. Warren SL. Preliminary study of the effect of artificial fever upon hopeless tumor cases. Am J Roentgenol 1935; 33:75-87.

46. Westermark F. Uber die behandlung des vecerirended cercixcarcinoms mittel konstanter warme. Zbl Gynak 1898; 1335-1309.

# ELECTROMAGNETIC FIELD FOCUSING (EFF) PROBE: APPLICATIONS IN ANEURYSM TREATMENT, ANGIOPLASTY, AND BRAIN TUMOR RESECTION

William S. Yamanashi, Arun A. Patil, Seppo J. Saksanen, Steven W. Phillips, Jimmie L. Valentine, Nabil A. Yassa, Benny D. Wagner, and Deborah L. Hill

The electromagnetic field focusing (EFF) probe described in this chapter is a new precision surgical tool (US patent pending file No. 906791). It has either a hand-held or an extended form such as a catheter. It produces an intense tissue vaporizing temperature in a precisely defined manner comparable to $CO_2$ laser. Unlike lasers, the EFF probe features tactile operation (i.e., action on touch only) such that the depth hazard associated with lasers is absent. The EFF probe simultaneously cuts (vaporizes) and cauterizes biological tissues except cortical bone. Cauterizing alone can be attained at a lower power setting.

The EFF system consists of a radio frequency (RF) generator, a solenoidal coil, and a hand-held or catheter probe (Fig. 1). The patient (animal, phantom) is placed near the coil. The electromagnetic field transmitted through the coil induces eddy currents in the subject. The eddy current is converged at the point of contact between the tissue and the probe which is connected to the return circuit. The effect of this convergence is further maximized by electrically tuning the probe to the resonance frequency of the coil.

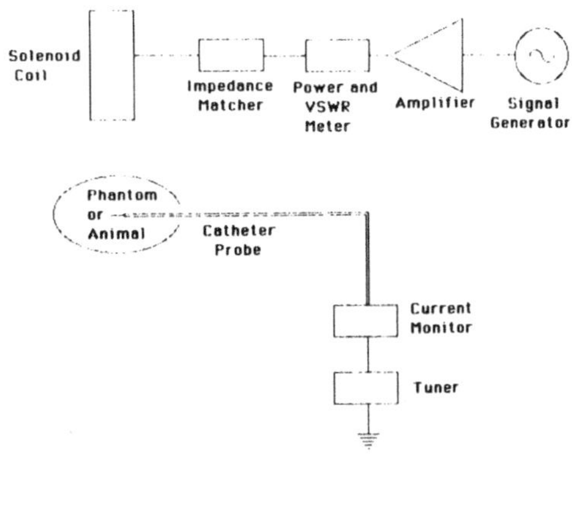


**Figure 1.** A block diagram of the essential components of the electromagnetic field focusing (EFF) probe.

Comparison between the EFF system and other electromagnetic heat generating methods such as inductive heating, capacitive heating, dielectrically loaded antenna, monopolar interstitial electrode, and bipolar cautery has been reported.[15] Because of its induced current convergence property the field focusing of the EFF system is more efficient in precision heating, cutting, or vaporizing. A probe tip temperature in excess of 1800°C was approximated from melting of platinum used as the probe tip in a tissue simulated phantom.

The EFF system can be used with the probe in a catheter form. Other currently available transcatheter surgical methods include lasers,[5] balloon catheters,[12] self expandable endoprosthesis or stents,[11,14] liquid and particulate embolization,[8,16] pulsed electrical spark erosion,[13] LF (Low Frequency) antenna,[7] and electrocautery.[1] Some of these are designed for angioplasty or endarterectomy, others are for thrombosis or occlusion.

Lasers can damage the arterial wall, causing perforation, dissection or aneurysm formation and stenosis is only partially relieved and reocclusion is common. Lasers are also expensive and difficult to manipulate. Late effects of lasers include re-stenosis, granulomas, and fistula formation.[9]

Balloon catheter angioplasty frequently develops re-stenosis. Endoprosthesis has a high incidence of thrombogenicity,[11] and the spark erosion technique often perforates the vessel wall.[13] Other techniques used in vascular lesions include nonlaser thermal devices such as monopolar and bipolar cauteries and sclerotherapy. Nonlaser thermal devices can coagulate but cannot cut or vaporize. Sclero-therapy has complications of toxicity, bleeding, necrosis with sloughing, and ulcer formation.[2]

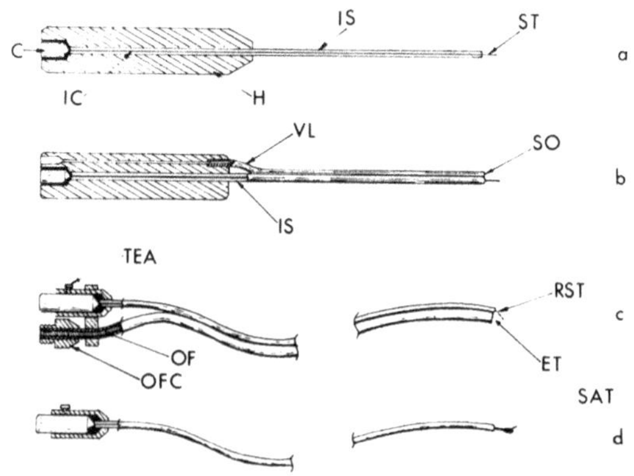


**Figure 2.** Examples of different probe tips: (a) microsurgical probe tip, (b) general surgical cutting tip with a parallel suction, (c) probe assembly attachable to an endoscope, and (d) catheter-form probe for angioplasty and endarterectomy.

Although the EFF probe is not claimed to be the solution to all of these problems, it is relatively inexpensive and is capable of producing well-circumscribed, intense focal heat at the point of contact with the tissue. The precision of the probe tip control, ease of use with catheters or endoscope, and its apparent perforation-free operation suggest that the EFF probe may offer distinct advantages over previous methods using surgical lasers, cauteries and other mechanical devices.

In the present report we summarize the preliminary studies using the EFF probe, phantom and various animal models including: I. Study of the properties of the probe; II. Aneurysm thrombosis; III. Angioplasty; and IV. Brain tumor resection. Various types of probe tips (Fig. 2) were used for different applications. For example, microsurgery probe tip (a), general surgical cutting tip with parallel suction (b), probe assembly attachable to an endoscope (c), and catheter form probe for angioplasty and endarterectomy (d).

## PROPERTIES OF THE EFF PROBE

### A. Materials and Methods

Figure 1 is a block diagram of the essential components of the RF field generating system and the EFF probe. A function generator (Hewlett-Packard HP-3314 A) was used to generate the sinusoid signal with frequency set at 13.65 MHz and the peak output voltage at 0.46 V. The signal generated was amplified first by a 50 W rms maximum, linear amplifier (Electronic Navigation Instrument [ENI] 350-L), then by a 1 kW rms maximum, linear amplifier (Ehrohorn Technological Operations [ETO], Alpha-76A). The impedance of the solenoidal network and that of the Alpha-76A amplifier was matched with a modified Nye Viking antenna matcher. The modification consisted of the removal of antenna-selector switches to prevent arcing among switching elements. The measurements of the RF output root mean square (RMS) power, and the voltage standing wave ratio (VSWR) were made by a power multimeter. The power measurement (in W) represents the total power dissipated by the resistive loads of both the coil and the tissue. The VSWR represents the amount of reflected power (VSWR=1.0 represents no reflected power). The RF output power RMS and the VSWR were measured with a Kenwood SW-2000 RF power multimeter. The output was calibrated by the use of a 1 kW, 50 ohm dummy load. The solenoid was tuned via a 50-500 pF, 15 kV, variable, vacuum bellow-type capacitor (ITT, Jennings). The rms RF output power is adjustable from 30 to 600 W by the use of the load adjustment in the Alpha-76A amplifier. (See specific power settings i the application section.) The measured RF output

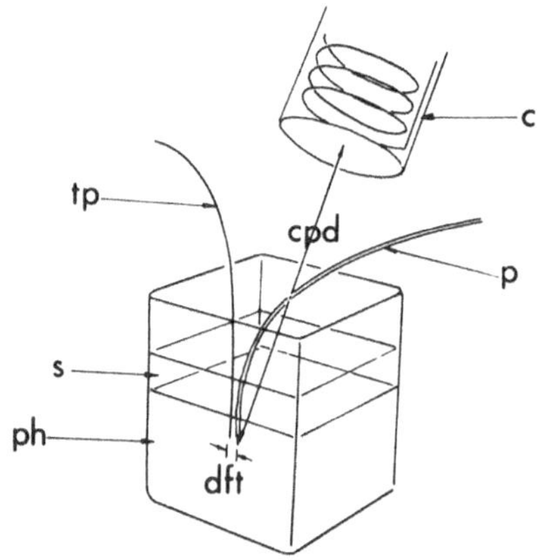


**Figure 3.** Experimental setting showing relative location of the coil, probe, phantom, and temperature probe. Symbols: C=coil, cpd=coil-to-probe tip distance, p=probe, tp=temperature probe, s=saline, ph=phantom, dft=distance from tip of the probe.

power in watts represents the total power dissipated by the resistive loads of both the coil and the biological tissue or phantom.

Experiments were performed with the RF RMS powers of 150, 200, and 250 W, and a VSWR of 1.2. Experiments were conducted on a uniform phantom (superstuff) developed by Guy, et al,[6] which simulates biological tissue electrical conductivity, dielectric constant, and magnetic permeability. A modified F2 polyethylene catheter was used as the probe insulator with its terminal hole permitting a 0.014 cm diameter, 175 cm long guide wire. A few mm of the guide wire protruding from the catheter were used as the probe tip. The solenoidal coil was placed above the phantom. Temperatures delivered and cooling curves were plotted by use of a fluoroptic temperature probe. Relative locations of the components of the experimental equipment are illustrated in Fig. 3.

Temperature vs. time and distance data were obtained for different power settings. Heating patterns were measured by using liquid crystal. In some experiments saline was added on top of the superstuff to approximate the in vivo conditions in blood vessels and brain, etc.

B. Results

Figure 4 demonstrates that the square wave form has the greatest heat dissipation, while the sinusoidal has the least. Figure 5 exhibits the effect of different coil-to-probe tip distances. It is seen from the curve that the closer the solenoidal coil to the probe tip, the higher the delivered temperature. There are

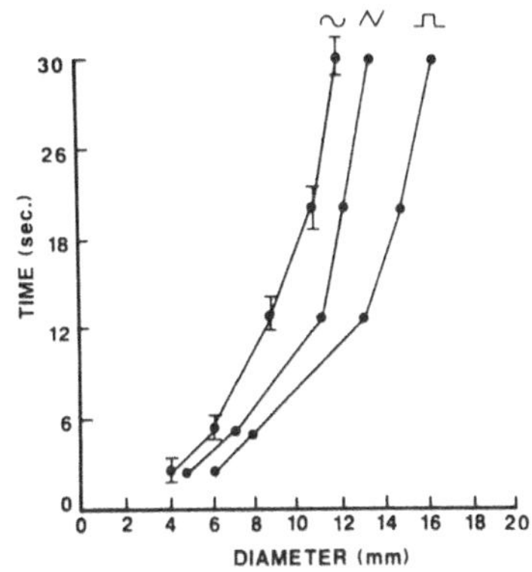


**Figure 4.** Diameter of hot spot over delta T=10°C as measured by liquid crystal when different wave forms were used at 2.5, 5, 10, 20, and 30 s. The power applied was 200 W.

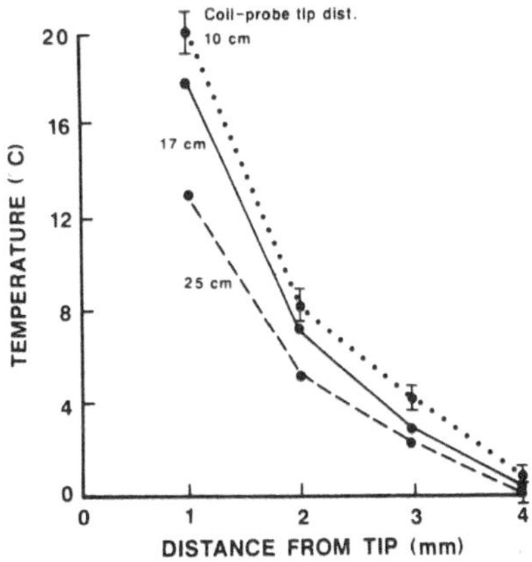


**Figure 5.** Difference in temperatures measured when coil-to-probe tip distances were varied (10, 17, and 25 cm). The probe tip was inserted 5 cm in the phantom with an uninsulated 5 mm portion of the tip exposed. A power of 200 W was applied for 5s.

no marked differences with coil-to-probe tip distances of 10 cm and 17 cm. In Fig. 6 temperatures delivered with different uninsulated probe tip lengths are compared. The shorter the probe-tip length, the higher the temperature delivered at a given power setting. Figures 7A, 7B, and 7C compare heat delivered with different powers at 2.5, 5, 10 and 20 seconds. Temperatures were measured at 1, 2, 3, and 4 mm from the probe tip. The coil-to-probe tip distance was 17 cm, and the experiment was conducted on the phantom under 128 $cm^3$ saline. For a given power setting, a broader thermal distribution is observed from the top of the probe with a longer time application. As the power level is increased, greater temperature is delivered with sharper thermal distribution as expected. It is apparent that temperatures delivered are related to the power applied.

**Figure 6.** The effect of different uninsulated probe tip lengths on the heating pattern. Temperatures were measured at 1 mm from probe tip, which was inserted 5 cm into the phantom and was 17 cm from the coil. The power applied was 150 W and temperatures were recorded at 2.5, 5, 10, and 20 s.

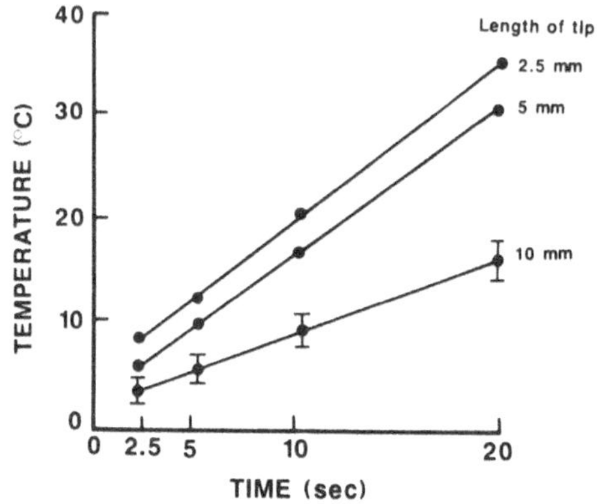


**Figure 7.(a-c)** The effect of different power settings on the heating pattern using 150 W (a), 200 W (b), and 250 W (c). The application time was 2.5, 5, 10, and 20 s. Temperatures were measured at 1, 2, 3, and 4 mm from the probe tip. Experiments were conducted under normal saline of volume=128 $cm^3$. The probe tip was inserted into the phantom 2 cm and had an uninsulated portion of 5 mm. The coil-to-probe tip distance was 17 cm.

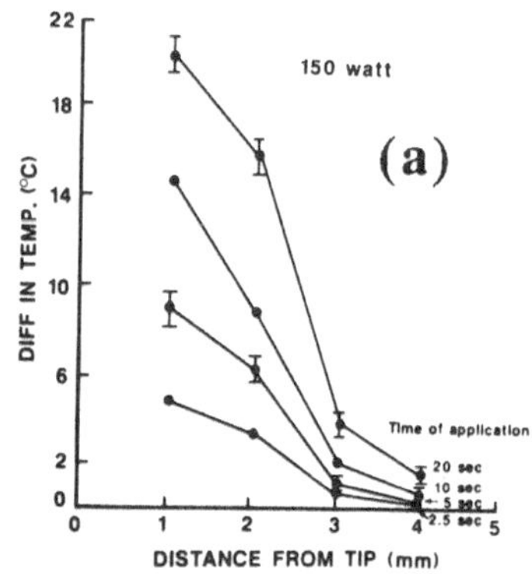


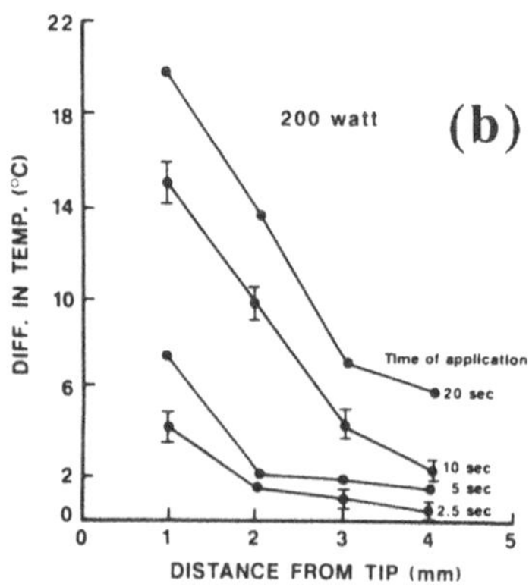


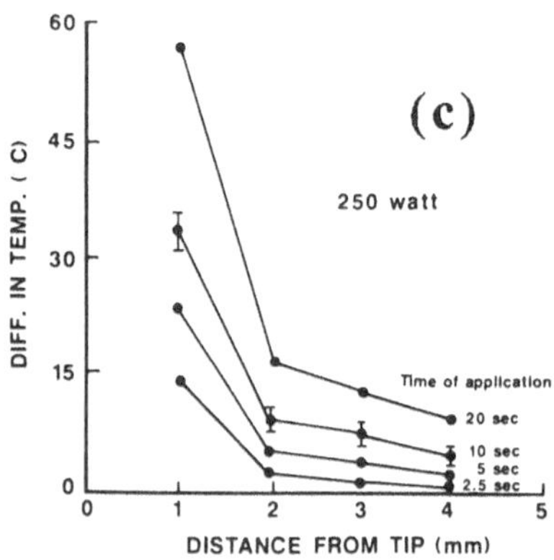

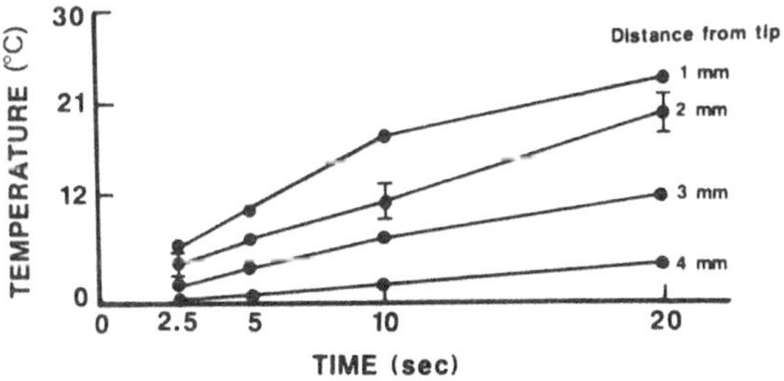


**Figure 8.** The time-temperature relationship measured with the use of a power of 200 W at 1, 2, 3, and 4 mm distance from probe tip.

**Figure 9.** The probe's cooling curves after 150, 200, and 250 W were applied for 5 seconds. Temperatures were measured at 3 mm distance from the probe tip. The probe tip was inserted in the phantom 2 cm and had an uninsulated portion of 2 mm. The coil-to-probe tip distance was 17 cm.

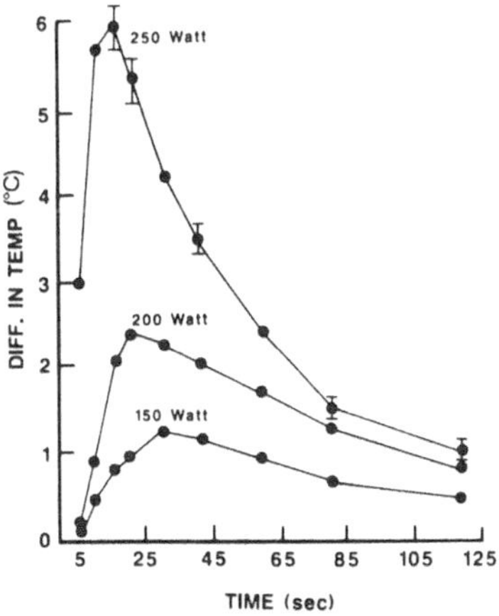


Figure 8 shows the time temperature heating exponential curve at varying distances from the probe tip. For a given coil-to-probe tip distance the temperature rises exponentially with the rate proportional to the distance of the thermal probe from the probe tip. It is apparent that temperatures delivered are related to the coil-to-probe tip distance. Figure 9 is the probe's cooling curve. It is seen that the higher the power delivered, the steeper the exponential cooling curve becomes.

C. Discussion

The EFF probe is an induced current convergence device. Upon contact of the probe with the tissue the eddy currents produced within the tissue are converged at the point of contact producing extremely high heat, which is very focal to the point of contact.

A laser is a coherent source of photons at a discrete frequency. At a particular frequency, a high enough temperature can be generated to cut, vaporize, and cauterize. It is, however, difficult to control the precise depth of penetration of the light beam front.[9] With the use of a metallic chamber, a hot tip laser acting as a light receptor has improved this problem to some extent, but the disadvantage of using a relatively rigid fiberoptics remains in the catheter applications.[3]

The EFF probe can be used with higher wattages with minimum danger of penetration, with an electrically insulated round ceramic tip followed by conducting metal. It can also be used through an endoscopic device with ease. Other heat probes do not reach the intense heat generation produced by the EFF probe, and they are difficult to operate endoscopically. The optimum probe-to-coil distance is from 15 to 30 cm. The coil can be located with different azimuthal angles relative to the surgical area, in order to accommodate such instruments as a fluoroscope. Sinusoidal wave forms are ideal for a very localized form of heat delivery since the heating pattern is more confined to the proximity of the probe tip than with other wave forms.

## AN ANIMAL MODEL FOR ANEURYSM TREATMENT

### A. Materials and Methods

A total of 97 rats were used for two types of aneurysm models with different geometry and extent of aneurysm. Fifty-six rats were used for the first model, and forty-one rats were used for the second model. In the first model, an abdominal aneurysm was created by side-to-side anastomosis between the abdominal aorta and the inferior vena cava in each rat. In the second model, abdominal aneurysms were created by an anastomosis between the abdominal aorta of one rat and the donor abdominal aorta from another rat. In the first model, prior to the anastomosis, the vena cava was ligated proximal and distal to the site of anastomosis. The length of the short and long axes of each aneurysm was measured a second time, 3 weeks later. All animals with the aneurysms were divided into treatment (88 rats) and control groups (9 rats). Those in the treatment group were treated at different time intervals after aneurysm formation and followed at different time intervals after treatment (see Table I for details). Nine animals in the treatment group had post-treatment angiograms performed by injecting 0.5-1.0 60% Conray (Mallickrodt Co., St.Louis, Missouri) by hand injection into the heart or aorta. The control group was followed for 3 weeks.

A microsurgical probe tip (Fig. 2a), 0.3 mm diameter and 7 mm long at the electrically conducting tip was used in the treatment of aneurysms. The

TABLE 1. *Animal (Rat) Experiment, Treatments, and Observations*

| | | | | | | | | | |
|---|---|---|---|---|---|---|---|---|---|
| Aneurysm model | II | I | II | I | II | I | II | I | II |
| No. animals | 9 | 48 | 12 | 8 | 11 | 0 | 6 | 0 | 3 |
| Average dimension | | | | | | | | | |
| Long axis (mm) | 4.0 | 7.0 | 4.0 | 5.0 | 4.9 | | 4.7 | | 5.6 |
| Short axis (mm) | 2.0 | 3.0 | 2.0 | 4.9 | 2.5 | | 1.5 | | 2.0 |
| No. angiogram | 1 | | | | | | 9 | | |
| Time of angiogram after aneurysm creation | 3 days | NA | | NA | | NA | | 3 days | NA |
| No. aneurysms histologically examined | 33 | | | | 23 | | | | |
| Time, follow-up after aneurysm creation | 3 weeks | Immediate | | 3 days | | 3 weeks | | 6 weeks | |
| Time, follow-up after treatment | NA | Immediate | | Immediate | | Immediate | | 3 weeks | |

length of the electrically insulated sheath was 10 cm. The average rms power used for the thrombosis of the aneurysms was 38.5 W; the average VSWR was 1.33. The average distance between the coil (inductive applicator) and the probe tip was 3.24 cm. The probe was first inserted using a direct transvascular approach into the caudal end of the aneurysm. Power was applied for a brief period (0.5-1.5 s) as the probe was gradually advanced toward the cephalic end. Average total treatment time was 11.2 s. During treatment, the distal end of the insulating sheath was held against the aneurysm wall, applying pressure to prevent bleeding from the puncture site.

A fluoroptic temperature probe (Luxtron model 1000-B, Mountain View, Calf.) was used to obtain time vs temperature data at several predetermined distances in the femoral artery and vein, as well as in dielectrically uniform phantoms. A dielectrically uniform phantom (superstuff) developed by Guy, et al,[6] which simulates the electric conductivity, dielectric constant, and magnetic permeability of biological tissue, was used to approximate the in vivo heating curve. The initial portion of the time-temperature curve approximates the in vivo situation where convective heat transport is minimal and absorption of applied energy maximal. In the in vivo plateau phase, absorption per unit volume is balanced by the local heat transfer. Surgery using this probe is done within the initial portion of the time-temperature curve. Figure 10 shows the heating pattern obtained near the probe tip inserted 1 mm into the superstuff phantom with 5 s of RF power, and with the tip inserted 5 mm with 20 s of RF power. The former represents a shallow and short exposure, and the latter, a deeper and longer exposure. Figure 11 shows the thermal distribution near the probe tip inserted into either the femoral artery or vein of the rat with the probe tip touching the vessel wall. The efficiency of heat conduction appears to be better in the artery than in the vein. Figure 10 illustrates that when the conduction tip has a small surface area, the same rms power results in more intense heating near the probe tip. As seen from Figs. 10 and 11, the shorter the tip length, the more discrete the heating pattern in the vicinity of the probe tip.

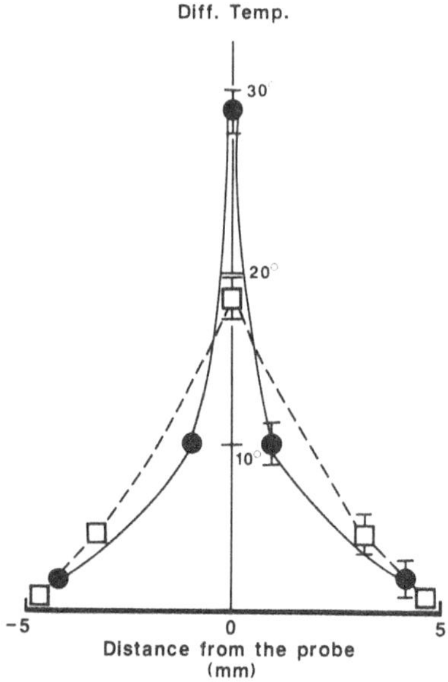


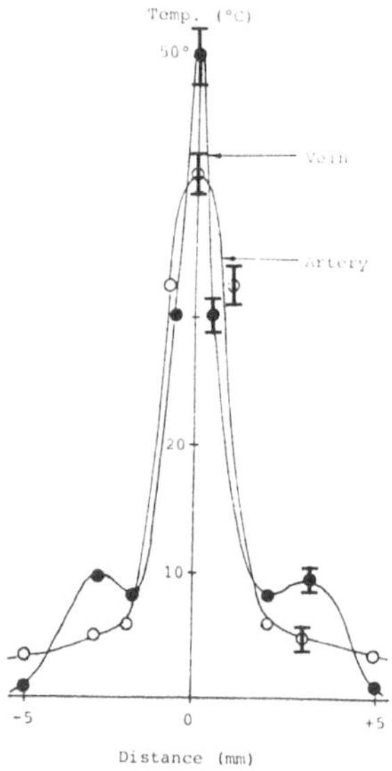


**Figure 10 (Left).** Thermal characteristics of the EFF probe at lower power: heating pattern near the probe tip inserted 1 mm into the phantom for 5 s and insertion to 5 mm with 20 s of 150 W rms power.

**Figure 11 (Right).** Thermal distribution near the probe tip inserted into the rat femoral artery or vein with the probe tip touching the vessel wall.

B. Results

As the first burst of power was applied, the blood in the immediate vicinity of the probe tip coagulated and the blood adjacent to this area became viscous, converting the "liquid-in-a-balloon" state into a raisin-like shriveled mass consisting of aneurysm fundus and coagulated blood. Several burst of focused RF were necessary during the linear manipulation of the probe to totally thrombose the aneurysm.

Aneurysm treatments and experimental animal models are summarized in Table I. Three weeks after formation of aneurysms, the nine control aneurysms were observed under an operating microscope. Swirling blood was seen through the thin, semitransparent vessel walls. The average dimensions of the aneurysms were 8.0 X 3.4 mm, compared with 4.0 X 2.0 mm when measured immediately after creation. One control aneurysm, 3 weeks after creation, was photographed and is shown in Fig. 12. A surgically created aneurysm before and

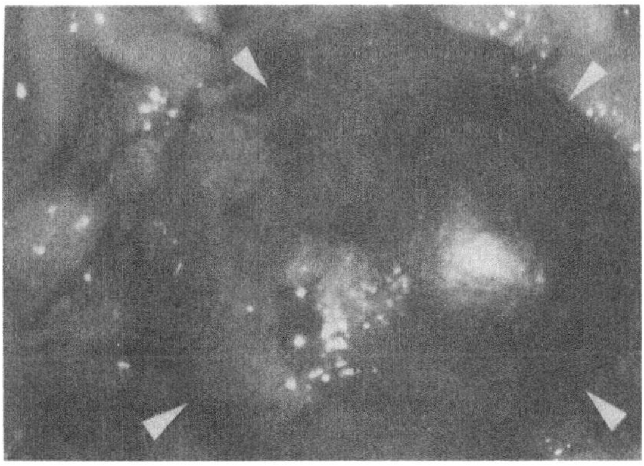

**Figure 12.** The control aneurysm 3 weeks after creation.

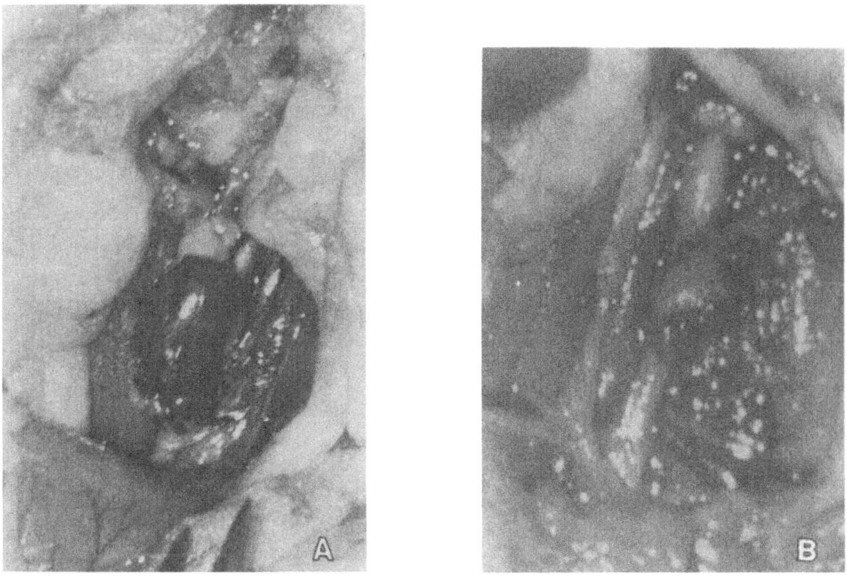


**Figure 13.** Surgically created aneurysm (a) before and (b) after EFF probe treatment.

after the EFF probe treatment is shown in Fig. 13, a and b, respectively. Histologic examination of an aneurysm shows the area of connection between the artery and vein (Fig. 14). All of the parent arterial vessels were patent, except two that were partially occluded. Histologic changes ranged from inflammatory infiltration of surrounding tissue to necrosis of the vessel wall. Grossly, the necrosis had the usual appearance of coagulation necrosis.

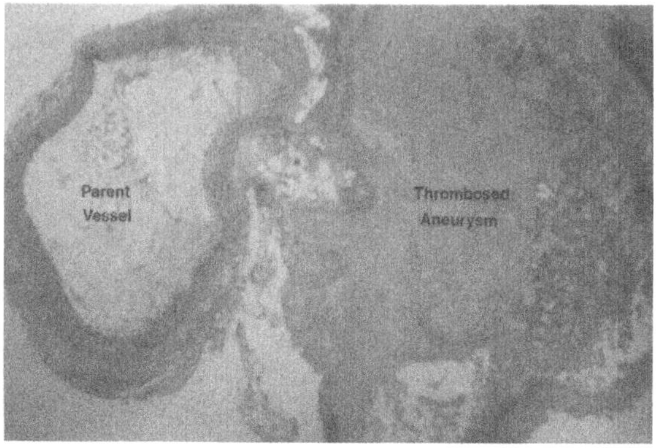


**Figure 14.** Histologic examination of an aneurysm showing surgical connections between arteries and a vein, and the coagulation of blood in the vein.

However, elongated nuclei typically seen in electrocution or high-temperature necrosis were not apparent, suggesting an appreciable difference in the mechanism of RF-induced pyrolytic thrombosis, although the absence of elongated nuclei is not sufficient evidence to rule out thermal damage completely. The success of the treatment was confirmed prior to histologic evaluation in eight animals by angiograms in which the aneurysm was not visible.

An aneurysm can be approached either transvascularly by inserting the probe directly into it or intravascularly by catheterization as illustrated in Fig. 15, a and b, respectively. Although all aneurysm treatments in this work were transvascular, a model for intraluminal approach is discussed in the section on angioplasty.

## ANGIOPLASTY: IN VITRO FEASIBILITY

### A. Materials and Methods

A modified polyethylene catheter was used as the probe insulator with its terminal hole permitting an 0.014 cm diameter, 175 cm long guide wire. Two mm of the guide wire protruding from the catheter was used as the probe tip.

Gas was analyzed by vaporizing phantom superstuff developed by Guy, et al,[6] that simulates biological tissue electrical conductivity, dielectric constant, and magnetic permeability.[13] A graduated 1 cc syringe was inserted into a 10.8 cm length of Tygon tubing half filled with phantom superstuff. The end of the Tygon tubing contained in a 1616 $cm^2$ Plexiglas box. This large phantom served as a model simulating the volume of tissue adjacent to the vessel from which induced current was converged. The graduated 1 cc syringe and half-empty

**Figure 15.** Two approaches to aneurysm (a) transvascular and (b) intravascular, using EFF probe, A=aneurysm, ST=sharp tip, IS=insulated sheath, and SAT=special alloy tip.

Tygon tubing were filled with normal saline. The EFF catheter probe was used to vaporize the superstuff in the Tygon tubing under normal saline at various powers (80-250 W), 16 cm coil distance, and 20 s duration (Table II). Immediately after the power was discontinued, the bubbles were tapped to the top of the graduated syringe and the volume in ml was measured. Equipment is illustrated in Fig. 16.

TABLE II

| Power* (W) | Coil Distance (cm) | Time (sec) | Volume Sum of Formed Bubbles (mL) |
|---|---|---|---|
| 80 | 16 | 20 | 0.08 |
| 150 | 16 | 20 | 0.10 |
| 200 | 16 | 20 | 0.14 |
| 250 | 16 | 20 | 0.14 |

*Total power dissipated in both the coil and the tissue (phantom).

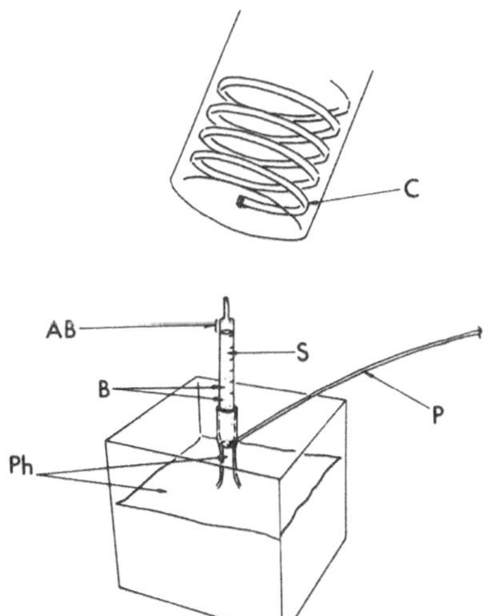


**Figure 16.** Experimental equipment. C=coil, P=probe, S=saline, AB=accumulated volume of bubbles, B=bubbles, Ph=phantom.

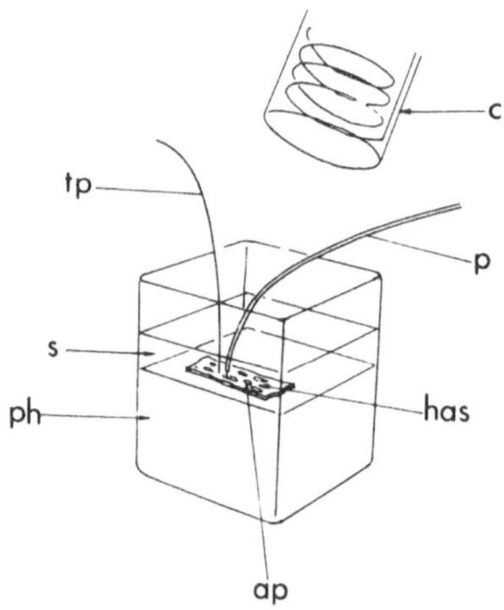


**Figure 17.**
Experimental equipment: c=coil, p=probe, tp=temperature probe, s=saline, has=human aorta specimen, ph=phantom, ap=atheromatous plaques.

Experiments on the human aorta specimens were done by using 150, 200, and 250 W and 2.5, 5, 10, and 20 s. The VSWR was adjusted to 1.2 and the coil-to-probe tip was fixed at 17 cm. Human cadaver abdominal aorta specimens less than 24 hours old and containing firm, raised, yellow atheromatous plaques were placed on phantom superstuff under normal saline. Different sizes of atheromatous plaques were vaporized and time of vaporization was recorded. Vaporized and control specimens were examined histologically. Temperatures delivered with different power applications and at different distances from the probe tip were measured by means of a fluoroptic temperature probe. Effects of changing the saline volume were also recorded. Relative location of the components of the experimental equipment are illustrated in Fig. 17.

B. Results

Table II exhibits the linear correlation between increase in RF power applied and the volume of microbubbles formed at the immediate probe tip. Within the power range of 80-250 W, the volume of bubbles varied from 0.08 to 0.14 ml and was directly proportional to the power applied.

Figure 18 shows the linear relationship between the size of atheromatous plaque and the time of vaporization. Figure 19 compares thermal profile in the vicinity of the probe delivered with the different wattage settings at each of the four time settings of the RF power application. Temperatures were measured at 1, 2, 3, and 4 mm from the probe tip. Figure 20 shows the effect of the volume of saline on the heat delivered. The smaller the saline volume, the greater the heat delivered during vaporization.

**Figure 18.** A plot of the atheromatous plaque area vs the time used for vaporization.

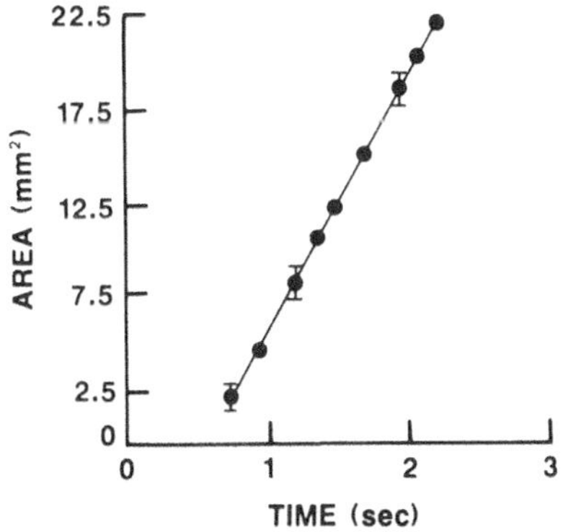


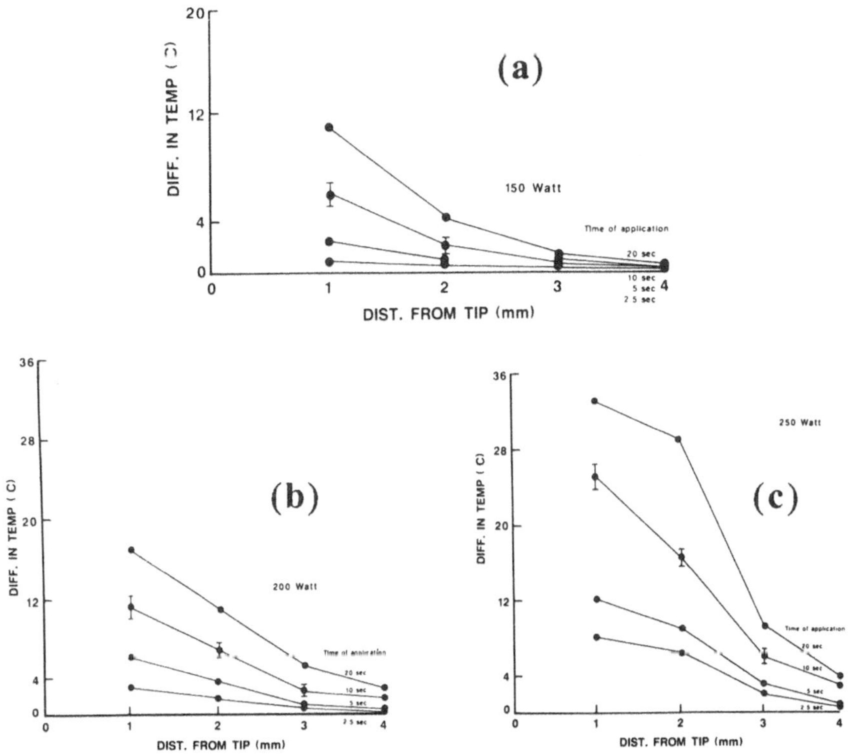


**Figure 19 A, B, and C.** Thermal distribution in vicinity of the probe tip at different powers and at 2.5, 5, 10, and 20 s.

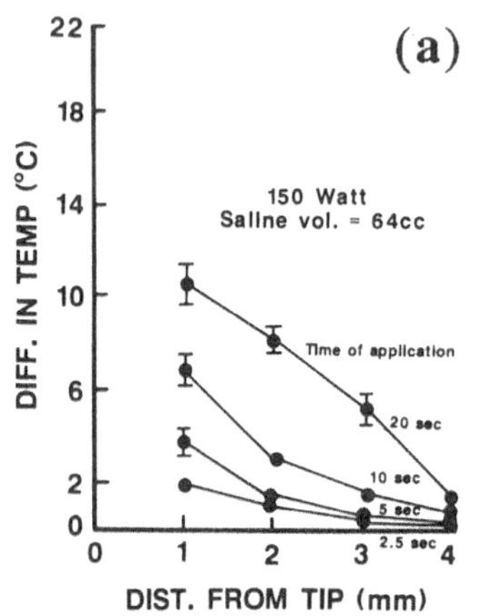


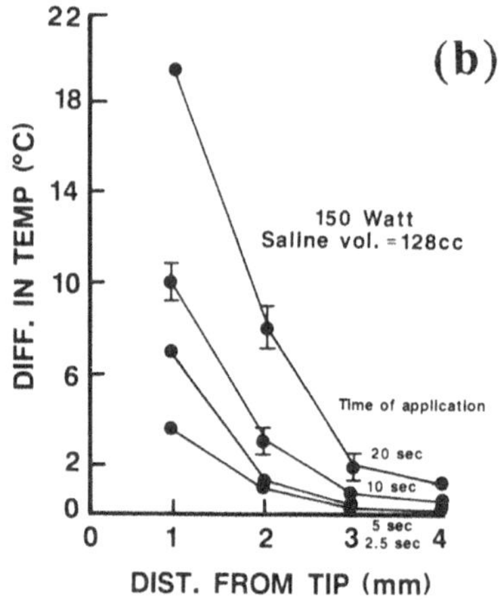


**Figure 20 A and B.** Dependence of temperature generated on the volume of saline.

Figures 21, 22, and 23 illustrate the histologic effects of the EFF application to a human atheromatous plaque with several shallow scalloped regions indicated by arrows. Clean vaporization with a thin layer of coagulum and viable cells are seen underneath the areas of the probe application.

C. Discussion

The EFF probe represents a new method that has potential use and advantages in angioplasty. The EFF probe produces high temperature at the point of contact with biological tissues resulting in clean vaporization. The probe is controlled with high precision and its heating pattern is maximum at the point of contact with the probe tip. The spread of heat to surrounding tissues is minimal.

This results in an advantage over laser angioplasty and avoids some of the latter's limiting factors[3,4] such as beam perforation, nonflexibility, and expensive fiberoptic catheters. The induced eddy current of EFF produces extremely intense heat at the probe tip, but this temperature is variable and easily controllable.[10] Vascular spasm is often observed in the use of lasers, but it is yet to be evaluated in the use of the EFF probe.

The coil must be positioned close to the operating field, but its orientation is not critically sensitive to the function of the probe. In general, the closer the

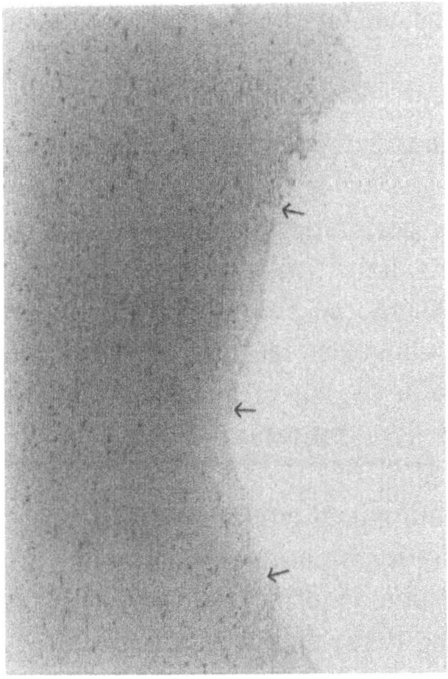

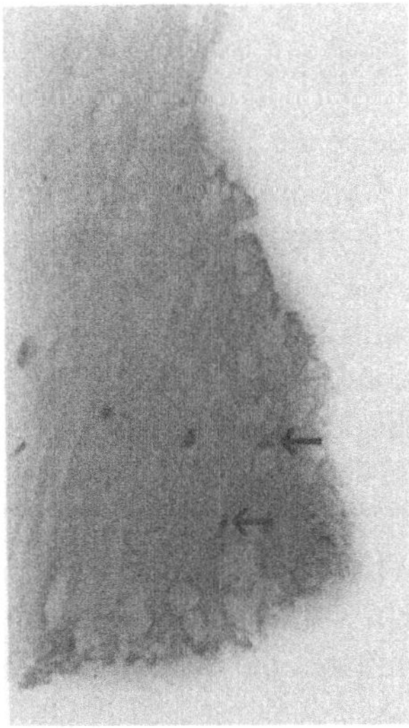

**Figure 21 (Left).** Histologic section of the EFF-treated cadaver aorta with atheromatous plaque. Notice several shallow scalloped regions (arrows).

**Figure 22 (Right).** EFF-treated cadaver aortic segment with atheromatous plaque and surviving cells (arrows). No adverse heat effects are immediately seen except in the regions most adjacent to the probe.

**Figure 23.** EFF-treated cadaver aortic segment with a relative smoothness representing the remnant of the tissue burned away by the heated probe.

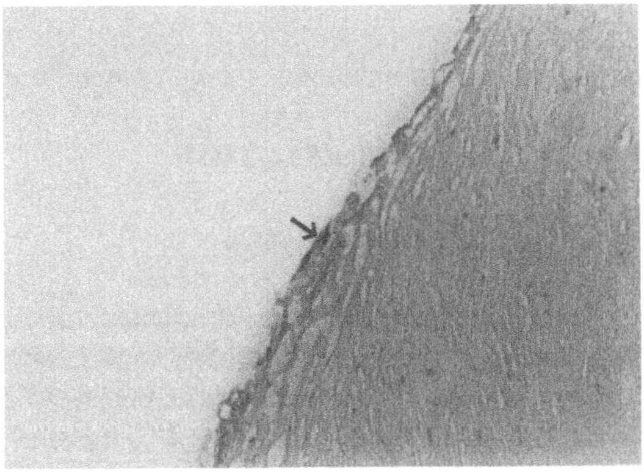

coil-to-probe tip, the higher the power delivered. Delivered power also depends on the magnetic flux density created by the coil, but the coil-probe distance gives the general guideline. Intermittent power application was found to be better than continuous power application. The latter decreases the probe's efficiency both electrically and thermally, probably due to carbon particles produced during vaporization. Heat delivered was found to be proportional to the power applied and to the duration of application. Measurements under smaller volumes of saline with high delivered temperatures indicated that smaller vessels need shorter application times or less power.

Since the EFF probe is a tactile device, there should be no perforation of a vessel wall unless sustained contact is made at the bend of a vessel when the power is applied. Although the issue of perforation is not addressed in this work, we have modified the probe tip so that it follows the electrically and thermally insulated flexible guide. This should substantially decrease the likelihood of perforation. This study demonstrates that gas bubbles produced with this technique are directly proportional to the power applied and, even at the highest wattage, the volume is not of embolic significance.

The distinct clear margin of vaporization path produced by the probe with the thin surface layer of coagulated tissue for normal endothelial growth demonstrates the relative lack of thermal trauma to the normal vessel. Viable cells were demonstrated in the surrounding tissue and were not affected by the adjacent heat.

### D. Conclusions

We believe the EFF probe has the potential as a surgical tool and for angioplasty with few complications. For safety, no grounding elements should be touching the patient, and surrounding monitor devices should be shielded from the electromagnetic field. As we have gained more experience with this device we have become more aware of its uses and relative safety. Further research into its advantages and limitations should expand its applications in preclinical trials.

## BRAIN TUMOR RESECTION

### A. Brain Tumor Treatment

The experiments were conducted to study heat distribution to the surrounding tissue in the brain after vaporization of intracerebral tumors in 19 rabbits. The follow-up period was as long as 47 days. Figure 24 shows a lesion produced in a rat by gently running a probe with a diameter of 0.3 mm over the cortical surface.

Using a fluoroptic temperature probe (Luxtron model 750), the following temperature change studies were performed.

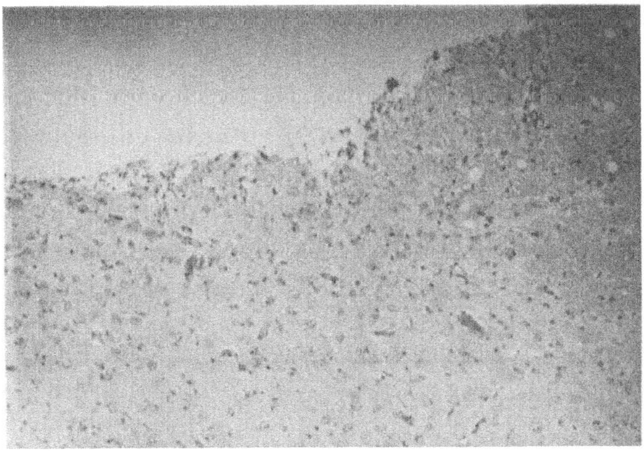

**Figure 24.** Lesion produced in a rat brain by gently running the EFF probe (0.3 mm diameter) over the cortical surface. The upper void represents vaporized tumor, and the rest represents the remaining normal cortical tissue. Notice virtual absence of thermal coagulation in the border.

1. Heating Curve as the Function of Distance from the Probe

At 130 W rms RF power, temperature rose at different distances from the probe tip at different time intervals, with the probe tip inserted to a depth of 2 mm in rabbit brains. The coil was 6 cm from the brain during this study.

2. Heating Patterns as the Function of RF Wattage

This study was conducted in a dielectrically uniform phantom (superstuff), which simulates the electric conductivity, dielectric constant, and magnetic permeability of biological tissue.[6] The solenoid is 17 cm long, with a diameter of 7.5 cm. The distance between the coil and the phantom was 6 cm. Total exposure time for each study was 5 s. Rises in temperature at different wattages and at different distances from the EFF probe tip were noted. The probe was inserted to a depth of 2 mm during this study.

3. Temperature as a Function of the Distance from the Probe tip

Using the superstuff phantom and the probe inserted to a depth of 2 mm, the temperature rise at different distances from the probe tip using 400 W of rms RF output power for 5 s of exposure time was studied.

4. Tumor Vaporization

Nineteen New Zealand rabbits were anesthetized with an intraperitoneal injection of ketamine hydrochloride and an intravenous injection of Nembutol.

At the level of the coronal suture on the right side, a buttoned 16 gauge Silastic catheter was introduced through a drill hole to a depth of 7 mm. The button served to hold the catheter in place. Four pieces of VX-2 tumor fragments 1 mm in diameter were introduced into the brain. The scalp was closed over the button.

After 3 weeks the incisions were opened, catheters removed, and craniectomies were preformed over the drill hole site. An operating microscope was used to view the catheter tract to its end. The tract was widened by making a cortical incision with the EFF probe. The mass at the end of the tract was vaporized, and the tissue for biopsy was obtained. Vaporization was done with an average rms power of 116 W and coil-to-probe tip distance of 16 cm.

Larger vessels on the cortical surface were coagulated prior to incision using the EFF probe at 40 W. The tumor vaporization was done by gentle, quick, sweeping movements of the probe over the tumor surface. Tumors that extended too deeply were only partially vaporized. The cranial defects were covered with Gelfoam slaps, and the wounds were closed. The animals were killed at varying intervals ranging from 3-47 days. The brains were removed, fixed in formalin for 5 days, and serial sections 1 mm thick were examined for hemorrhage and edema using the operation microscope. Histologic study of the tumor bed was preformed using hematoxylin and eosin stain.

The heating curve as a function of distance from the probe tip (Fig. 25) showed minimal increase in temperature at 2 mm from the probe tip with only a 4°C rise in temperature after 5 s of sustained exposure. The temperature changes were less at 3 mm and absent at 4 mm. The heating pattern as a function of RF output wattage (Fig. 26) showed a temperature rise of 250°C at the probe tip, but a sharp drop as the distance increased. Higher temperatures were noted with higher wattages. The heating pattern as a function of the coil-to-probe tip distance (Fig. 27) showed that the temperature rise increased with decreasing coil distance and sharply decreased with increasing probe distance.

Vaporization of the tumor was bloodless and quick. The vaporization procedure resembled wiping the tumor from the brain surface. The procedure could be performed layer by layer, leaving the brain surface unharmed. The probe tip had to be cleaned from time to time. Less charring was seen with higher wattages. The procedure was found to be similarly efficient even under the flow of saline solution and cerebrospinal fluid. The average diameter of the tumors was 6 mm. On gross examination, the brain surface showed no signs of edema. Similarly on serial section, no edema or hemorrhage was seen. In histologic studies on 14 of the 19 rabbit brains, no changes adjoining or distal to the tumor bed were seen. Of the remaining five brains, two showed edema measuring 0.2 mm and 1 mm thick adjoining the tumor bed; two showed

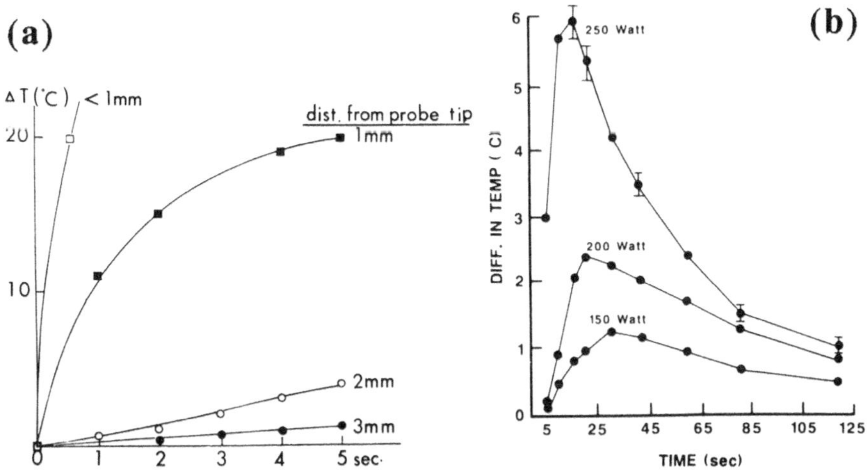


**Figure 25 a and b.** (a) Heating curve as a function of distance from the probe tip. (b) Cooling curve as a function of distance from the probe tip after 5 s heating with power settings 250, 200, and 150 W. Uninsulated probe tip was 2 mm long and temperature was measured 3 mm from the tip of the probe, which was inserted 2 cm into the phantom and 17 cm from the coil.

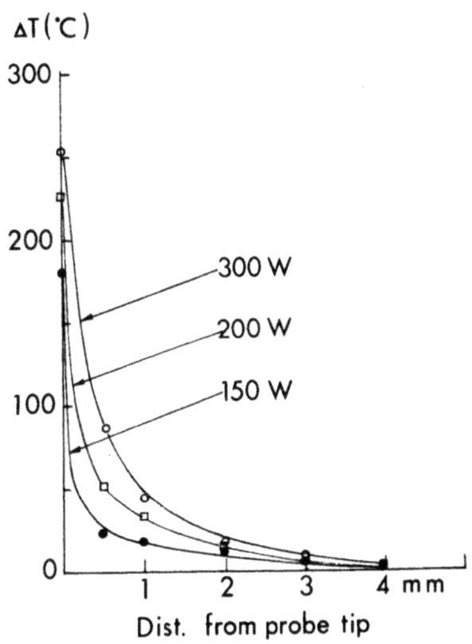


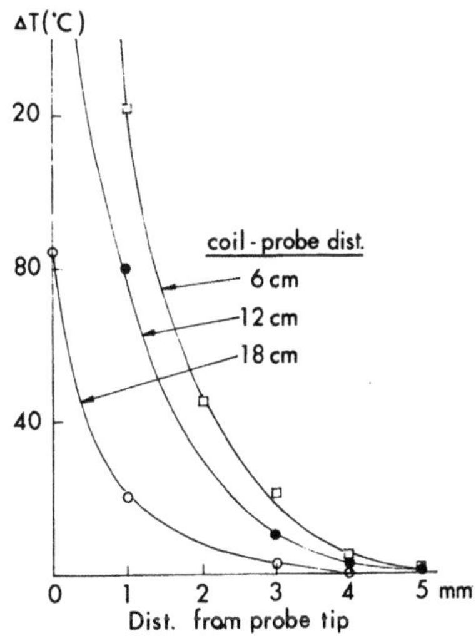


**Figure 26 (Left).** Thermal distribution in a phantom in the vicinity of the probe tip with high RF power (400-600 W rms).

**Figure 27 (Right).** Dependence of heating on the distance between the probe and coil.

hemorrhage 1.3 mm and 2 mm thick, and one showed both hemorrhage (0.3 mm) and edema (0.4 mm).

The EFF probe enables bloodless removal of tumors without disturbing the surrounding brain tissue, which is probably the most important factor in reducing morbidity and mortality. In addition, it produces intense and circumscribed pinpoint heating. Because the heat produced is very focal, even with a rise in temperature up to 250°C at the probe tip, there is minimal rise in temperature at a distance of 2 mm from it, even after sustained exposure for 5 s (Figs.. 24 and 25). Intense heat is produced during vaporization only when the probe touches the tissue. As the result of the quick, sweeping technique, each contact time is less than a fraction of a second. The histologic studies were remarkable in their paucity of observation of edema and hemorrhage. Contact vaporization also gives the surgeon tactile feedback, which is lacking with lasers.

Because heating is produced in tissue irrespective of consistency or pigmentation, the EFF probe can simultaneously coagulate vessels and vaporized tissue, making the procedure bloodless. Because the system works efficiently even under flowing saline solution and cerebrospinal fluid, intraventricular procedures could be performed with ease and allow irrigation with saline solution during use.

Adequate shielding of pacemakers against the RF electromagnetic field produced by the EFF probe should be investigated before use on paced patients with tumors and atheromatous plaques.

As long as the coil is directed to the operative field, its position, such as the angle at which it is pointed toward the patient (phantom, animal), is relatively unimportant. It can be placed anterior, posterior, or lateral to the head and thus can be kept outside the surgical field. The coil distance is an important factor. In general the closer the coil is to the tissue, the higher the power delivered (Fig. 27). For tumor resection in rabbits, the coil distance of 16 cm was ideal.

The probe can be constructed of thin wire; therefore, it could also be used through endoscopic devices. A prototype of the EFF system is shown in Fig. 28.

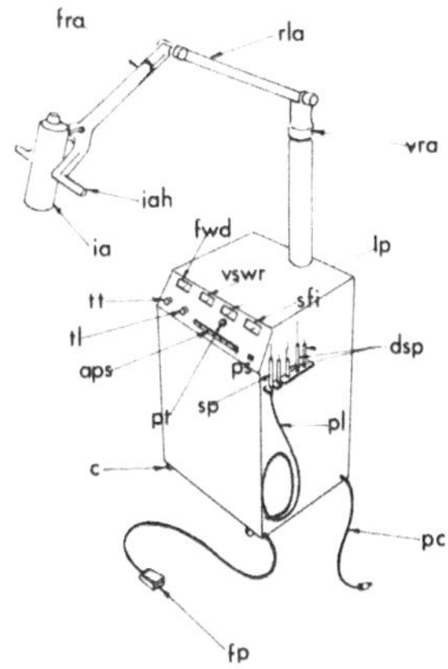


**Figure 28.** External view of EFF unit. ia=inductive applicator, iah=inductive applicator handle, fra=front rotating arm, rla=rear leveling arm, vra=vertical rotating arm, fwd=forward RF power meter, vswr=voltage standing wave ratio meter, lp=probe current meter, pt=probe tune, tt=transmitter tune, tl=transmitter load, ps=power switch, sfi=synthesizer frequency indicator, aps=application/power setting (preset), sp=surgical probe, pl=probe line, dsp=detachable spare probe, fp=foot power pedal, pc=power cable.

## CLINICAL STUDY

A clinical study was conducted to assess the effectiveness of the EFF system in neurosurgery for cutting of tissue, coagulation of vessels and vaporization of tissue. The safety of the system and convenience were also assessed.

The protocol consisted of neurosurgical patients with tumors. Patients with pacemakers were excluded from the study. Initial approval for the Institutional Review Board and FDA were obtained to use this system as an investigative device. Patients were invited to enter into the study. The final decision to use the system was made at the time of the operation.

### A. Materials and Methods

The study period was over sixteen months. It included patients with intracranial and intraspinal tumors. One patient had a supraclavicular neurofibroma. The intracranial lesions included lesions in the supratentorial as well as the infratentorial space. Those tumors that were soft enough to be easily sucked out using ordinary suction were not treated with this probe. Wattage between 125 and 150 were used during the surgery. The probe was used for coagulation of surface vessels, cutting adhesions and arachnoidal bands, incision into cortex and tumor surface, and vaporization of tumor.

Vaporization of the tumor was done using quick rapid movement over the surface of the tumor. Simultaneous suction was used to suck the gas that was generated by the vaporization.

Special precautions were also taken. The unipolar cautery was disconnected during the use of the system and a special effort was made to avoid any metal object that might serve as a grounding device from touching the patient.

B. Results

Table III gives the breakdown of the different types and locations of tumors that were excised using this system:

TABLE III

| LOCATION OF TUMOR | TYPE OF TUMOR | No. OF CASES |
|---|---|---|
| Intracranial | | |
| a. supratentorial | glioma | 6 |
| | meningioma | 3 |
| | neurocytoma (intraventricular) | 1 |
| b. infratentorial | acoustic neuroma | 1 |
| | metastatic tumor | 1 |
| Intraspinal | meningioma | 1 |
| | metastatic tumor | 1 |
| Supraclavicular | neurofibroma | 1 |
| **TOTAL** | | 15 |

The system was found to be effective for incising cerebral cortex and tumor. Most vessels coagulated during cutting except those vessels which were larger than a millimeter that tended to bleed after cutting. Larger vessels were coagulated with bipolar cautery and then cut using the EFF probe. Arachnoidal adhesions could be easily cut. Similarly tissues that were extremely firm could also be cut with the EFF probe. The EFF probe was particularly useful for vaporization and this was most evident for very firm tumors where the ultrasound suction device failed to perform. This was evident in the case of neurofibroma where we initially attempted to suck the tumor using the ultrasound suction device, and failed to do so. However, using the EFF probe it could be easily debulked by vaporization. During vaporization of tumors that were vascular, very minimal bleeding was observed. The vaporization could be achieved rapidly and precisely. Only the tissue that was touched vaporized. As the probe diameter is extremely small (0.5 mm), this instrument could be inserted through narrow openings into deep wounds. Minimal to no charring was seen on the probe tip except when the tumor bed had blood or necrotic material.

No post-operative complications were observed as a result of use of the system. Post-operative scans reveal no evidence of damage to surrounding structures (Figs. 29a, 29b, 30a, 30b overleaf). Electrocardiogram and evoked potential monitoring and TV camera function was interfered with during the use of the system but returned to normal when the system was not in use. None of the patients appeared to have any burns even though sharp pinpoint fixation pins were used for craniotomy procedures.

C. Discussion

The EFF system was found to be an effective tool for cutting, coagulation and vaporization. The capability of the system to produce sharp focusing of the electromagnetic field produces a safe way of vaporizing tumors. This capability reduces the risk of damage to surrounding structures and also reduces blood loss during vaporization of the tumor. The critical part of the system is the fine tuning of the probe, which enables sharp focusing, so that any other sharp objects are not tuned and do not produce the heating pattern. This made the use of standard surgical instruments, including the three point fixation of the head, safe. As a precaution, the unipolar cautery was disconnected and the patient was kept free of grounding.

Several advantages were observed with the use of this system. The advantages included the fact that during cutting and coagulation, minimal to no bleeding was observed. This reduced the need to coagulate vessels as the tumor was being debulked. This is an advantage over the $CO_2$ laser or the ultrasonic suction device, where bleeding can be a problem. The EFF probe offers the capability of tactile feedback which is lacking with the $CO_2$ laser. The feedback enables the surgeon to vaporize structures precisely and even discriminate

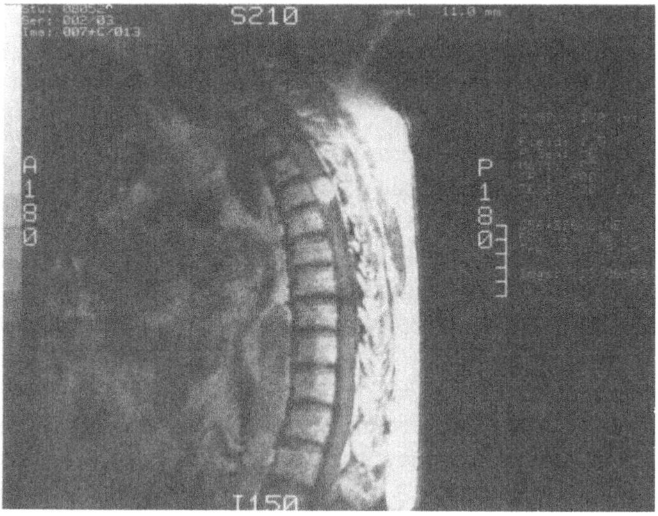


**Figure 29 (a).** Scan of a 57 year old male showing spinal cord soft tissue at scan section number 210.

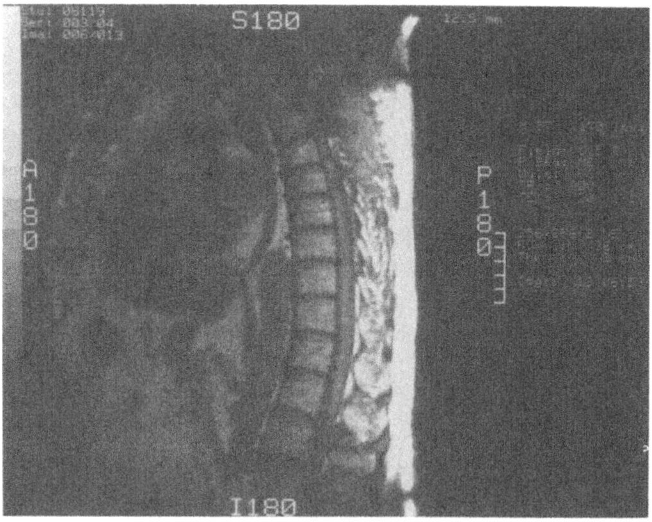


**Figure 29 (b).** Scan of the same patient showing spinal cord tissue at scan section number 180.

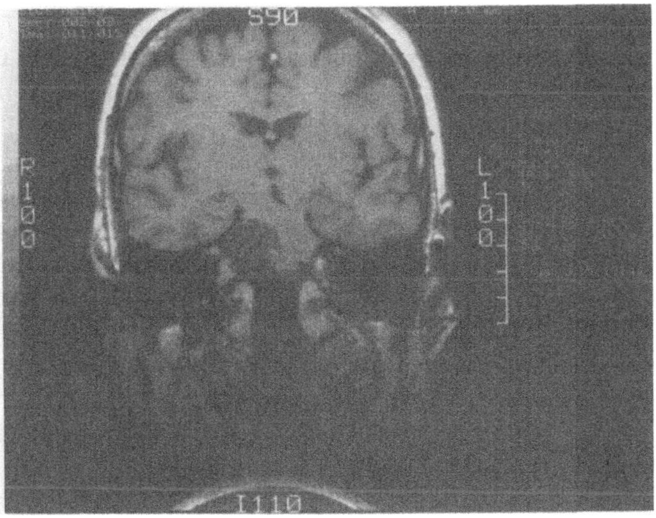


**Figure 30 (a).** Scan of a 73 year old female showing tumour site at scan section number 90.

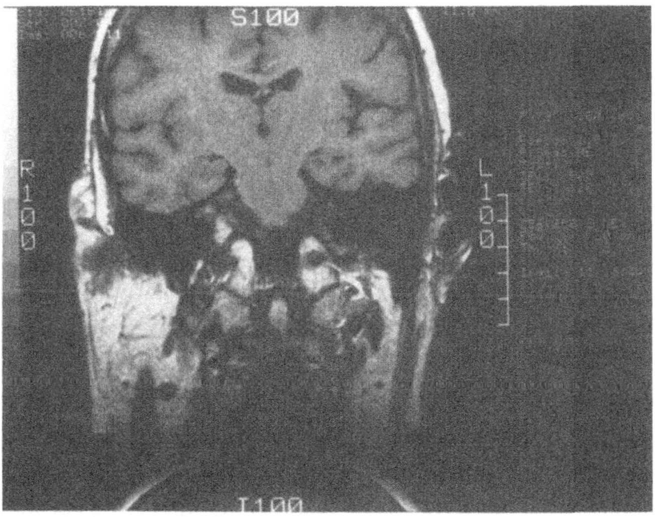


**Figure 30 (b).** Scan of same patient at scan section number 100.

between normal and abnormal tissue. The small size of the probe enables one to insert the probe through narrow openings and into corners of deep wounds, which would be impossible to do with an ultrasonic suction device. The system was superior to the ultrasonic suction device while vaporizing firm tumors. In the case of neurofibroma, the ultrasonic suction device failed to work and this was easily vaporized by the EFF system.

One of the biggest problems with the system is the inability to use evoked potential monitoring during the actual procedure because of the interference from the EMF with this system. The use of special shields on the coil might prevent this problem. Although this is only preliminary data, and a large patient group would be needed to document effectiveness of this system over other systems, the present data suggests that this would be a valuable tool for, not only neurosurgical, but other surgical disciplines as well.

## SUMMARY

The EFF probe cuts, vaporizes, and cauterizes biological tissue, except bone. Cutting and vaporization is accomplished by cauterizing. Cauterizing alone can be attained at a lower power setting. Studies using the EFF probe on in vivo and in vitro models included treatment of aneurysms, brain tumor vaporization, angioplasty, and endarterectomy. These studies demonstrated the strengths of the EFF probe, such as tactile operation (action on touch only), precision in cutting, ease in locating the probe to the lesion, and negligible trauma in surrounding tissues, including absence of perforation hazard in intravascular procedures. The probe, with its simplicity, lower cost, and wide range of applicabilities, challenges the $CO_2$, YAG, and other surgical laser devices.

### References

1. Barry KJ, Kaplan J, Connolly RJ, et al. The effect of radiofrequency-generated thermal energy on the mechanical and histologic characteristics of the arterial wall in vivo: Implications for radiofrequency angioplasty. Am Heart J 1989; 117:332-341.

2. Barsoum MS, Mooro HA, Boulous FI, et al. The complications of injection sclerotherapy of bleeding esophageal varices. Br J Surg 1982; 69:79-81.

3. Cumberland DC, Sanborn TA, Taylor DI, et al. Percutaneous laser thermal angioplasty: Initial results with a laser probe in total peripheral arterial occlusions. Lancet 1986; 1:1457-1459.

4. Geschwind H, Boussignac G, Teisseire B, et al. Laser angioplasty: Effect on coronary artery stenosis. Lancet 1983; 2:1134.

5. Geshwind HJ, Teisseire B, Boussignac G, et al. Laser angioplasty of arterial stenosis. Cardiovasc Intervent Radiol 1986; 9:313-317.

6. Guy AW, Webb MD, McDougall JA. Critical comparison RF field delivery methods to thermal responses of models and experimental animals. USAF/SAM contract 1976; final report.

7. Jackman WM, Kuck KH, Naccarelli GV, et al. Radiofrequency current directed across the mitral anulus with a bipolar epicardial-endocardial catheter electrode configuration in dogs. Circulation 1988; 78:1288-1298.

8. Kricheff II, Berenstein A. Simplified solid-particle embolization with a new introducer. Radiol 1979; 131:794-795.

9. Lee G, Seckinger D, Chan MC, et al. Potential complications of coronary laser angioplasty. Am Heart J 1984; 108:1577-1579.

10. LePage JR, Yamanashi WS, Patil AA, et al. Induced current convergence (ICC), a new method for angiopyroplasty, 73rd Scientific Assembly and Ann Meeting, Radiological Society of North America, Chicago, Illinois. 1987; p 128.

11. Rousseau H, Puel J, Joffre F, et al. Self expanding endovascular prosthesis: An experimental study. Radiol 1987; 164:709-714.

12. Sanborn A, Haudenschild CC, Gaber GR, et al. Angiographic and histologic consequences of laser thermal angioplasty; comparison with balloon angioplasty. Circulation 1987; 75:1281-1286.

13. Slager CJ, Essed CE, Schuurbiers JCH, et al. Vaporization of atherosclerotic plaques by spark erosion. J Am Coll Cardiol 1985; 5:1382-1386.

14. Sigwart U, Puel J, Mirkovitch V, et al. Intravascular stents to prevent occlusion and restenosis after transluminal angioplasty. N Eng J Med 1987; 316:701-706.

15. Yamanashi WS, Yassa NA, Hill DL, et al. Properties of electromagnetic field focusing probe. Angiolo 1988; 39:953-959.

16. Zanetti PH, Sherman FE. Experimental evaluation of a tissue adhesive as an agent for the treatment of aneurysms and arteriovenous anomalies. J Neurosurg 1972; 36:72-79.

# BIOLOGICAL EFFECTS OF PULSED HIGH PEAK POWER ELECTROMAGNETIC ENERGY USING DIAPULSE®

Jesse Ross

## INTRODUCTION

After meeting the inventor of Diapulse, the late Dr. Abraham J. Ginsberg, in 1955, I became intrigued with the medical possibilities of non-thermal, high frequency, pulsed, high peak power electromagnetic energy. I proceeded to establish research in the United States and at international universities and hospitals to explore the beneficial biological and biophysical effects of this technology. The specific parameters of Diapulse equipment are: 27.12 MHz (11 meter band); pulse repetition rate of 80 to 600 pulses per second; 65 microsecond pulse width; power range per pulse between 293 and 975 watts; duty cycle between 1/2 of 1% to 3.9%; square pulse with rise and fall time less than 1%; and ability to tune the patient to resonance at the applicator.

There are three basic processes I consider necessary before healing can occur: elimination of edema, absorption of hematoma, and increased blood-flow.[21] A review of the beneficial effects reported by researchers utilizing Diapulse found that Diapulse accelerated these processes with no reported contraindications. The research on Diapulse from the early 1930's to 1981 includes laboratory, animal, and control blind and double-blind clinical studies on the acceleration of wound healing.[22].

Diapulse has been reported to have beneficial effects for the prevention and relief of acute and chronic pain.[23] When trauma is induced, surgically or accidentally, three processes appear to be triggered. The first response to trauma is electrical, the second is electrochemical, and the third is chemical. I believe that Diapulse accelerates the return of the electrical charge to damaged tissue, and hastens the electrochemical and chemical responses. Therefore, it may be able to accelerate the healing processes by as much as 50%.[22] Several standard medical texts refer to the value of the application of Diapulse [5,17,25,30] for accelerated reduction of edema and pain.

This chapter provides an updated review containing additional research demonstrating the beneficial effects of Diapulse in reducing edema and pain,

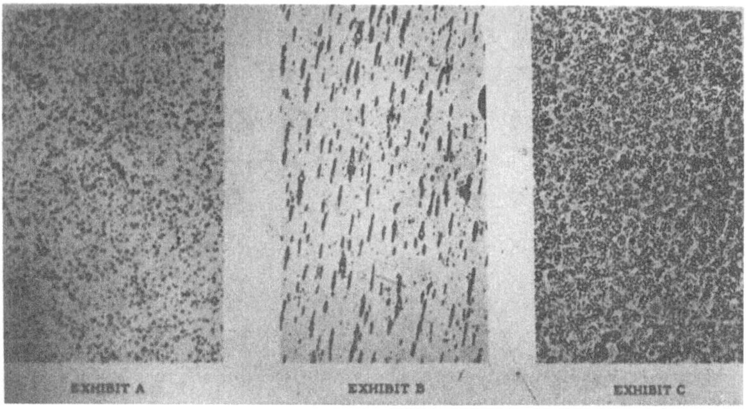


**Figure 1.** Pearl Chain Phenomenon. (Courtesy of Ginsberg.[6])

thereby accelerating healing in surgical and traumatic injuries, and reducing the requirements for analgesics.

## PEARL CHAIN PHENOMENON

An introduction to our present understanding of the non-thermal effect of Diapulse energy should include a description of the Pearl Chain Phenomenon representing some of the earliest data on Diapulse. In order to demonstrate this non-thermal effect, fat globules of milk were exposed to Diapulse. Electrodes were placed on either side of the slide, and the energy introduced. A Pearl Chain effect was observed. When the energy was eliminated, the globules returned to their original random positions. When exposed to any measurable heat, the globules formed an agglomerated pattern which was irreversible.[6] As shown in Fig. 1, the same effects are observed in vitro on fat globules of milk and on a variety of cells.

In 1961 a control study demonstrated the effect of Diapulse on the acceleration of wound healing in dogs.[4] The author observed; (a) the acceleration of transverse alignment, and (b) that collagen is deposited in the transverse alignment-line of stress, as compared to the controls.

It has also been reported that, macroscopically, Diapulse treated nerves of rats post operatively appeared less swollen and that the ease of removal of treated nerve specimens was marked.[19] There was less intraneural edema and more orderly arrangement of endoneural collagen fibers in treated animals as compared to the controls. Another study on nerves illustrated the accelerated cell

**Figure 2.** is a photograph of the regenerated spinal fibers traversing the site of injury, showing lack of scar tissue formation in the treated nerve (B), as compared to the control (A). (Figure from Ghista et al,[5] with permission.)

organization in treated tissue as opposed to the untreated tissue in sacrificed animals.[5] The study reported the ease of dissection in the treated group of nerves, and observed there was less fibrosis and scarring around and within the nerve from the Diapulse treated animals (Fig. 2).

## PERIPHERAL NERVE TISSUE REGENERATION

Early research also indicated accelerated peripheral nerve tissue regeneration[28] as shown in Table I. Expansion of this research confirmed the findings.[29] These works and others[18] on accelerated healing of nerve injuries were reviewed earlier.[22] These findings have been confirmed including the accelerated recovery of injured limbs, the degeneration, regeneration,[19] and maturation of myelinated axons, and the reduction of fibrosis using Diapulse energy.[20]

The clinical effect of Diapulse after microsurgical nerve repair on 640 patients, evaluated according to international criteria,[3] showed that sensory and

**Table I.**

**Fiber Density in Normal Rat Nerves. Average Fibers/mm$^2$ in Nerves from Ten Rats:**

| | Range | Average |
|---|---|---|
| Less than 8.3 | 4.88-8.54 | 6.71 |
| More than 8.3 | 3.60-4.42 | 3.94 |
| Total | 8.98-12.48 | 10.64 |

Nerve Fiber Density/mm$^2$ 30 days after section and resuture

| Treated with Diapulse | Controls |
|---|---|
| 5.63* | 0 |
| 10.26 | 0 |
| 10.35 | 3.00 |
| 10.91 | 4.00 |
| 11.12 | 5.00* |
| 12.70 | 5.59 |
| 13.10 | 5.59 |
| 14.22 | 6.96 |
| 15.28 | 7.50 |
| 15.33 | 13.00* |

*Regeneration limited to part of the section

Nerve fiber density/mm$^2$ 30 days after section and resuture, Diapulse treated, and controls. (Table from Ghista et al,[5] with permission.)

motor recovery is achieved two to three months earlier in the patients treated with Diapulse as compared with the patients in the control group where Diapulse was not applied.[13,14] These results showed that fascicular sutures with Diapulse produced better motor and sensory recovery. Three cases of median nerve artery suture in combination with fascicular suture and Diapulse resulted in total recovery with a two point discrimination test of 0.4, four months after surgery. Diapulse indicated a clearly positive effect, with the Tinel sign appearing much more distal than usual, 48 hours after surgery.

## SPINAL CORD CRUSH INJURIES

The earliest control study using Diapulse was on spinal cord regeneration following hemicordotomy in cats.[29] When the animals were sacrificed three months post-surgery, the extent of the scar tissue of the Diapulse treated animals was significantly less than in the untreated animals. There were abundant regenerating neurons traversing the region of the cordotomy of the treated group.

Functional recovery and altered calcium was observed following standard crush injury of the cat spinal cord.[32] The Diapulse field applied to spinal cords shortly after injury reduced calcium accumulation in the cord adjacent to the injury site. This change was consideıed to be associated with improved recovery and motor and sensory function in severe spinal injury. Additional reports on the use of Diapulse to alter calcium in spinal cord injury gave additional evidence of Diapulse effect to alter calcium in spinal cord injury.[33,34] In another study, these same investigators discussed an objective test of descending motor function which supported the efficacy of Diapulse to restore motor function following spinal injury.[8]

Another study on Diapulse alteration of calcium and functional recovery in spinal cord injury was reported.[35] This study stated that the calcium concentration increases several fold and that these changes were consistent with massive calcium influx into cells, believed to cause axonal damage in spinal cord injury. Diapulse applied for 2 hours, commencing 45 minutes post-injury, reduced tissue calcium accumulation by 30% in cats' spinal cord, at 3 hours after severe contusion, without altering concomitant shifts in tissue, sodium, potassium, and water contents. These data indicate that the effects of Diapulse are not due to general changes in cell membrane ion permeability, metabolism, or ionic transport. Nor do they appear to be due to thermal mechanisms. It would appear therefore, that very little or no edema develops, especially in conditions of extreme stress within 45 minutes. The immediate application of Diapulse restores the electrical charge, eliminating the necessity for the sodium and water to accumulate.

A clinical study performed on 97 patients with complete or deep spinal cord lesions which were treated with Diapulse reported a pronounced neurological recovery observed in 38 patients (ie, 40%).[16] It should be noted that the data on the 97 patients included 10 that died of complications mainly due to respiratory causes in the early post-traumatic period. The authors stated that, remarkably, in 28 individuals the recovery had substantial functional value. The patients were discharged from the neuro-orthopedic department with paresis slightly impairing the function of the extremities. It is noteworthy that all patients who presented with paralysis of the extremities with traces of deep sensibility at the time of admission experienced a marked neurological recovery.

## TRAUMA AND BURN WOUNDS

The application of Diapulse also appears efficacious in the treatment of hand burn injuries, severe hand trauma, finger transplants, nerve repairs, and facial surgery.[10] Results over a five year period on over 5,000 injuries demonstrated accelerated wound healing with Diapulse in all categories.

Diapulse has also been reported to affect the dynamics of enzymes in traumatic and burn wounds in humans.[11] Data gathered on the amount of proteins and some principle enzymes in normal, traumatic, and burned tissue, before and after Diapulse, showed enzymatic changes in epidermal cells with samples taken from; (a) the traumatized and burned areas before treatment, (b) nearby undamaged area (the control section), and (c) taken 24 hours after the injury and after two applications of Diapulse. This report suggests that compared to normal skin (control), the activities of protein, lactate dehydrogenase (LDH), glutamic-oxalacetate (GOT), glutamic-pyruvic transaminase (GPT), and alkaline phosphatase were significantly modified after Diapulse treatment. It appeared as if the earlier this therapy was applied, the more rapidly the normal enzymatic activates were restored. The treatment was effectively applied through bandages and surgical dressings. The data demonstrate statistically significant accelerated healing with the utilization of Diapulse on traumatized and burned tissue.

Accelerated healing of over 2,000 patients during a nine year period with the application of Diapulse has been reported with some patients with traumatic lesions, burns, inflammatory diseases, severe evoluted necrotic lesions, and radionecrosis.[12]

## PLASTIC SURGERY

There are reports on accelerated healing of rhinoplasty,[9] and donor sites[7] treated with Diapulse.

Reduction of capsular contracture after two-stage augmentation mammaplasty was reported on 231 patients (462 implants) over a period of three years.[27] At the time of the second surgical stage, 216 cavities were found to have hematoma. After removal of the hemotoma and completion of the second stage, only 41 breasts developed capsular contraction. Treatment of the 41 with Diapulse, massage, and closed capsulotomy completely eliminated all capsular contracture. The period of observation ranged from three months to two and a half years.

## HAND INJURIES

Diapulse has also been used on hand injuries, in particular on tenosynovitis and tendovaginitis.[31] Diapulse accelerated reduction of pain and swelling and was considered especially valuable because it can be administered through strapping or through plaster casts, and is unattended therapy.

In another study of hand injuries 230 randomly selected patients presenting a variety of injuries, 115 were treated with Diapulse and 115 were used as controls.[1] These patients showed that hand injuries treated with Diapulse within 36 hours of injury had a marked reduction of swelling in three days, compared with the controls which had increased swelling until the seventh day. The reduction of swelling had a statistically significant effect on function and pain. From the larger group, 30 pairs were matched for age, weight, sex and degree of trauma. All patients were examined, measured and assessed on alternate days for swelling, pain and disability. The results showed that the application of Diapulse had a biological effect on recently injured soft tissue, especially in the reduction of swelling, the diminution of pain, and the improvement in function. No adverse effects of any kind were observed either during or after the trial as illustrated in Fig. 3.

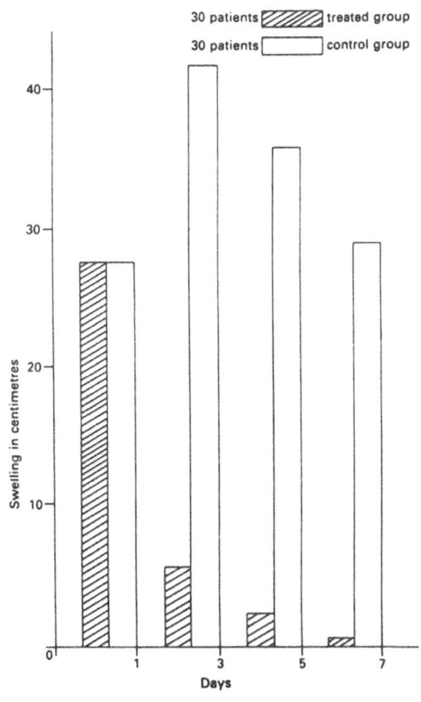


**Figure 3.** Normalized histogram, combined swelling of 30 matched pairs. (Figure from Barclay et al,[1] with permission.)

## RHEUMATOLOGY AND DECUBITUS ULCERS

Diapulse has been used in treatment of 200 rheumatic patients.[2] Diapulse constituted an important method in rheumatology that gave favorable results in relief of pain and inflammation in degenerative and non-articulated arthritis, with an improvement of 80 to 100% in each patient. Another study reported on use of Diapulse on infected pressure ulcers.[24] In a study of a group of eleven patients, seven were completely healed and surgery for grafts was successfully accomplished in the remaining four cases.

## BRAIN EDEMA, HEAD TRAUMA, COMA

The effect of Diapulse on cerebral edema in rats has also been reported.[15] A control study found that Diapulse treated rats showed less cerebral edema. Intravital dye studies showed seepage of dye at the injury site of non-treated animals. The authors postulated that the energy: (a) potentiates cell membrane integrity, (b) augments the efficiency of the sodium pump, (c) prevents accumulation of fluid by regulating calcium channels and protects piaglial cell membranes as well as vascular capillary endothelium so that the accumulation of interstitial fluid is minimized or prevented, (d) protects against accumulation of

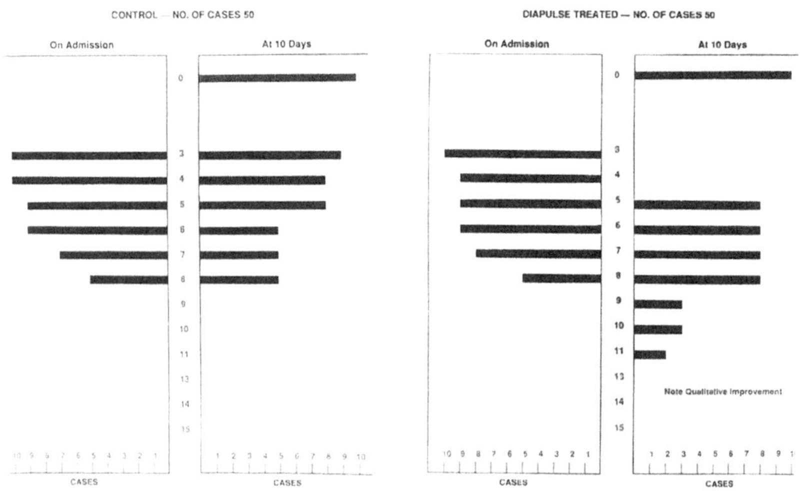


**Figure 4**. Clinical Assessment, GCS, Control, 50 Cases (left), Treated, 50 Cases (right). (Figure from Jayakumar et al,[15] with permission.)

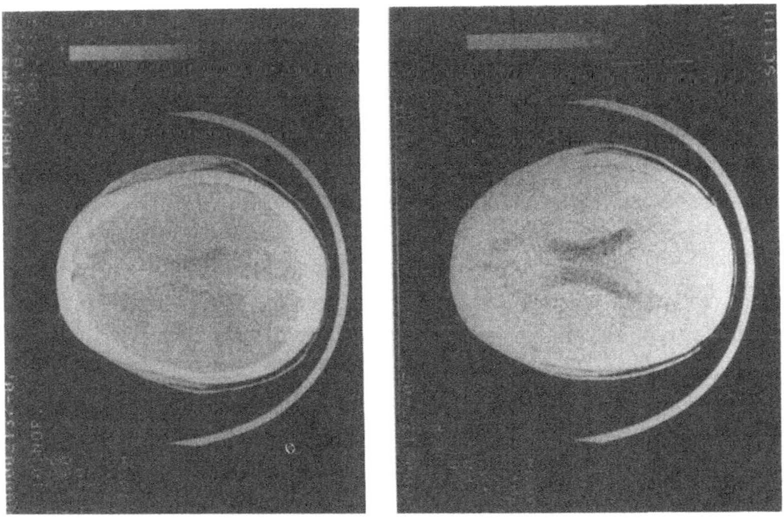

**Figure 5.** CT Brain Scan on admission (left image), and after Diapulse treatment (right image). (Figure from Jayakumar et al,[15] with permission.)

interstitial fluid, (e) augments venous return, (f) improves movements of interstitial fluid to absorption sites.

Human studies have also been performed with the use of Diapulse in the management of head injuries.[26] One hundred patients with severe head injuries with similar Glasgow Coma Scale (GCS) scores of eight or less were included in the study. Fifty patients were treated with Diapulse and fifty were used as controls. Only patients with edema and contusions were included. Cases with surgically treated lesions such as hematomas were excluded. Alternate cases were treated with Diapulse for 30 minutes twice daily at 12 hour intervals. The untreated patients showed little or no improvement on the GCS (Fig. 4), whereas the Diapulse treated patients showed significant improvement on the Scale (Fig. 4). Clinical observations, as well as CT scans (Fig 5), showed improvement in the Diapulse treated cases compared to the untreated group. In contused brain syndromes, the use of Diapulse, a non-invasive procedure, eliminated the need for dehydration or steroid therapy. The exposure to Diapulse seemed to prevent and/or reduce cerebral edema. Cases with similar frontal craniotomies were also exposed to Diapulse and studied. Facial and eyelid edema resolved faster than the untreated cases.

## CONCLUSION

The accumulated clinical research during the past thirty-four years with control, double-blind studies performed in universities and medical centers in many parts of the world demonstrate the beneficial biological and biophysical effects of non-thermal, high frequency, pulsed high peak power electromagnetic energy as delivered by the Diapulse apparatus.

Current research has been in progress for several years in universities and medical centers in various countries. Preliminary observations indicate that these studies will expand the applications of Diapulse.

Diapulse technology has emerged as a valuable tool in clinical medicine today. Current use provides physicians and surgeons with a means for adjunctive patient care in cases where edema is a complication to the healing process. Implications for future research and applications include:

1) enhanced recovery time of injuries which normally require long-term care;

2) reduced use of analgesics and medication without compromising the patient by side effects;

3) reduction of nursing care due to the ease of application of this non-invasive, unattended therapy with no detriment to the operator; and

4) a reduction in hospital stay with associated reduction in hospital costs for patients, the government, hospitals, and insurance companies.

### References

1. Barclay V, Collier RJ, Jones A. Treatment of various hand injuries by pulsed electromagnetic energy (Diapulse). Physiotherapy 1983; 69(6):186-188.

2. Borrachero S. Experiencia con diapulseterpia en el tratamiento de enfermos reumaticos. Neuvas Tecnicas Diagnosticas y Terapeuticas en Patologia del Aparato Locomotor 1986; MAPFRE 157-163.

3. Bowden REM. Factors and influencing functional recovery. In: Peripheral Nerve Injuries (Ed. HJ Seddon). MTC Special Report Series No 282, London; HMSO. 1954; 298-353.

4. Cameron BM. Experimental acceleration of wound healing. A J of Ortho 1961; 3(11):336-343.

5. Ghista DN, Frankel HL. In: Spinal Cord Injury Medical Engineering. Charles C. Thomas, Springfield, IL. 1986; 125-129.

6. Ginsberg AJ. (abstract) Pearl chain phenomenon. 35th Annual Meeting of the American Congress of Physical Medicine and Rehabilitation. 1958; 36:112-115.

7. Goldin MB, Broadbent NRG, Nancarrow JD, Marshall T. The effects of Diapulse on the healing of wounds: a double-blind randomised controlled trial in man. Br J of Plast Surg. 1981; 34:267-270.

8. Gruner JS, Young W. (abstract) An objective test of descending motor function to study the efficacy of pulsed electromagnetic fields and ganglioside treatments on spinal injury. American Association for Neurological Surgeons. 1984; 53:220.

9. Herrera RP. (film) Application of Diapulse Therapy in Otorhinolaryngology. 1966 Azucarera Hospital (Mexico City, Mex.).

10. Ionescu A, Milicescu S. (abstract) Five years of Diapulse therapy (DT) in the department of burned patients. Fifth International Congress of Burn Injuries. 1978; 5:86.

11. Ionescu A. (abstract) Study of efficiency of Diapulse therapy on the dynamics of enzymes in burned wound. Sixth International Congress on Burns. 1982; 6:25-26.

12. Ionescu A, Talnar E. (abstract) The effect of Diapulse therapy on burned wounds: A clinical and experimental study. The Medical Congress of the Balkan Countries. 1988; 8:96.

13. Ionescu D, Ionescu A. Results of microsurgical suture in 200 nerves. Acta Chirurgiae Plasticae 1984; 26 3: 166-183.

14. Ionescu D. (abstract) The effect of Diapulse therapy after microsurgical nerve repairs. Eighth Symposium of the International Society of Reconstructive Microsurgery. 1986; 8:45.

15. Jayakumar K, Sambasivan M, et al. Effect of pulsed electromagnetic field (PEMF) in cerebral oedema. Neurology India 1986; 34:241-247.

16. Kiwerski J, Chrostowski T, Weiss M. Clinical trials of the application of pulsating electromagnetic energy in the treatment of spinal cord lesions. Narz. Ortoped. Pol. 1980; 45 3:273-277.

17. Linkow LI, Chercheve R. Theories and Techniques of Oral Implantology. C.V.Mosby, St.Louis, MO. 1970; 279-280.

18. Meissl G. Nerve regeneration under the influence of an intermittent electromagnetic energy; An experimental study on rabbits. Handchirurgie 1979; 11:31-35.

19. Raji ARM. Effects of high peak pulsed electromagnetic field on degeneration and regeneration of the common peroneal nerve in rats. Lancet 1982; 444-445.

20. Raji ARM, Bowden REM. Effects of high-peak pulsed electromagnetic field on the degeneration and regeneration of the commonperoneal nerve in rats. The J Bone and Joint Surg 1983; 65B 4:478-492.

21. Ross J. (abstract) Utilization of pulsed high peak electromagnetic energy (Diapulse therapy) to accelerate healing processes. Digest, International Symposium of Antennas and Propagation Society. 1977;

22. Ross J. (abstract) Results, theories and concepts concerning the beneficial effects of pulsed high peak power electromagnetic energy (Diapulse Therapy) in accelerating the inflammatory process and wound healing. The Bioelectromagnetics Society Third Annual Conference. 1981; 3:13.

23. Ross J. Evolution, prevention and relief of acute and chronic pain with the application of Diapulse therapy (pulsed high peak power electromagnetic energy). Sonderdruck aus: Schmerz, Anschrift des Verfassers 1984; 9-16.

24. Rodrigues R, Cot A. Estudio anatomopatologico de las ulceras por decubito y sus modificaciones con las corrientes electromagneticas pulsatiles de alta frecuencia. Nuevas Tecnicas Diagnosticas y Terapeuticas en Patologia del Aparato Locomotor, 1986; MAPFRE:173-180.

25. Rutherford WG, Nelson PG, Weston PAM, Wilson DH. In: Accident and Emergency Medicine. Pitman Medical, London, 1980; 253, 258-261.

26. Sambasivan M. Pulsed electromagnetic field in the management of head injuries. Presented 1988: Eighth European Congress of Neurosurgery.

27. Silver H. Reduction of capsular contracture with two-stage augmentation mammaplasty and pulsed electromagnetic energy (Diapulse therapy). Plast and Reconstr Surg 1982; 69:5.

28. Wilson DH, Jagadeesh P, Neuman PP, Harriman DGF. The effects of pulsed electromagnetic energy on peripheral nerve regeneration. Ann of the N Y Acad of Sci 1974; 238.

29. Wilson DH, Jagadeesh P. Experimental regeneration in peripheral nerves and the spinal cord in laboratory animals exposed to a pulsed electromagnetic field. Paraplegia 1976; 14:12-20.

30. Wilson DH, Hall MH. Casualty Officer's Handbook, Butterworths, London, 4th Edition 1981; 94, 107-109, 139-141, 167, 182.

31. Wilson DH. Tenosynovitis, tendovaginitis and trigger finger. Physiother 1983; 69:10.

32. Young W, Ransohoff J. (abstract) Pulsed electromagnetic fields alter calcium (Ca) and functional recovery in spinal cord injury. The American Association of Neurological Surgeons Annual Meeting. 1984; 67:127.

33. Young W. Pulsed electromagnetic fields alter calcium in spinal cord injury. Presented 1984: The Society of Neurological Surgeons, 75th Meeting.

34. Young W. Pulsed electromagnetic fields (Diapulse) alter calcium in spinal cord injury. Presented 1984: The First International Symposium on Central Nervous System Trauma.

35. Young W. (abstract) Pulsed electromagnetic fields alter $Ca^{++}$ and functional recovery in spinal cord injury. Eighth Annual Meeting Bioelectromagnetics Society. 1986; 8:51.

# ELECTROTHERAPY APPLIED TO IRRADIATED HEAD AND NECK CANCER PATIENTS

Gordon E. King and Jan Scheetz

## INTRODUCTION

Dentists were first attracted to auriculotherapy and bioconductive therapy as a noninvasive, painless, and rapid means of relieving myofacial pain caused by temporomandibular joint problems and whiplash trauma. Physical therapists at the University of Texas M. D. Anderson Cancer Center have long been using electrotherapy in the traditional, recognized ways, and they also have been using auriculotherapy for pain control for at least 12 years. Auriculotherapy and bioconductive therapy gained further attention for their usefulness in treating sports injuries during the 1984 Olympic Games in Los Angeles. Oleson and others[4] at the University of California at Los Angeles School of Medicine showed how auriculotherapy could be used to detect a person's physical problems by scanning his or her ear points with an electric probe. If an abnormal condition existed the corresponding ear point was hyperactive and elicited an electrically detectable response. Fenn showed that applying an electromagnetic field to an injured area, results in rapid dispersion of edema and hematoma.[2] In the treatment of sports injuries, electrodes are placed on either side of an injured area to direct the electromagnetic field through the area.

The use of electrotherapy to treat sports injuries, myofacial pain and whiplash injuries appeared to have had dramatic success. Therefore, we thought that applying electrotherapy to head and neck cancer patients suffering from post-radiation complications might provide similar results, and the treatment would be painless and noninvasive.

Post-radiation problems of patients treated for head and neck cancers have challenged dentists since radiation was first used as an anticancer therapy. Radiation therapists have greatly improved their techniques to minimize the exposure of healthy tissues to the radiation beams, but the post-radiation complications are still with us. The undesirable side effects are considered to be the price paid for curing the disease and are accepted by both patients and physicians. The sequelae and problems resulting from the progressive ischemia of irradiated tissues are treated symptomatically, as are the acute problems that occur during radiation therapy. We are essentially dealing with treatment of irradiated wounds and radiation wound healing.

The major clinical problems dentists try to resolve or prevent are radiation caries, osteoradionecrosis, xerostomia, loss of oral opening, and loss of range of head and neck motion. Traditionally, we have managed these problems from a purely dental perspective, and our results have not been satisfactory.

## EXPANDED TREATMENT USING ELECTROTHERAPY

### A. Physical Therapy

During the past four to five years, members of my department have been working closely with our physical therapists, recognizing their superior training and skills in restoring patients to a normal range of motion and neuromuscular health. We have challenged them with the problems of our irradiated patients, hoping that together we might achieve better results. We are now employing traditional physical therapy techniques and electrotherapy.

Physical therapy techniques, per se, offer little that is new. Manual stretching and massaging and home exercises are employed to free adhesions, restore free-moving facial planes, and improve circulation. In an irradiated field, physical therapy alone is minimally effective at improving circulation. When the physical therapy techniques are supplemented with electrotherapy, however, we seem to get some encouraging results.

### B. Hyperbaric Oxygen

We have also used hyperbaric oxygen to improve healing in irradiated tissues, and this has proven extremely beneficial. Hyperbaric oxygen therapy is well established as a modality for treating irradiated tissues. Revascularization of irradiated tissues can be achieved by following specific hyperbaric oxygen regimens, and we have seen dramatic instances of irradiated wound healing[3]. Interestingly, we have observed similar results with the use of electrotherapy alone. Unfortunately, we cannot prove our results histologically, nor can we prove that the patients did not heal independently of our electrotherapy. The application of electrotherapy has been empirical, based on clinical observations. Also, we have varied our techniques from patient to patient in order to make use of what appeared to be increased clinical efficacy.

### C. Auriculotherapy

The first irradiated patient we treated with auriculotherapy for a condition other than pain had received 70 Gy of external-beam radiation bilaterally for metastatic disease in the neck following a marginal maxillectomy for squamous cell carcinoma of the gingiva. Her face and neck were darkly pigmented, and had the typical hard, alligator-hide texture where she had been irradiated. She received 10 auriculotherapy treatments over two weeks from our physical therapist, as well as the normally prescribed head and neck exercises.

At the end of the two weeks, we noticed a dramatic change in the texture and color of her irradiated tissues. We also noticed that during auriculotherapy her neck and lower face became quite warm, as if hot compresses had been applied. It seemed plausible that marked increases in blood flow and cellular metabolism were occurring within this area. We also assumed that the combination of increased blood flow and increased cell-membrane permeability, as described by Pilla,[5] was allowing tissues to repair and perhaps revascularize in an area that would otherwise experience progressive ischemia and fibrosis. We would like to demonstrate this histologically.

D. Electromagnetic Therapy

Soon after treating this patient, we learned of work done by Bauer and Bosworth.[1] They reported that auriculotherapy for irradiated patients could be enhanced by directing electromagnetic fields in microamperage currents across the fields being treated. They used Alpha Stim 350 (Electromedical Products, Inc., Hawthorne, CA 90250) units to do this. After acquiring these same units we routinely treated irradiated areas by crossing them with electromagnetic fields. After some empirical adjustments, we found a 50 microamp current at 0.5 Hz provided the best results. Stronger currents appeared to make the area tighter, suggesting we were stimulating fibrous tissue regeneration. Although we are still uncertain what, specifically, occurs physiologically and histologically, we believe we are getting some degree of angiogenesis or revascularization in the soft tissues.

## ATTEMPTED LABORATORY DEMONSTRATION

We tried to demonstrate revascularization by treating a group of miniature pigs that had been irradiated with the same dose-rate as our head and neck cancer patients (200 cGy per day, five days per week, for six to seven weeks). Two of three sites (shoulders, sides, or flanks) had been irradiated; this was done tangentially to avoid radiating vital organs. The objective of the radiation therapists who supplied the study animals was to measure skin contracture following radiation; we thought the animals would provide an opportunity to demonstrate revascularization. We biopsied the irradiated sites and then initiated bioconductive therapy. We applied the electrodes of Alpha Stim 350 units to crisscross the irradiated area with fields of 50 microamps at 0.5 Hz. After six weeks of 30-minute treatments, five days per week, we repeated the biopsies of the areas.

Our initial biopsy specimens appeared normal because of the pigs' large subdermal fat layer. Our post-treatment biopsy specimens showed no increase in vascularity but did show a decrease in the amount of collagen. All we could determine was that either the pigs were not a good model for the study or our treatment was not effective in increasing angiogenesis in irradiated pigs.

## CLINICAL STUDIES

Despite the disappointing results in the pig study, our clinical efforts with patients encouraged us to do a study to document specific results in humans. We had been using auriculotherapy to relieve the discomfort of mucositis and the pain of swallowing while patients received radiation therapy. We thought we were getting good results in loosening up the stiff, fibrotic neck tissues of previously radiated patients. Several patients obtained their own Alpha Stim 350 units so they could treat themselves at home.

To obtain a measurable clinical result, we selected the restoration of lost oral opening as an end point. We monitored 37 patients who had received ablative head and neck surgery and radiation therapy and suffered from the usual

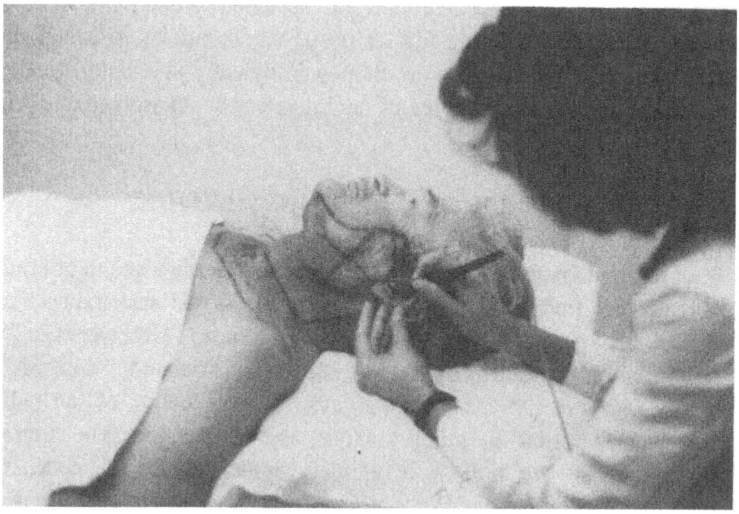

**Figure 1.** An irradiated patient receiving auriculotherapy.

unpleasant sequelae. These patients were sent to our physical therapists to receive exercise instruction, physical therapy, and both auriculotherapy and bioconductive therapy (Fig. 1). Because of the wide variety of diseases and defects, and in the treatments these patients received, we limited our study to 16 patients who had received a maxillectomy followed by radiation therapy and had decreased oral openings. They regained varying amounts of their range of opening after receiving physical therapy and electrotherapy.

Fifteen treatments were initially planned for each patient: one 30-minute treatment, five days per week for three weeks. However, the patients actually received from 1 to 57 treatments; only six received the planned 15 treatments. Six patients received more than 20 treatments; one had two separate courses of 33 and 36 treatments each. The patient who received 57 treatments was being treated palliatively and died of his disease. Four patients received one, three, seven, and nine treatments and then, for various reasons, discontinued participation in the study. The patient who had only one treatment learned to use the Alpha Stim 350 by himself and obtained one for home use. The patient who received three treatments had residual disease and discontinued therapy. The patient who received seven treatments increased her oral opening from 11 mm to 16 mm, had no evidence of disease, and quit the study. The patient who received nine treatments died of his disease.

Of the six patients who received their 15 treatments in three weeks, four had increases in their range of opening (from 12 to 20 mm; 10 to 19 mm; 14 to 20 mm; and 10 to 20 mm.) The other two patients died of residual disease; one had no increase in oral opening; the other lost another 4 mm of opening.

Three patients developed osteoradionecrosis. Two received hyperbaric oxygen therapy, and one received electrotherapy only. In each of these patients, the osteoradionecrosis resolved.

Although each of the 16 patients in the study group had had a maxillectomy and radiation therapy, their conditions and treatments varied so much that a valid comparison of their treatment modalities could not be made.

## DISCUSSION

Our efforts to restore the range of lost vertical oral opening have been disappointing. Patients who did not regain their original oral opening through exercises between the time of surgery and the beginning of radiation therapy saw their range of opening cut in half, and the likelihood of regaining their preradiation oral opening was poor, no matter what treatment was received or for how long. Patients who had necrosis or residual or recurrent disease in the pterygoids, continued to lose their range of oral opening.

Generally speaking, patients who have a maxillary malignancy and receive postmaxillectomy radiation therapy will have a presurgical oral opening, measured between central incisors, of 40-50 mm. Patients who do not strive to regain their normal opening through exercises frequently find their opening range reduced to about 20 mm by the time radiation therapy begins. By the time radiation therapy has been completed, that range will have been reduced to about 10 mm. The best we can expect for these patients, with exercises and electrotherapy, is to increase their opening back to about 20 mm. Patients with residual disease or osteoradionecrosis will not likely regain more than 2-3 mm, if anything.

Clinically, it is evident that the combination of auriculotherapy and bioconductive therapy has a beneficial effect on the recovery of irradiated tissues when applied soon after radiation therapy. The increased cell membrane permeability and the increased flow of tissue fluid through the treated area evidently prevent long-term fluid stasis and allow many damaged cells to recover from what would otherwise be certain cell death and increased fibrosis.

It is recognized that cells respond differently to different levels of electric current. Our results with electrotherapy have been achieved with equipment that provides a very limited range of current. We obtained the best results with 50 microamps at 0.5 Hz. If the current could be fine-tuned from 0 to 100 microamps and pass unimpeded through the tissue field, I believe we could be far more successful in enhancing the repair of irradiated tissues.

## CONCLUSION

It is clear that electrotherapy by itself, as we have applied it, is ineffective in overcoming the progressive or established fibrosis and contracture of irradiated soft tissues that can limit oral opening. We have been most successful when we have treated compliant, disease-free patients shortly after radiation therapy. What is most impressive is the restoration of normal color, texture, and suppleness of the irradiated soft tissues.

It would appear that regaining lost oral opening is related to the compliance of the patient in performing exercises at home, as instructed, while healing between surgery and radiation therapy. Electrotherapy, as we are applying it, can be more profitably directed to assist in wound healing and pain relief.

## References

1. Boswell NS, Bauer W. Non-invasive electrical stimulation for the treatment of radiotherapy side effects. Am J of Electromedicine 1985; 1:5-6.

2. Fenn JE. Effect of pulsed electromagnetic energy (Diapulse) of experimental hematomas. Can Med Assoc J 1969; 100:251-254.

3. Marx RE, Johnson RP, Kline SN. Prevention of osteoradionecrosis: a randomized prospective clinical trial of hyperbaric oxygen versus penicillin. J Am Dent Assoc 1985; 3:49-53.

4. Oleson ID, Kroening RJ, Bresler DE. An experimental evaluation of auricular diagnosis: the somatotopic mapping of musculoskeletal pain at ear acupuncture points. Pain 1980; 8:217-229.

5. Pilla AA. Electrochemical information transfer at living cell membranes. Ann NY Acad Sci 1974; 238:149.

# SAFETY ISSUES IN ELECTROMAGNETIC MEDICINE

Mary Ellen O'Connor

The technology and developments introduced in the preceding chapters indicate that we are on the verge of a revolution in medicine. A revolution that is predicated upon the recognition of the electrical side of biology. There is substantial evidence that low level electromagnetic fields (EMF) do indeed interact with the electrical signals and mechanisms in the body. It is exciting to contemplate the beneficial diagnostic and therapeutic applications of this interaction. However, there also is concern that the changes induced at the cellular level can be harmful to living organisms.[1] This chapter presents some of the research that raises the question of risk. In addition, some of the general concepts and concerns associated with risk/benefit analysis are discussed.

## INTRODUCTION

Any new development in medicine, particularly one that promises increased healing power or reduction in human discomfort, is met with varying levels of excitement. Patients for whom available treatments failed, or were only partially successful, are most excited about a new procedure. A physician in a speciality area where success is limited and prognosis is generally poor, is also likely to be excited. Physicians and health professionals in areas where current techniques and available medications provide good or professionally acceptable patient prognosis are likely to be skeptical, or simply less enthusiastic. The scientific community, who are trained and rigidly socialized to be skeptical of any new development, ordinarily express some curiosity, but no excitement unless the underlying mechanism of action is known.

It is not unusual for different elements of society to disagree over the promise offered by a new technology. The responsibility of the physician, manufacturer or producer, as well as regulatory commissions or agencies are all different. Each of these groups, in turn have their own external or internal review or overview system whose role is to ensure that they behave responsibly. However, most individuals do not receive information about developments in medical treatment directly from any of the above sources. Rather, the information is conveyed via the popular media. The role of the media is to make information known to the public in as timely a manner as possible. It is important to recognize that the media in the US is not merely an information system, it is a for-profit business. Being the first to discover and inform the public about a new beneficial invention, or on the other hand a new hazard, is part of the status and reward structure in the field of communications. As such, the information that reaches the general public via the media is often based on preliminary evidence. In current times, even scientists have short circuited the established peer review system and announced new discoveries via the popular media.[24,25]

It is not surprising that the general public has difficulty recognizing the difference between information based on anecdotes, testimonials and controlled clinical trials. Educators in the US repeatedly have expressed their concern that the general public is not well versed in science issues or scientific methodology. Recent surveys tend to support this concern.[5]

No treatment, regardless of how revolutionary it may be, is a panacea. However, it is beneficial for patients to believe that a treatment will work. Scientific evidence supports the idea that patients have a better prognosis if they believe that they can get well and if they have a positive attitude regarding their physicians and treatments.[9,17] Also, simple treatments, cure-alls and miracles will always be popular with a certain segment of the population. There are members of the general public who seem to be particularly susceptible to information that scientists consider preliminary, biased or, at worst, 'hype'. And, there seem to be ample numbers of media representatives who are willing to deliver the 'hype'.

Next to the popular media, the most likely source of information about medical developments is the personal physician. Several other contributors to this volume[6,11,31] discussed the current reliance in medicine on pharmacological treatment. The emphasis on pharmacology permeates the training of physicians and our entire society. Thousands of legal and illegal drugs are available for medical and recreational use.[22,23] Most of these drugs have some non-beneficial side-effects. Of the thousands available, only a limited number are used at any point in time. New drugs are added but the older drugs that they replace are not removed necessarily from the market. Physicians tend to rely on a relatively smaller number of products, with which they have encountered success in the past. The current practice of medicine in the US, particularly general medicine, places untoward demands on the time of the physician. It is, thus, not surprising that their information about improved efficacy for a product is received from a representative of a pharmaceutical manufacturer rather than articles in peer reviewed pharmacology journals.[16,23] An additional difficulty is encountered in the introduction of electromagnetic technology because the physician is not likely to be as well versed in biophysics as in biochemistry. An introduction to the basic mechanisms by which electromagnetic (EM) devices most probably function are not part of current medical education.

The US Congress has charged the Food and Drug Administration with determining the efficacy and the safety of all medical devices including those that use EM energy.[18] However, all of the key terms in the above statement are open to interpretation. What, exactly has been legislated? How do you determine efficacy? What constitutes safety? The last question requires definition of risk and hazard and in some instances a formal or informal risk/benefit analysis. The fundamental methodology and techniques used by FDA were developed in the context of reviewing food, food additives and drugs as opposed to technological devices. There is also the problem of statistically determining either the efficacy or the safety of a treatment that is intended to be adjunctive to other forms of already established efficacious treatments.

The data associating EM devices with health risk developed in the context of safety of EMF encountered in the general environment rather than in a medical setting. The research generating this data was aimed at determination of any biological effects that might result from exposure of occupational groups or the general public to fields encountered in the environment. Very few of the statistical techniques or models that are used by experts in the relatively neophyte area of risk/benefit analysis have been applied to the general question of safety of EMF. Existing safety guidelines for EMF[3,15] specifically exclude medical devices and applications from coverage.

## DEFINITION OF TERMS

### A. Hazard

A hazard is an intrinsic property of a device, agent or event. A device, agent or event is thus termed hazardous when it is known to result in harm to the environment, instruments, or organisms with which it comes into contact. A hazard represents the potential for injury or danger.[6] Since the properties of a hazard are intrinsic they can not be changed. An event can be hazardous without being toxic. Toxicity may develop at some level of exposure to the hazard.

### B. Risk

Risk, on the other hand, is not intrinsic and, as such, can be manipulated. The concept of risk adds the notion of probability and consequences to the concept of hazard. The magnitude of the consequences can range in seriousness including damage to property or the environment, loss of money, injury, or fatalities. Risk associated with a known hazard can be reduced by manipulating the environment, (eg, by wearing protective devices and or clothing and by limiting the duration of exposure). Avoiding a hazard reduces risk. Limiting the amount of time that a hazard is experienced can reduce the risk. Even the risk associated with natural hazards often can be manipulated. For example, electrical energy produced in the form of lightning during electrical storms is hazardous. However, the risk to an individual is variable depending on whether the person is in the open environment, under a tree, in a building, or in an automobile.

### C. Assessing and Quantifying Risk

The first step in assessing and quantifying risk is to identify what level over a specified time period constitutes a hazard. Before the issue of risk can be addressed adequately, it should be established scientifically that an event is a hazard at some level for some period of time. With regard to low level electric and magnetic fields, the question of hazard is still debated. Dose and dose response relations can not be quantified until some level and some duration of exposure is deemed hazardous.

Such circumstances and difficulties in determining hazard are not unique to electric and magnetic field research. These questions have also plagued researchers regarding ionizing forms of radiation and radiation exposure. In cases where hazard can not be clearly identified, these other areas of radiological concern have adopted the concepts of 'no observed effects level' (NOEL) and 'no observed adverse effects level' (NOAEL). In addition, occupational exposure guidelines have relied on the concept of 'as low as reasonably achievable (ALARA)' and '*de minimus*,' indicating that it is a level that has been determined to be below regulatory interest. Researchers and regulators in the nuclear radiation area have considered the perception of risk in assessing the health hazard of a nuclear reactor selection site.[21,30] These concepts were developed following concensus that there is no threshold dose for hazardous ionizing radiation exposure; as such the concepts are not encountered frequently in the non-ionizing EM effects or exposure standards literature.

## CELLULAR CHANGES FOLLOWING EXPOSURE TO EM FIELDS

Among the responses found in laboratory studies of cells and tissues exposed to EM fields are modulation of ion flows, interference with DNA synthesis and RNA transcription,[14] and interaction with the biomedical kinetics of cancer cells.[8] With the exception of some of the work with pulsed electromagnetic fields (PEMF) and bone cells, almost all of the studies were initiated in relation to questions of the safety of working with or living near electric power lines.

A. Calcium Efflux

One of the earliest reports of cellular effects resulting from exposure to low level EM fields was that of calcium efflux from neonatal chick brain.[4] The effects were induced at particular frequency windows. The observations were replicated by Blackman et al.,[7] and were extended to in vivo demonstrations in awake cat brain.[2]

B. Other Nervous System and Cellular Effects

Other cellular changes in central nervous system amines have been demonstrated. Rats exposed to low level fields at 60 Hz showed phase shifts in diurnal rhythms of brain biogenic amines[32] and similar field exposures in monkeys increase levels of biogenic amines in cerebrospinal fluid.[28]

Effects of ELF at the tissue level have been reported for the past 20 years. However, the mechanism at the molecular or ionic level have not been established. For example, exposure of bone and bone cells in vitro to PEMF causes a membrane-mediated desensitization of the osteoblast to parathyroid hormone.[19] The magnetic sensitivity of the pineal gland has also been reported.[29,34] Changes in both calcium and sodium transport in chick tibia exposed to PEMF have been reported.[10] The effect of PEMF on chondrogenesis and osteogenesis adenylate cyclase activity in skeletal tissues[26] and other membrane activities have been reported.

## WHOLE BODY ANIMAL STUDIES

Most of the whole body animal studies have been performed using the radio frequency (RF), particularly microwave, portion of the EM spectrum. The threshold for the ANSI recommendation[3] as well as subsequent recommendations[15] was actually based on studies showing behavioral disruption at low levels of exposure. Fields in the ELF region have been associated with circadian rhythm disruption and some disorientation.[12]

## EPIDEMIOLOGY STUDIES

An area of intense concern with regard to EM exposure is the possible link between exposure to EM fields and childhood cancer.[13,20,27,33] The initial studies were criticized for the rigor of the experimental design and the lack of attention to variables other than EM fields. However, the subsequent studies have offered some support to the suggested relationship and continue to be reported, quoted and used as the emphasis for additional research funds in this area. Definitive statements can not be made until more data are available.

The epidemiology studies are for cases of chronic, not acute exposure to EM. The two groups that have been studied repeatedly are power line workers and persons living in residences near power lines. Most medical exposure, on the other hand, is for a limited time with durations of exposure on the order of a few hours a day for a defined period of treatment. If EM were limited to diagnostic purposes, the exposures would be of even more limited duration. One of the themes repeated in this volume is the fact that the use of EM in medicine is indeed both diagnostic and therapeutic.

## SUMMARY

Raising the issue of safety and risk in the area of EM in medicine results in many unanswered questions. As noted, the terms hazard and risk are not well defined for this area. The data suggesting concern for safety comes from a literature that developed outside of the area of EM devices in medicine and is based primarily on whole body exposure over extended periods of time. Some of the effects mentioned in this chapter as well as some of the cellular effects discussed earlier (eg, electroporation) argue for caution in tampering with such basic cellular function.

The safety issues, particularly with regard to risk/benefit, are obviously different if one is concerned about the patient or the operator of the device. There are a number of individuals who are excited about the prospects of EM in medicine. New devices should be viewed with caution and the questions of benefit versus risk should continue to be asked. This caution should not prevent each device from being evaluated on its own merits for its clinical efficacy.

### References

1. Abelson PH. Effects of electric and magnetic fields. Science 1989; 245:241.

2. Adey WR, Bawin SM, Lawrence AF. Effects of weak amplitude modulated microwave fields on calcium efflex from awake cat cerebral cortex. Bioelectromagnetics 1982; 3:295-307.

3. American National Standard Safety Levels with Respect to Human Exposure to Radio Frequency Electromagnetic Fields, 300 kHz to 300 GHz: ANSI C95.1-1982. IEEE, New York. 1982.

4. Bawin SM, Kaczmarek LM, Adey WR. Effects of modulated VHF fields on the central nervous system. Ann NY Acad Sci 1975; 247:74-81.

5. Bell TE. Managing risk in large complex systems. IEEE Spectrum 1989; 26:22-27.

6. Bentall RHC. Electromagnetic energy: A historical therapeutic perspective; it's future. In: Emerging Electromagnetic Medicine (Eds. ME O'Connor, RHC Bentall, JC Monahan). Springer-Verlag, New York, 1990; 1-17

7. Blackman CF, Benane SG, Joines WT, Hollis MA, House DE. Calcium-ion efflex from brain tissue: power-density versus internal field-intensity dependencies at 50 MHz RF radiation. Bioelectromagnetics 1980; 1:277-283.

8. Byus CV, Lundak RL, Fletcher RM, Adey WR. Alterations in protein kinase activity following exposure of cultured lymphocytes to modulated microwave fields. Bioelectromagnetics 1984; 5:34-51.

9. Cohen SI. Voodoo death, the stress response, and AIDS. In: Psychological, Neuropsychiatric, and Substance Abuse Aspects of AIDS (Eds. TP Bridge, AF Mirsky, FK Goodwin). Raven Press, New York. 1988; 95-109.

10. Collacicco G, Pilla AA. Chemical, physical and biological correlations in the CA-uptake by embryonal chick tibia in vitro. Biochem and Bioenerget 1983; 10:119-131.

11. Cook IA. Cellular physiology and electromagnetic fields: Cellular communication, human communication. In: Emerging Electromagnetic Medicine (Eds. ME O'Connor, RHC Bentall, JC Monahan). Springer-Verlag, New York. 1990; 59-65

12. Ehret CF. (abstract) The actions of high strength 60-Hz electric fields on circadian rhythms in small rodents. Department of Energy Annual Contractors Review. 1985.

13. Fulton JP, Cobb S, Preble L, Leone L, Forman E. Electrical wiring configurations and childhood leukemia in Rhode Island. Am J Epidemiol 1980; 111:292-296.

14. Goodman R, Abbott J, Henderson AS. Transcriptional patterns in the X chromosome of Sciara coprophila following exposure to magnetic fields. Bioelectromagnetics 1987; 8:1-7.

15. Guy AW, Adey WR, Alpen EL, Justesen DR, O'Connor ME, Phillips RD. Biological Effects and Exposure Criteria for Radiofrequency Electromagnetic Fields. National Council on Radiation Protection and Measurements, Bethesda MD, 1986.

16. Julien RM. Drugs and the Body. WH Freeman and Co, New York, 1988.

17. Levy SM. Behavior as a biological response modifier: The psychoimmunoendocrine network and tumor immunology. Behav Med Abstr 1985; 6:1-4.

18. Liska K. Drugs and the Human Body. Macmillan, New York. 1986; 62-85.

19. Luben RA, Cain CD. Use of hormone receptor activities to investigate the membrane effects of low energy electromagnetic fields. In: Nonlinear Electrodynamics in Biological Systems (Eds. WR Adey, AF Lawrence). Plenum, New York. 1984; 23-34.

20. McDowall ME. Mortality of persons resident in the vicinity of electricity transmission facilities. Br J Cancer 1986; 53:271-279.

21. Okrent D. The safety goals of the US Nuclear Regulatory Commission. Science 1987; 236:296-300.

22. Physician's Desk Reference: 43rd Edition. Medical Economics, Oradell, NJ, 1989.

23. Physician's Desk Reference for Non-Prescription Drugs: 10th Edition. Medical Economics, Oradell, NJ, 1989.

24. Pool R. How cold fusion happened -- Twice! Science 1989; 244:420-423.

25. Pool R. Cold fusion: End of act I. Science 1989; 244:1039-1040.

26. Rodan GA, Bourret LA, Norton LA. DNA synthesis in cartilage cells is stimulated by oscillating electric fields. Science 1978; 199:690-692.

27. Savitz DA, Calle EE. Leukemia and occupational exposure to electromagnetic fields: Review of epidemiologic surveys. J Occup Med 1987; 29:47-51.

28. Segal RF. (abstract) Exposure to 60-Hz electric and magnetic fields alters biogenic amine metabolite concentrations in non-human primate cerebrosopinal fluid. Soc Neurosci Ann Meeting Abstr 1985; 11(1):443.

29. Semm P. Neurobiological investigations on the magnetic sensitivity of the pineal gland in rodents and pigeons. Comp Biochem Physiol 1983; 76A:683-689.

30. Slovic P. Perception of risk. Science 1987; 236:280-285.

31. Tyler P. EMR and the brain: A brief literature review. In: Emerging Electromagnetic Medicine (Eds. ME O'Connor, RHC Bentall, JC Monahan). Springer-Verlag, New York, 1990; 145-158

32. Vasquez BJ. (abstract) Effects of 60-Hz electric fields on the daily rhythms of biogenic amines. Bioelectromagnetics Society Eight Annual Meeting Abstracts 1986; 65.

33. Wertheimer NW, Leeper E. Electrical wiring configurations and childhood cancer. Am J Epidemiol 1979; 109:273-284.

34. Wilson BW, Anderson LE, Hilton DI, and Phillips RD. Chronic exposure to 60-Hz electric fields: Effects on pineal function in rats. Bioelectromagnetics 1981; 2:371-380.

# SUBJECT INDEX